Recent Progress and Development in Nanostructures

Edited by Mohammed Muzibur Rahman

Published in London, United Kingdom

Recent Progress and Development in Nanostructures
http://dx.doi.org/10.5772/intechopen.1004676
Edited by Mohammed Muzibur Rahman

Contributors
Abdullah M. Al-Mayouf, Akamu J. Ewunkem, Ananda Kumar Chettupalli, A'lyiha F. Beard, Brittany Justice, Dinesh K. Singh, Dmitry Evgenyevich Milovzorov, Feng Yang, Henry Agbe, Himabindu Peddapalli, Hongdong Li, Ilunga Tshimanga, Jeffery Meixner, Kavitha Marati, Lei Zhao, Mabrook S. Amer, Maged N. Sh, Maram A. AlGhamdi, Mounika Kuchukuntla, Niore Johnson, Noha Taymour, Prabhakarn Arunachalam, Rajendra Kumar Jadi, Samuel Kwofie, Shahab Ahmadi Seyedkhani, Shaik Abdul Rahaman, Shawn Y. Wu, Tao Peng, Wessam S. Omara, Xiaoxi Yuan, Zahirah J. Williams

First published in London, United Kingdom, 2025 by IntechOpen
IntechOpen is the global imprint of INTECHOPEN LIMITED, registered in England and Wales, registration number: 11086078, 167-169 Great Portland Street, London, W1W 5PF, United Kingdom

For EU product safety concerns: IN TECH d.o.o., Prolaz Marije Krucifikse Kozulić 3, 51000 Rijeka, Croatia, info@intechopen.com or visit our website at intechopen.com.

British Library Cataloguing-in-Publication Data
A catalogue record for this book is available from the British Library

Recent Progress and Development in Nanostructures
Edited by Mohammed Muzibur Rahman
p. cm.

This title is part of the Nanotechnology and Nanomaterials Book Series, Volume 11
Topic: Nanomaterials and Nanostructures
Series Editor: Jung Huang
Topic Editor: Alberto Jiménez-Suárez
Associate Topic Editor: Gonzalo Seisdedos

Print ISBN 978-0-85014-826-8
Online ISBN 978-0-85014-825-1
eBook (PDF) ISBN 978-0-85014-827-5
ISSN 3029-0538

IntechOpen Book Series
Nanotechnology and Nanomaterials
Volume 11

Aims and Scope of the Series

Humans face growing challenges in environmental sustainability. We need better materials, more effective devices, and higher computing power to conquer these challenges. Researchers have long explored the realm of science at the nanometer scale for developing novel problem-solving technologies. In the past decades, breakthroughs in multiscale computer simulation, novel hierarchical material and device structures, nanometrology, and new functionalities from interfacial and quantum size effects have enhanced our ability to utilize nanotechnology to solve real-world problems. This book series addresses important advancements in nanotechnology and nanomaterials. It showcases how nanotechnology is being continually developed and implemented in a variety of domains of science and technology. The two main topics this series covers are Nanotechnology and Nanodevices, and Nanomaterials and Nanostructures.

Meet the Series Editor

Jung Y. Huang, a university educator and researcher, has been working to unravel the structures and functional properties of materials and cellular events in living cells with varying optical methodologies. He has co-authored hundreds of journal papers and five book chapters. He also holds tens of patents in laser techniques and single-molecule/hyperspectral imaging and has developed architectural photonics based on hierarchically structured materials. Currently, his research focuses on the integration of artificial intelligence methodology with optics to automatically discover meaningful information from optical sensing/imaging data cubes. He has been an editorial board member and reviewer for several scientific journals. As a member of the global scientific community, he sincerely supports and endeavors to promote the spread of scientific knowledge.

Meet the Volume Editor

Dr. Mohammed Muzibur Rahman, a professor at the Center of Excellence for Advanced Materials Research (CEAMR) and Chemistry Department, King Abdulaziz University in Saudi Arabia, has been a leading researcher since 2011. He obtained his Ph.D. from Chonbuk National University in South Korea in 2007, followed by academic roles in South Korea, Japan, and Saudi Arabia. Dr. Rahman has authored over 550 research articles, 44 book chapters, 23 books, and holds 11 US patents. With over 26,500 citations and an H-index of 81, his research focuses on nanomaterials, sensors, catalysis, and electrochemistry. He also serves as editor and guest editor for international journals, significantly contributing to advancements in nanotechnology and environmental solutions.

Contents

Preface

Nanostructures, with their unique properties and diverse applications, have emerged as a cornerstone of modern science and technology. This book, *Recent Progress and Development in Nanostructures*, comprehensively explores the fundamental principles, innovative synthesis methods, and groundbreaking applications of nanostructured materials. It serves as a vital resource for understanding the interdisciplinary nature of nanotechnology, bridging the fields of chemistry, physics, biology, engineering, and medicine.

This book discusses the primary phases of nanostructure development, covering synthesis techniques, growth mechanisms, advanced characterization methods, and the innumerable properties of nanoscale materials. A significant focus is placed on functionalized approaches and their relevance to various applications, including sensing, detection, catalysis, optoelectronics, and environmental remediation. Furthermore, the book delves into cutting-edge areas such as drug delivery, disease diagnosis, antimicrobial properties, and biomedical therapies, highlighting the transformative potential of nanostructures in healthcare and sustainability.

This work also emphasizes the importance of porous nanostructured materials, noble metal and carbon nanocomposites, and new facile synthesis techniques. It aims to provide valuable insights into the latest advancements in nanostructures, equipping scientists, researchers, academics, and postgraduate students with an up-to-date understanding of this rapidly evolving field.

The increasing significance of nanostructured materials in technology-driven economies underscores the need for such a comprehensive resource. By compiling advanced developments in nanostructures, this book aspires to inspire further innovation, fuel interdisciplinary collaboration, and address pressing global challenges through the power of nanotechnology.

We hope this book will guide and inspire readers, offering a window into the immense possibilities of nanostructures and their role in shaping a sustainable and technologically advanced future.

Mohammed Muzibur Rahman
Center of Excellence for Advanced Materials Research (CEAMR)
and Chemistry Department,
King Abdulaziz University,
Jeddah, Saudi Arabia

Section 1

State-of-the-Art Synthesis

Chapter 1

Carbon Nanomaterials for Neural Interfaces: Synthesis, Properties, Applications

Shahab Ahmadi Seyedkhani

Abstract

Carbon nanomaterials have become integral to the advancement of neural interfaces due to their exceptional electrical conductivity, biocompatibility, and mechanical properties. In this chapter, we provide a detailed examination of various carbon-based materials, including glassy carbon (GC), carbon nanotubes (CNTs), carbon nanofibers (CNFs), graphene, and synthetic diamond, in the context of neural interfacing. We review the synthesis methods, electrochemical properties, and applications of these materials, emphasizing their roles in neural stimulation, high-resolution recording, therapeutic agent delivery, and biosensing. We discuss the interactions between neural cells and these materials, highlighting their biocompatibility and potential to influence cell differentiation. Additionally, we address the fabrication methods for diamond electrodes and microelectrode arrays (MEAs) and their applications in neurochemical sensing and neural stimulation. Finally, we consider the challenges and future prospects of carbon nanomaterials in neural interfaces, aiming to provide a comprehensive overview for researchers in the field.

Keywords: neural interface, carbon nanotube (CNT), graphene, glassy carbon, nanofiber, diamond, brain-computer interface (BCI)

1. Introduction

Carbon nanomaterials, encompassing a diverse range of materials such as carbon nanotubes (CNTs), graphene, graphene oxide, and diamond, have garnered significant attention for their potential in neural interface applications. These nanomaterials, defined by their nanoscale dimensions and unique carbon atom arrangements, exhibit superior electrical conductivity, mechanical strength, and chemical stability, making them ideal candidates for interfacing with the nervous system [1–3]. Historically, the exploration of carbon nanomaterials began with the discovery of fullerenes in 1985 [4], followed by the identification of CNTs in 1991 [5], and the isolation of graphene in 2004 [6]. These milestones marked a pivotal shift in material science, leading to extensive interdisciplinary research aimed at harnessing the properties of carbon nanomaterials for various applications, including biomedical engineering and neural interfaces. The integration of carbon nanomaterials into neural interfaces emerged as a promising field in the early 2000s, driven by the need for materials that could overcome the limitations of traditional neural interface materials like metals and silicon [7].

The importance of carbon nanomaterial-based neural interfaces lies in their ability to provide improved biocompatibility, electrical conductivity, mechanical flexibility, and surface functionalization capabilities compared to conventional materials [8, 9]. These attributes enable the development of neural interfaces that are not only more effective in recording and stimulating neural activity but also more durable and adaptable to the dynamic environment of neural tissues. Applications of carbon nanomaterials in neural interfaces include high-resolution neural recording and stimulation electrodes, neural prosthetics, drug delivery systems, tissue engineering scaffolds, and optogenetic and photothermal therapies [2, 9, 10]. The exceptional properties of carbon nanomaterials facilitate high signal-to-noise ratios (SNRs) in neural recordings, enhance the longevity and performance of neural prosthetics, allow for targeted therapeutic delivery, support neural tissue regeneration, and enable precise control of neural activity through light-based methods [11–13]. As research and development in this field continue to advance, the integration of carbon nanomaterials into neural interfaces holds great promise for bionanoelectronics and neuroscience, along with improving the quality of life for individuals with neurological disorders.

2. Carbon nanomaterials for neural interfaces

Neural interfacing systems commonly employing rigid metal electrodes can lead to harm in the surrounding tissue and compromise the long-term performance of implanted systems. While recent approaches, such as nanowire field-effect transistor (FET) sensors and polymer fiber probes, have not completely overcome these challenges, carbon-based electrodes offer a promising alternative. They have demonstrated the capability to mitigate immune responses during implantation, possess larger cross-sectional profiles, improve SNR for neural recording, and provide enhanced resolution for neural stimulation. This section will delve into the characteristics and distinctive advantages of devices utilizing carbon-based nanomaterials in this context.

2.1 Glassy carbon

Glassy carbon (GC), which is also known as glass-like carbon or vitreous carbon, is a unique form of carbon that combines the properties of glass and graphite. The GC is a highly appealing material for neural interfaces due to its proven excellent electrical properties, chemical inertness, electrochemical stability, and adaptability for various geometries [14–20]. However, its consideration for such applications is relatively new, primarily attributed to advancements in technologies enabling pattern transfer and the integration of GC electrodes into thin film polymeric substrates [21–23]. This holds significant promise in the field of neural prosthetics. The utilization of thin film and ultra-flexible polymer substrates in device fabrication addresses the mechanical mismatch between the device and tissue, thereby mitigating the inflammatory reactive response. Recently, studies have indicated a significant breakthrough by designing a neural interface that utilized GC microelectrodes (sensing pads) placed on flexible and thin film polyimide substrates [24]. The study revealed that GC microelectrodes exhibit an electrochemical window over 70% wider and a charge transfer capacity (CTC) over 70% higher than Pt microelectrodes of comparable geometry. Furthermore, it was demonstrated that GC microelectrodes can endure at least 5 million pulses at a charge density of 0.45 mC/cm^2 with an impedance change of less than 7.5%, whereas Pt microelectrodes experienced delamination after only 1 million pulses. In addition, the researchers applied electrodeposition of poly(3,4-ethylenedioxythiophene)-poly(styrenesulfonate) (PEDOT-PSS) to both sets of devices (GC and Pt) to selectively reduce impedances for diameters less than 60 μm. They observed superior adhesion of PEDOT-PSS to GC compared to Pt, enabling a significant reduction in electrode size while maintaining a consistent level of delivered current. For *in vivo* characterization in rats, the study demonstrated that GC microelectrode arrays recorded somatosensory evoked potentials (SEP) with an almost two-fold signal-to-noise ratio (SNR) compared to Pt arrays [24]. Despite the widespread use of GC in neural interfaces, there are scalability limitations [24]. The challenge lies in producing carbon-based electrodes with both high conductivity and uniform configurations for large-scale manufacturing. Research efforts are actively addressing this challenge, aiming to develop efficient production methodologies for the consistent fabrication of high-quality electrodes [24]. **Figure 1** illustrates a GC neural interface MEAs. MEAs fabricated on ultra-thin and finger-like polyimide substrate [25].

2.2 Carbon nanotubes (CNTs) and carbon nanofibers (CNFs)

Carbon nanotubes (CNTs) are nanostructures with a cylindrical morphology, comprising carbon atoms arranged in a hexagonal lattice. These structures possess notable characteristics, including a high aspect ratio, robust mechanical strength, and excellent electrical conductivity. These distinctive properties make CNTs highly advantageous in diverse neuroscience applications [1]. In neural engineering, CNTs offer unique advantages due to their nanoscale size, surface characteristics, and porosity, making them an ideal interface for precise neural signal recording [26]. Moreover, their biocompatibility and bioactivity can be enhanced through surface modifications such as covalent functionalization, noncovalent wrapping, and

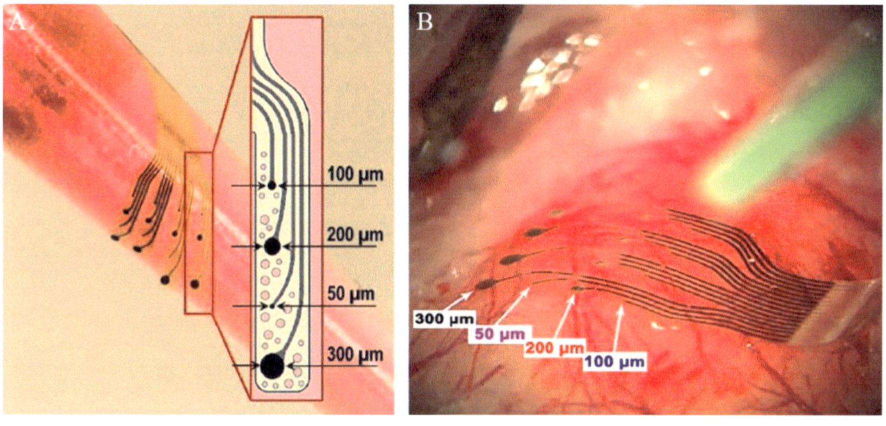

Figure 1.
A GC electrocorticography (EcoG) electrode on ultra-thin and finger-like polyimide substrate [25].

attachment of bioactive molecules [27, 28]. CNTs also exhibit different electrical states, promoting anisotropic electrical propagation and potentially aiding in cell alignment and migration [29, 30].

2.2.1 Biocompatibility of CNTs and cell differentiation

CNT biocompatibility is complex, with varying outcomes in different studies. Early research in animal models uncovered potential toxicity near mucous membranes and upon intravenous injection [31, 32]. Cellular and molecular effects have yielded inconsistent findings, with some studies suggesting benignity and others indicating toxicity [33, 34]. Recent research has focused on specific cell types, including neural cells, and strategies to enhance compatibility and mitigate cytotoxicity, such as noncovalent wrapping and covalent functionalization [35]. On the other side, CNTs influence the differentiation of stem cells toward neural lineages, which is crucial for neural engineering and regeneration [36]. Mesenchymal stem cells (MSCs) exposed to CNTs adopt neural-like characteristics [37]. Neural stem cells (NSCs) can differentiate into neurons and glial cells, influenced by CNTs [38, 39]. Embryonic stem cells (ESCs) can differentiate toward the neural lineage, with CNT materials playing a significant role [40, 41]. Surface chemistry and electrical stimulation further modulate differentiation [42, 43]. Nevertheless, concerns about the toxicity of CNTs in medical applications have prompted extensive research, as biocompatibility depends on factors like dosage, administration method, and exposure type [44, 45]. **Table 1** shows different types of toxicities related to CNTs.

Source	Challenges	References
General effect of CNTs	Toxic effects of CNTs have been reported	[46–49]
	The dispersion of CNTs is associated with toxicity in the brain, lungs, and skin.	[31, 46, 50]
	CNTs have the potential to induce aneurysms, especially in the cerebellum, which may suggest impairment of the blood-brain barrier (BBB).	[51]
	In cell culture, neurons, and astrocytes cultivated on certain CNT surfaces exhibit diminished cell viability and adhesion relative to standard growth surfaces. In some instances, this reduced viability persists even after the cells are transferred from the CNT-coated substrate.	[52, 53]
	The surface charge or polarity of CNTs may contribute to cytotoxicity, potentially by affecting metabolic processes.	[52]
	Soluble CNTs can impair the capacity of neurons to transport Ca^{2+} ions in response to membrane depolarization. This impairment is associated with the formation of fewer but longer neurites, accompanied by a reduction in neurite branching.	[54]
	CNTs have the potential to induce epigenetic changes that persist long after the CNTs have been physically removed. This concern is particularly significant because CNTs are highly resistant to biodegradation and may remain in the body indefinitely.	[55]
	CNT films have been demonstrated to modify neurite branching, neurite length and density, and the number of growth cones.	[52, 53, 56]

Source	Challenges	References
Effects attributed to CNT composition	CNTs exhibit a range of sizes and surface properties.	[57]
	Different types of CNTs possess markedly distinct electrical properties, leading to varying behaviors in bulk versus nanophase contexts.	[47, 58]
	CNTs structurally resemble graphite, which can induce toxic responses, especially in the lungs.	[47]
	The lung is a particular concern for CNTs because their manufacturing processes often result in their easy release into the air.	[55]
	In the context of lung toxicity or effects on alveolar cells, CNTs typically induce the formation of granulomas.	[31, 47, 59]
	Reactive oxygen species (ROS) are likely generated by the photoexcitation of CNTs, which can produce free electrons. The production of ROS appears to be dose-dependent and is probably mediated through a Fenton-type reaction, where hydrogen peroxide (H_2O_2) is decomposed to yield hydroxyl (OH^-) radicals. This process may be associated with contamination from iron catalysts.	[46, 59–62]
	CNTs may interfere with ROS scavenging pathways, such as glutathione, rather than directly generating ROS.	[59]
	Observed ROS-related effects of CNTs include reduced cell viability through both apoptotic and necrotic pathways.	[33, 59, 63, 64]
	Exposure to CNTs disrupts mitochondrial membrane potential, though it remains uncertain whether this effect is due to ROS production or direct penetration of the mitochondrial membrane.	[62, 65]
	Unmodified hydrophobic CNTs may intercalate between the hydrophobic lipids of the cell membrane, leading to membrane disruption.	[65–68]
	CNTs can act as a physical barrier, disrupting cytoskeletal proteins and organelle placement, as well as impairing external mass transfer. This disruption has been linked to the upregulation of apoptotic proteins and a reduction in the expression of various cell adhesion proteins, which may affect neurite branching. It is hypothesized that these cellular responses are an attempt to encapsulate CNTs by surrounding them with detached apoptotic cells. This reaction maybe triggered by the similarity in size between CNTs and viruses, leading to the observed upregulation of several proteins associated with antiviral responses in CNT-exposed cells.	[33, 54, 62, 63, 65, 69, 70]
	Hydrophobicity of CNTs can be modified through oxidation processes, such as acid treatment or plasma etching, resulting in COOH-terminated CNTs. Some oxidized CNTs have been observed to exhibit greater toxicity compared to their untreated counterparts.	[64, 69, 71]
	CNTs are often contaminated with trace amounts of nickel or iron used as synthesis catalysts. These contaminants may contribute significantly to reactive oxygen species (ROS) production and oxidative stress.	[55, 62, 69, 70, 72]

Source	Challenges	References
Effects related to Surface Functionalization of CNTs	Surface functionalization of CNTs, including both the type and degree of functional groups, significantly influences their toxicity. Specifically, the charge of CNTs has been shown to affect neurite extension, density, and length.	[57]
	Cells exposed to CNTs with more positively charged surfaces exhibit increased neurite branching and a higher number of growth cones.	[56, 73]
	The polarity of CNTs seems to influence toxicity, with polar surfaces generally causing less toxicity and inducing higher neurite densities compared to nonpolar surfaces.	[52, 74, 75]
	The number of functionalization units on CNTs affects their toxicity, with an increasing number of units generally leading to reduced toxicity.	[49]
	The toxicity of CNTs may be attributed to an increased tendency for aggregation.	[57]
Toxicity related to CNT's Dimensions	Studies in the lung have established that, on a mass basis, the general trend of toxicity is SWNTs > MWNTs > C60 (fullerenes). The theory suggests that materials with the smallest surface area tend to be the most toxic.	[70, 76]
	In toxicity responses, longer CNTs are more likely to elicit an inflammatory reaction and are less readily engulfed by macrophages.	[48]
	The similarity in size and shape between CNTs and viruses may influence the toxicity response, potentially triggering immune responses. Small hydrophobic particles, resembling viruses, may provoke similar immune reactions.	[33, 70]
	CNT aggregates seem to loosely adhere to cells, potentially through interactions with extracellular proteins.	[77]
	CNT aggregates may affect the transport of nutrients and waste products across the cell membrane.	[69]

Table 1.
Different types of toxicity related to the CNTs in biomedical applications.

Owing to their distinctive properties, CNTs have found extensive applications in biotechnology, particularly in neuroscience. The ensuing discussion outlines key properties and capabilities inherent to these nanotubes.

2.2.2 CNT substrates for neural cell culture

CNTs show potential as substrates for promoting the growth of neural cells, potentially replacing traditional methods that require surface modifications with charged groups or biomolecules to facilitate neuron adhesion. Conductive polymers (CPs) like poly(3,4-ethylenedioxythiophene) (PEDOT) and polypyrrole (PPy) are considered neural interface coatings due to their superior electrical conductivity. However, their complex and occasionally toxic fabrication processes limit their effectiveness compared to naturally occurring factors [78–81]. Various techniques can modify surfaces with CNTs, such as direct growth, spray coating, and film deposition [82–84]. CNT-enhanced surfaces promote the adhesion and proliferation of various cell types, including neural cells [85], with the topography of the surface influenced by the length scales of CNTs playing a critical role in cellular responses [86]. Pristine CNTs, when purified but unprocessed, can accommodate various cell types, including neurons [87, 88]. However, poor solubility and potential toxic impurities in unmodified CNTs present challenges in cell-contact applications [89]. Furthermore, the patterning of substrates significantly influences neural cell behavior, including adhesion, spreading, and growth [90]. Diverse configurations of multi-walled carbon nanotube (MWCNT) arrangements, including cross-hatch, square, and spiral patterns, have undergone scrutiny regarding their influence on the extension and guidance of neurites [91]. Neurons exhibit an affinity for MWNT patterns owing to their nanoscale surface irregularities, fostering interactions and facilitating migration. Substrates based on CNTs possess the capability to both stimulate and record signals from neural cells, thereby maintaining the normal resting membrane potentials [92]. The intentional development of cellular networks with precise control finds applications in high-throughput assays, biosensors, and the creation of synthetic neural networks [93–95]. Patterned CNT substrates offer valuable insights into potential mechanisms of control [95]. Furthermore, CNTs exhibit remarkable electrical conductivity, rendering them adept at transmitting and receiving signals in neural applications [96]. Notably, researchers leveraged films of single-walled carbon nanotubes (SWCNTs) to concurrently culture and electrically stimulate neuroblastoma cells [97]. Such systems enable investigations into the interactions between neuronal cells and other cell types within co-culture environments.

2.2.3 Cell-CNT interactions

Interactions between cells and CNTs play a critical role in replicating normal neural cell functions, such as attachment, growth, maintenance of resting membrane potentials, and cell signaling. While natural extracellular matrix (ECM) proteins support the growth of neural cells, they lack electrical conductivity which is a crucial factor for many neural devices. CNTs have the potential to not only support cell growth but also to electrically integrate cells with their substrate, exhibiting conductivity when seeded with cells [98]. The conductive nature of CNTs makes them appealing coatings for neural culture substrates; however, the complete impact of conductivity on neural cell growth remains not entirely understood [99]. Surface modifications of CNTs using bioactive molecules can induce specific responses from neural cells [75, 100]. CNT-based substrates enable the manipulation of stimuli through nanotube properties, including nanoscale topography, surface charge, and bulk electrical conductivity. Various techniques for shaping and designing CNT substrates have been explored [101, 102]. In summary, the intricate interplay between CNTs and neural cells, encompassing electrical integration and surface modifications, holds significant promise for advancing neural culture substrates and devices.

2.2.4 Neural electrodes based on CNTs and CNFs

Due to the unique properties of CNTs and CNFs, including appropriate electrical conductivity, flexibility, small dimensions, and electrochemical properties, the development of neural electrodes based on these materials has also been considered. In recent studies, researchers constructed MEAs by incorporating CNTs into gold-based arrays (Au-MEAs) [103, 104]. The outcomes revealed a 50% reduction in overall electrode impedance at 1 kHz for CNT-modified-Au MEAs compared to bare Au-MEAs.

Moreover, the researchers showed the favorable characteristics of CNT-modified-Au MEAs, including high stability, durability, and scratch resistance, rendering them suitable for prolonged utilization in neural interfaces [103]. Recently, an ultrathin, flexible shape-changing electrode array (SCEA) was developed for large-scale electrocorticography (ECoG) mapping with minimal invasiveness [104]. **Figure 2** shows the prepared CNT-based SCEA for intracranial brain activity mapping.

The fabrication of large-scale CNT-based microelectrodes has also been undertaken for biomedical applications. In a recent study, a continuous dip-coating method was employed to coat 4.8 m long CNT fibers, ranging in diameter from 25 to 66 μm, with hydrogenated nitrile butadiene rubber (HNBR) [105]. The resulting HNBR-coated CNT fibers displayed uniformity, absence of pinholes, and compatibility with biological systems. Microelectrodes, utilizing these flexible HNBR-coated CNT fibers, demonstrated consistent electrochemical performance, featuring a specific impedance of approximately 27.0 MW μm^2 at 1.0 kHz and a cathodal charge storage capacity of around 487.6 mC cm^{-2} [105]. These electrodes exhibit attributes that render them suitable for integration into implantable medical devices for enduring *in vivo* applications.

In another work, researchers created flexible CNF electrodes designed for intraspinal recording [106]. They performed ongoing *in vivo* impedance monitoring and histology studies to investigate the key elements influencing the durability of the recording and to depict the development of the interface between the electrode and tissue. Using these CNF electrodes, they effectively captured single neuronal and local field potential activities in response to somatic mechanical stimulation from the spinal dorsal horns. Histological findings showed the flexible CNF microelectrodes' capability to establish enhanced intraspinal interfaces, exhibiting significantly diminished gliosis compared to rigid metal counterparts. Also, the CNTs and CNFs can be

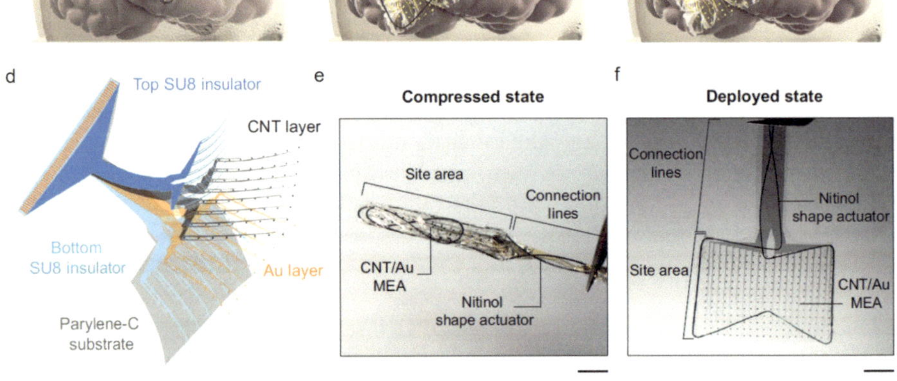

Figure 2.
(a-c) Schematic representation for prepared CNT-based neural interfaces for intracranial brain activity mapping. (d) expanded configuration of the prepared neural interface. (e, f) Optical images of the fabricated CNT-based neural interfaces in the compressed and deployed states [104].

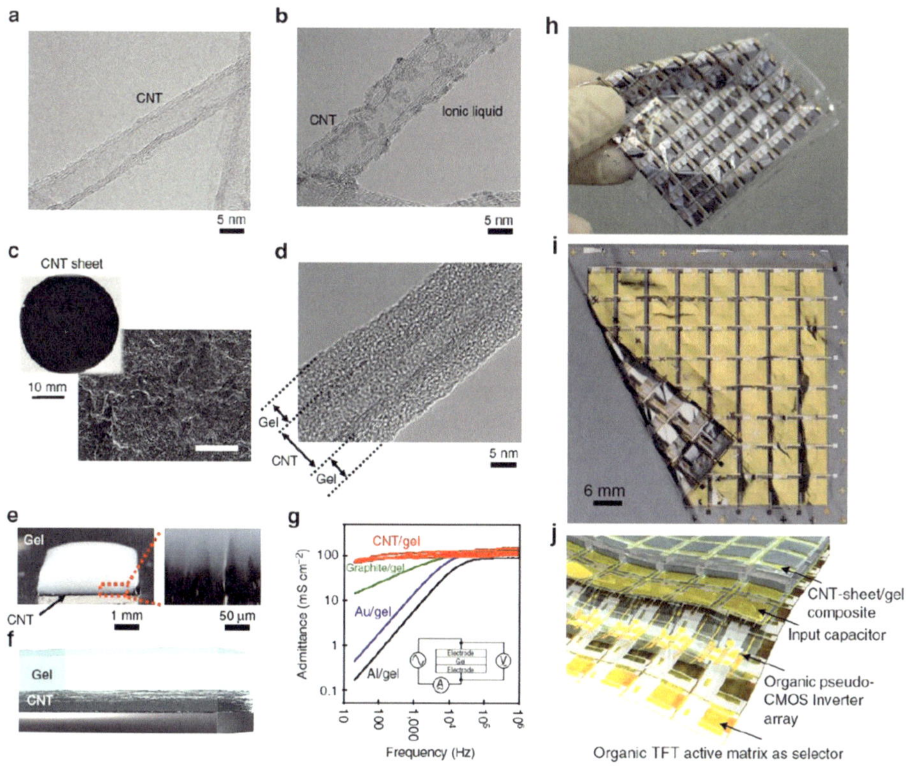

Figure 3.
(a-f) Neural electrode comprising of CNT/gel interface. (g) Admittance (mS cm^{-2}) of CNT/polyrotaxane gel in the vertical direction as a function of frequency, represented as a red line. (h, i) Optical images, and (j) schematic illustration of fabricated neural interfaces based on CNT/polyrotaxane gel [107].

appropriately integrated with gels to make ultra-flexible and highly conductive neural interfaces [107]. **Figure 3** shows an organic amplifier comprising CNT/gel interfaces developed to obtain biosignals with high spatiotemporal resolution and sensitivity.

2.3 Graphene neural interfaces

Graphene is a single layer of carbon atoms arranged in a hexagonal lattice, exhibiting exceptional electrical conductivity due to its sp^2 hybridized carbon-carbon bonds. Its remarkable mechanical strength, coupled with flexibility, positions graphene as a promising material for applications in areas such as flexible electronics and composite materials. The two-dimensional (2D) nature of graphene, with its high surface area, contributes to its potential in diverse fields, from energy storage to biomedical applications. Graphene-derived materials exhibit remarkable electrical and electrochemical characteristics, presenting potential solutions for overcoming challenges in the design of neural interfaces [108]. Thorough investigations have been conducted to assess their compatibility with neural cells and their capacity to interact with neuronal tissue. Experiments involving neural cell cultures on 2D graphene substrates have revealed substantial enhancements in adhesion, viability, neurite sprouting, and outgrowth [109, 110]. The evaluation of biodegradability is a critical aspect of implantable device design, and certain studies propose the enzymatic biodegradation of graphene nanomaterials [111]. Conducting long-term studies is imperative to comprehensively evaluate the biocompatibility, biodegradation, and safety of neural devices based on graphene.

Planar graphene substrates, synthesized *via* chemical vapor deposition (CVD), possess conductive properties that facilitate the electrical stimulation of human neural stem cells. This stimulation guides their differentiation toward a neuronal phenotype [112]. Moreover, three-dimensional (3D) scaffolds based on graphene induce stimulation in adult hippocampal neural stem cells, prompting their differentiation into astrocytes and neurons [113]. The implementation of porous hydrogels made of graphene oxide (GO) aids in the differentiation of embryonic neural progenitor cells into neurons and glial cells, thereby supporting the development of dendrites, axons, and synaptic connections [114]. These distinctive attributes empower the creation of devices for both electrical recording and stimulation. Additionally, there is potential for integrating the release of therapeutic molecules to alleviate the formation of fibrotic capsules around electrodes.

2.3.1 Graphene for neural stimulation

Efficient neural tissue stimulation is essential for various electrode applications, including deep brain stimulation (DBS), motor brain-computer interfaces, cochlear and retinal implants, and spinal cord stimulation. Electrical stimulation requires a specific charge injection capacity (CIC) depending on the target tissue, typically ranging from tens of $\mu C\ cm^{-2}$ to several $mC\ cm^{-2}$. Graphene-based materials, while exhibiting slightly lower interfacial capacitance than noble metals, show improved performance when used in porous thin films or polymer composites [115, 116]. Porous GO electrodes, achieved through laser treatment or Langmuir–Blodgett methods, offer exceptional charge injection values between 1 and 3 $mC\ cm^{-2}$ [117, 118]. Emerging materials like free-standing porous graphene and GO foams show potential for charge injection capabilities, though they require further testing [119, 120]. Structured or 3D graphene-based films and composite films hold promise for stimulation applications. **Figure 4** shows a fabricated graphene neural interface for neural imaging and optogenetic applications [10]. Moreover, neural interfaces hinge on electrodes' capacity to convert stimuli into electrical patterns sent to the brain. Beyond sensitivity to stimuli, electrodes must exhibit stability in operating conditions, efficient charge transfer to neurons, and refrain from altering the physiological properties of the target tissue. Consequently, researchers evaluated the impact of graphene-based substrates on brain cell cultures, assessing the functional and synaptic integrity of developing neuronal networks [121]. The findings demonstrated that graphene substrates served as permissive interfaces, maintaining unaltered neuronal signaling properties even without cell adhesion layers, making them suitable for carbon-based neural prosthetic devices.

2.3.2 Graphene for high-resolution neural recording

Monitoring the electrical activity of the brain demands high spatial and temporal resolution. Graphene's electrochemical capabilities for neural recording rival traditional materials like platinum (Pt) and gold (Au). In recent investigations, researchers have documented local field potentials originating in the rat cortex by employing graphene [10, 122]. Nevertheless, diminutive microelectrodes featuring single layer or few-layer graphene may encounter elevated impedance and thermal noise. The efficacy of recording can be improved by utilizing structured or 3D graphene films, which augment specific surface area. This enhancement in recording performance

DOI: http://dx.doi.org/10.5772/intechopen.1006603

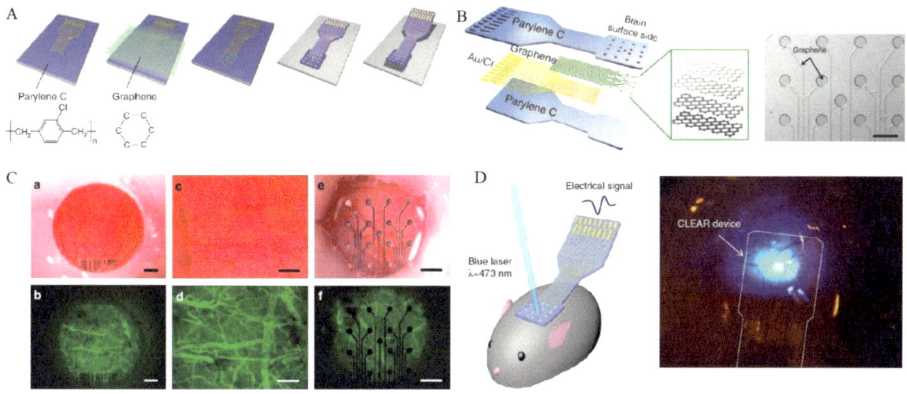

Figure 4.
(A) Fabrication of graphene neural interface, (B) expanded interface depicting Au connections, parylene C insulating layers, and graphene sensing pads. (C) Applying the prepared graphene neural interface on living tissue, and (D) Optogenetic stimulation and neural imaging using graphene neural interfaces [10].

has been demonstrated in the context of *in vivo* brain activity monitoring through the utilization of porous graphene-related materials [117, 118]. Graphene's advantage lies in its field-effect transistor (FET) configuration, which enables signal amplification and higher integration density compared to traditional electrodes. Flexible graphene FETs have demonstrated recording capabilities with a high SNR [123, 124].

To ensure the reliable functionality of bioelectronic devices, it is imperative to establish interfaces with biological tissues that are conformal, stable, and conductive. Traditionally, the integration of such devices with tissues has relied on physical attachment or surgical suturing. However, these approaches present challenges such as non-conformal contact, unstable fixation, tissue damage, and the potential for scar formation. In response to these challenges, researchers have innovatively developed an electrical bioadhesive (e-bioadhesive) interface. This interface is constructed using a thin layer of a graphene nanocomposite [125]. The unique properties of this e-bioadhesive facilitate rapid adhesion formation (within ~5 seconds), coupled with robust interfacial toughness exceeding 400 J m^{-2}. Importantly, this interface allows for on-demand detachable integration of bioelectronic devices onto various wet dynamic tissues. The electrical conductivity of the e-bioadhesive interface surpasses 2.6 S m^{-1}, enabling bidirectional bioelectronic communications. The researchers demonstrated the biocompatibility, applicability, mechanical and electrical stability, as well as recording and stimulation functionalities of the e-bioadhesive interface using *ex vivo* porcine and *in vivo* rat models [125]. These comprehensive findings present a promising and innovative strategy to enhance tissue–device integration, ultimately improving the overall performance of biointegrated electronic devices.

2.3.3 Therapeutic agent delivery by graphene

Graphene and its derivatives serve as nanoplatforms for controlled delivery of therapeutic compounds to specific cell populations. GO interacts with therapeutic compounds like doxorubicin and nucleic acids, enhancing versatility [126]. When integrated with neural interfaces, graphene flakes act as nanocarriers, enabling the controlled release of agents like dopamine, microglial inhibitors, and anti-inflammatory substances [127]. Electroresponsive graphene-based hydrogels for stimulation-triggered drug release have been proposed [128]. In these strategies, graphene sheets are incorporated into a hydrogel matrix to enhance mechanical, electrical, and thermal properties. Neural tissue engineering faces a significant challenge in finding suitable supporting materials that not only control cell behaviors but also facilitate the development of functional neural networks. Graphene, with its distinctive electrical and mechanical properties, emerges as a promising candidate for neural interfaces. Regenerative strategies in therapy aim to harness and enhance the inherent ability of neurons to reorganize by promoting controllable plasticity mechanisms. In this regard, researchers have explored how graphene influences the formation and functionality of neural networks in cultures of neural stem cells (NSCs) [129]. Through techniques such as calcium imaging and electrophysiological recordings, they have illustrated graphene's capacity to promote the growth of functional neural circuits, enhance neural performance, and improve electrical signaling within the network.

2.3.4 In situ biosensing in graphene neural interfaces

Graphene's electronic properties and its favorable surface-to-volume ratio make it an excellent candidate for biosensing in neural interfaces. Enzyme-modified graphene FETs are capable of detecting neurotransmitters like acetylcholine. Additionally, functionalized graphene and graphene/polymer composites are employed for the detection of dopamine and other biomolecules [130, 131]. Quantitative assessments of neurotransmitter functions in their natural environment offer valuable insights into the mechanisms governing stem cell differentiation, the establishment of neuronal networks, and the progression of neurodegenerative disorders. Presently, existing methods for detecting neurotransmitters face limitations such as inadequate spatial resolution, nonspecific detection, and a lack of *in situ* analytical capabilities. To overcome these challenges, researchers have devised a GO-hybrid nanosurface-enhanced Raman scattering (SERS) array specifically designed for dopamine (DA) detection. Through the utilization of this GO-hybrid nano-SERS array, they achieved swift and reliable measurements of a broad spectrum of DA concentrations ranging from 10^{-4} to 10^{-9} M [132]. Integrated graphene-based biosensors can enable closed-loop systems for adjusting stimulation or therapeutic molecule release based on biosensing data. Achieving specificity in target molecule recognition involves controlled chemical modifications of graphene for selective recognition of neurotransmitters and biomarkers.

The rising importance of electrophysiology in 3D neuronal cultures lies in its crucial role in uncovering cellular communications related to neurodevelopment and neurological diseases within a brain-like 3D environment. Despite the inherent modular architecture essential for cortical processing in the brain, interfacing a network consisting of multiple 3D neuronal tissues has proven to be a formidable challenge. In response, researchers have innovatively developed a self-folding graphene-based electrode array, facilitating the reconstruction of modular 3D neuronal tissue and the exploration of firing dynamics among these modules [133]. **Figure 5** illustrates the self-folding graphene-based interface fabricated for brain-like modular 3D tissue.

This approach involves culturing encapsulated cells within the folded graphene, leading to the spontaneous formation of 3D cell aggregates and ensuring a robust contact between the graphene surface and encapsulated cells. The use of the inner graphene surface as an electrode enables reliable, long-term electrical recording from multiple 3D aggregates, thanks to the secure cell–electrode contact. Significantly, the modular network formed by these multiple 3D aggregates displays more diverse firing patterns compared to a conventional homogeneous 2D network. This observation

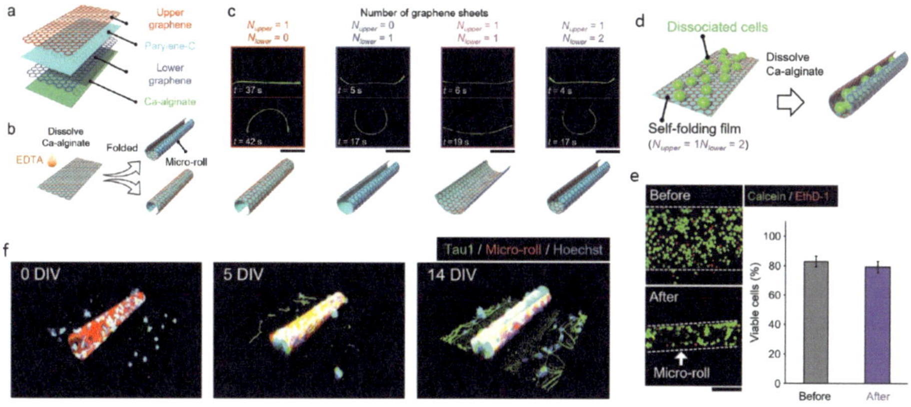

Figure 5.
(a) Schematic images of film composition, (b) folding process, (c) horizontal views of the self-folding graphene neural interface, (d) schematic images of cell encapsulation by self-folding graphene-sandwiched trilayer film, (e) viability assay by staining neurons with dyes of calcein (live; green) and ethidium homodimer-1 (EthD-1; dead; red) before and after folding, (f) 3D reconstructed images of encapsulated neurons at 0, 5, and 14 DIV [133].

underscores the effectiveness of the approach in enabling measurements of firing dynamics within intricate 3D neuronal networks. In fact, the development of a deformable graphene electrode serves as a potent platform for exploring the complexities of cellular communications in brain-like 3D cultures, highlighting its potential to advance our understanding of neurobiology. **Figure 6** shows MEAs fabricated based on the self-folding graphene neural interfaces.

2.4 Neural interfacing based on diamond

Human fascination with diamonds, rooted in their exceptional optical and physical properties and rare occurrence in nature, has persisted for millennia [134]. The 1950s marked a turning point with the onset of synthetic diamond production, notably through High-Pressure High-Temperature (HPHT) methods, enabling large-scale

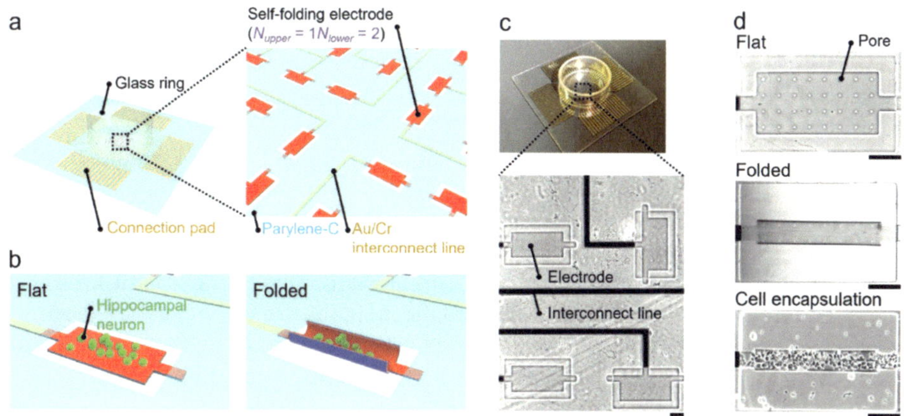

Figure 6.
Neuronal cell culture on self-folding graphene-based electrode array: schematic illustrations show (a) the configuration of the self-folding graphene MEA and (b) cell culture on the fabricated electrodes. (c, d) Optical images of the fabricated MEAs based on self-folding graphene [133].

manufacturing [135]. Affordable synthetic diamonds gained prominence for their ultra-hardness in cutting tools. In the 1980s, Japanese scientists optimized chemical vapor deposition (CVD) for diamond production, offering versatility in growing microcrystalline or nanocrystalline films with applications such as wear-resistant coatings due to diamond's low chemical reactivity [136, 137]. Medical applications emerged with diamond-coated artificial hip joint components, cardiovascular stents, and dentistry tools, backed by extensive investigations demonstrating negligible cytotoxicity of diamond-like coatings [138]. Nanodiamonds explored for drug delivery exhibit biocompatibility, with surface termination influencing suitability [139]. Diamond-like coatings extend to diverse fields like food packaging and plastic materials [140]. The electronic properties of diamond, from its insulating nature to conductivity achieved through boron doping, expanded its applications [141, 142]. Recognizing diamond's potential in neuro-modulation devices, despite challenges due to extreme hardness, has grown over the past decade. Initially, as a growth substrate for various cells, diamonds consistently showed non-cytotoxicity and enhanced cell viability [143–145]. Recent advancements include prototype implantable devices, often featuring diamond electrodes, notably in retinal prostheses where diamond's suitability for hermetic encapsulation of implanted electronics is under exploration [146, 147].

2.4.1 Synthesis methods and products

The High-Pressure High-Temperature (HPHT) synthesis method replicates the natural diamond formation process. In this procedure, seed diamonds are positioned within a chamber containing high-purity carbon and a metal alloy solvent. The chamber undergoes heating to approximately 1400°C, causing the metal solvent to undergo melting. Subsequently, hydraulic pistons apply high pressure to the chamber, resulting in the dissolution and diffusion of carbon from the high-purity source through the molten metal solvent. Upon the release of pressure, the dissolved carbon precipitates onto the diamond seed, adopting its crystal structure. A common treatment for HPHT diamonds involves irradiation, inducing lattice defects, and generating distinct color effects [148]. Despite the thermodynamic stability of graphite being higher than that of diamond, the elevated atomic density of diamond, particularly in tetrahedral stacking, favors its formation under extremely high pressure. HPHT techniques extend beyond the realm of synthetic diamonds, finding application in the modification of natural diamonds to enhance clarity or alter color. Furthermore, the intentional incorporation of dopants, such as boron, into the carbon source facilitates the production of electrically conducting HPHT diamonds [149].

The Chemical Vapor Deposition (CVD) technique, similar to HPHT, employs diamond seeds. Activated carbon originates from a carbon source onto these seeds, leading to the growth of the crystal. The CVD needs a diamond seed, a feed gas containing carbon, and a means to activate the mixture of feed gas. Atomic hydrogen in the gas mixture plays a crucial role in determining diamond formation over graphitic forms of carbon. Different carbon sources, from mundane methane to exotic ones, have been used for CVD diamond growth. Two common CVD methods are hot-filament CVD and plasma-enhanced CVD (PECVD), each with its specific approach [150, 151].

2.4.2 Electrochemical properties of diamond and diamond films

When evaluating a novel electrode material for electrochemical purposes, essential factors encompass corrosion resistance, background current, and the usable potential window. Boron-doped diamond (BDD) demonstrates exceptionally elevated corrosion resistance. The background current, which makes a significant noise source, is low in BDD due to its low electrochemical double-layer capacitance. BDD also offers a wide electrochemical window in aqueous electrolyte solutions, making it suitable for detecting analytes across various voltages without interference from faradaic reactions. The low capacitance of BDD contributes to obtaining low-noise electrodes [152].

Nitrogen-included ultra-nanocrystalline diamond (N-UNCD), consisting of small diamond crystals in a non-diamond carbon matrix, exhibits electrochemical properties influenced by nitrogen concentration during growth. Despite containing non-diamond carbon, N-UNCD shows diamond-like electrochemical behavior. The electrochemical capacitance of N-UNCD can be altered by growth conditions, such as nitrogen concentration and substrate temperature. Careful control of growth conditions is crucial for achieving repeatable material properties in CVD-grown N-UNCD [153, 154].

2.4.3 Growth of neurons on diamond (neuron-diamond interactions)

Diamond's chemical inertness renders it biocompatible, making it appealing for various biomedical applications, including coating medical devices and supporting cell growth. Notably, diamond has proven effective as a substrate for cultivating diverse cell types, such as neurons, fibroblasts, and osteoblasts, without inducing cytotoxicity [155–157]. For electrodes to function successfully, establishing a nontoxic interface with local neurons is vital, particularly for prolonged use. Thus, exploring the diamond's role as a growth substrate becomes crucial for maintaining electrical and chemical functionality *in vivo*. As we know, surface properties, which are influenced by fabrication methods and post-processing steps, play a pivotal role in neural interfacing. However, the sp^2/sp^3 content, as a variable affecting cell viability on diamond-like carbon, has not been extensively explored for neurons [158]. Surface termination, such as hydrogen and oxygen termination, significantly impacts neural adhesion and proliferation [159–161]. Oxygen-terminated surfaces often exhibit better neuronal attachment, while conflicting results exist regarding the necessity of protein coatings [160–162]. Moreover, the introduction of elements like phosphorus, boron, and nitrogen into diamond alters conductivity and impacts neuronal growth [161, 163]. While phosphorus-doped diamond supports better survival of cortical neurons, boron-doped and nitrogen-doped diamonds show positive effects on cell survival, neurite formation, and apoptosis levels [161, 163, 164].

Surface morphology affected by factors like diamond microcrystal size, nanorod structures, and 3D nanostructured substrates, influences neuronal attachment and behavior [165, 166]. Neurons exhibit preferences for surfaces mimicking their biological environment, highlighting the importance of adhesion promoter coatings [163, 166]. Neurons respond to biomaterial properties, and the debate persists on whether adhesion promoter coatings such as poly-D-lysine, poly-DL-ornithine, or laminin are essential for neuron attachment and proliferation on diamond films. Conflicting evidence exists, with studies reporting varied outcomes [165, 167, 168]. The role of surface roughness, especially in homoepitaxial diamonds, is intricate. While solid nanocrystalline diamond (NCD) films may not promote neuronal adhesion, isolated nanodiamonds exhibit favorable interactions, suggesting subcellular interactions are crucial [168]. Studies highlight that responses vary across diamond types, including microcrystalline, NCD, and UNCD. The need for adhesion promoter proteins depends on the specific type of diamond, with conflicting evidence on the survival and functionality of neurons on bare diamond surfaces [165, 167, 169].

2.4.4 Fabrication methods for diamond electrodes and diamond microelectrode arrays (MEAs)

Diamond, renowned for its extraordinary hardness, chemical inertness, and resistance to melting, poses significant challenges in the fabrication of microelectrodes and microelectrode arrays (MEAs) for neural interfacing applications. While the production of flat diamond films is relatively straightforward, the demands of neural interfacing require intricate microscale structures. Several methods have been developed and extensively reviewed [170, 171]. Here, we highlight a select few examples to elucidate the most commonly employed techniques.

Among different configurations of diamond-based neural interfaces, standalone diamond electrodes and diamond-based Multi-Electrode Arrays (diamond-MEAs), exhibit considerable potential for diverse neuroscientific applications. Standalone diamond electrodes are frequently crafted through the CVD of diamonds on a high melting temperature metal substrate. Researchers [172] have conducted significant work, showing the application of boron-doped polycrystalline diamond (PCD) on an electrochemically sharpened platinum wire. Tungsten wire is also commonly used as a substrate for these diamond electrodes [173]. The formation of MEAs typically involves selectively removing material from a continuous conductive diamond film, leading to discrete islands or channels [174]. Researchers have introduced an innovative technique employing a laser milling apparatus to produce a nano-ultrananocrystalline diamond (N-UNCD) microelectrode array [147]. In that work, individual electrodes were isolated by selectively removing sections of the film through laser milling. The insulating polycrystalline diamond (PCD) serves as the substrate, and the vias created by laser milling act as hermetic electrical feed-throughs for the electrode array. This design facilitates direct bonding to microelectronics. Additionally, they proposed an alternative method involving the growth of a diamond film in an etched silicone template to mold diamond spikes [175]. Another noteworthy approach to fabricating diamond-based neural electrodes utilizes laser ablation [176]. In this method, a standalone film of boron-doped diamond (BDD) is laser ablated through a mask, generating an array of BDD columns. Subsequently, insulating PCD is grown over the structure, and a final polishing step reveals the BDD columns penetrating through the PCD [176]. An unconventional yet remarkable fabrication method involves patterning the diamond seed film before CVD. This results in a patterned diamond film directly from the CVD chamber [146, 177–179].

2.4.5 Neurochemical sensors based on diamond

Diamond's broad electrochemical window and low background current make it a compelling material for neurochemical sensing. Neurotransmitters, which are crucial for signal transmission in neural networks, are central to normal neural function, and abnormalities in their function are linked to various diseases. Common targets for electrochemical detection include biogenic amines such as catecholamines (dopamine, norepinephrine, epinephrine) and serotonin. Dopamine, abundant among catecholamines, plays a key role in brain function and is associated with conditions like schizophrenia and attention deficit hyperactive disorder (ADHD). Serotonin, derived from tryptophan, is linked to depression. Electrochemical detection involves oxidizing these molecules to their corresponding ketones, a process often hindered by the formation of insoluble dimers, reducing electrode sensitivity. Studies have revealed that Boron-doped Diamond (BDD) outperforms glassy carbon but lags behind carbon nanotube (CNT) films in serotonin detection sensitivity [180]. However, BDD electrodes demonstrate superior resistance to fouling over multiple scans, which could be a critical advantage. Researchers establish a low detection limit of 44 nM for dopamine using BDD microelectrodes, even in the presence of interfering ascorbic acid. Surface modifications, such as carboxy functionalities, enhance sensitivity by preventing fouling. N-UNCD has been also reported as highly sensitive to dopamine detection. Some works showed a diamond field-effect transistor (FET) as a sensor for acetylcholine [181], highlighting the diverse applications of diamond in neurotransmitter detection. Furthermore, the diamond's chemical stability and biocompatibility position it as an ideal material for long-lasting implantable sensor electrodes. In living systems, BDD electrodes effectively map the exocytosis of catecholamines from single cells, illustrating its potential for real-time monitoring in neural environments. Researchers have demonstrated *in vivo* detection of norepinephrine in rat intestine sections, emphasizing the antifouling performance of BDD electrodes over extended periods. They detected nitric oxide released by neurons using the fabricated electrodes, showing the versatility of BDD electrodes in detecting various species. In addition, researchers used BDD electrodes successfully to detect dopamine in different animal models [182–184]. The studies advanced to awake-behaving Japanese monkeys, demonstrating BDD's effectiveness in recording dopamine expression in response to stimuli [185]. This approach strengthens the reliability of BDD electrodes in neurochemical detection.

2.4.6 Neural recording and stimulation

Recording electrodes necessitates overcoming electrochemical capacitance and electrical impedance. Electrode parameters like size, shape, and material intricately affect recording quality. For example, smaller electrodes, although have higher impedance values, can sample nearby firing neurons, facilitating single-unit recording. On the other side, large electrodes, which show lower impedances and are influenced by a broader electrical environment, can record from a larger neuron population, termed local field potentials (LFPs). Diamond electrodes, predominantly boron-doped diamond (BDD), are commonly used for neural recording [186]. Researchers showed high-quality action potential recordings using high-purity, single-crystal, hydrogen-terminated diamonds. Despite the diamond's electrical insulation, hydrogen termination provides limited conductivity *via* surface transfer doping [187]. In addition, researchers developed a diamond transistor array for recording membrane potentials [181, 188]. Halpern et al. demonstrated a BDD electrode in a Sea Hare, exhibiting improved signal-to-noise ratio (SNR) and resistance to fouling compared to stainless steel [173]. Piret et al. used a diamond microelectrode array (MEA) to record hindbrain activity, demonstrating enhanced performance with nanostructured BDD [166]. **Figure 7** shows a diamond-based electrode for electrochemical sensing and neural recording.

For neural devices employing diamonds, the scope for advancements goes beyond conventional electronic and electrochemical methods. Innovations in optical techniques utilizing inherent defect centers in diamonds show promise for developing multi-magnetometer arrays (MMAs). These MMAs, rooted in noninvasive high-resolution technology, offer a compelling avenue for visualizing magnetic fields

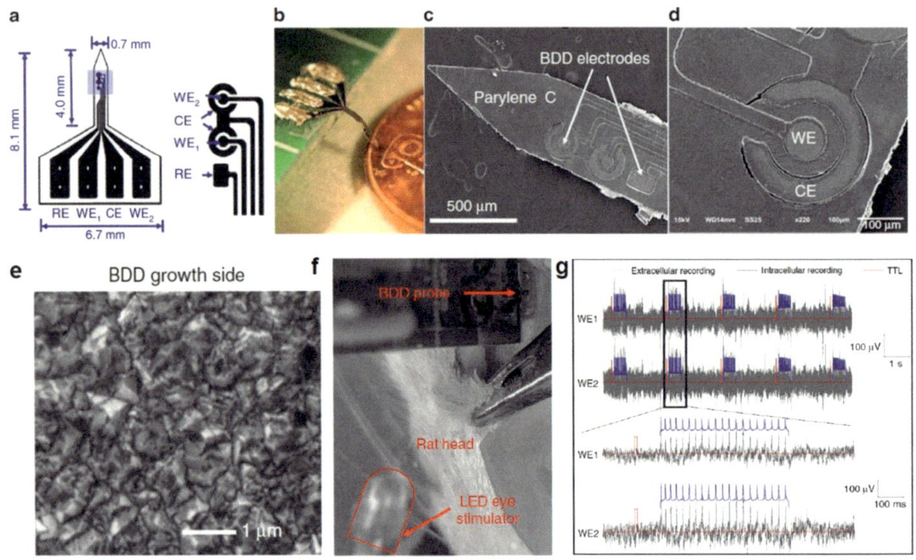

Figure 7.
(a-d) Detailed configuration and FESEM images of the diamond-based neural electrode, (e) FESEM image of diamond layer microstructure. (f) Application of the prepared diamond electrode to an animal model, and (g) recorded electrophysiological signals [189].

arising from transmembrane potentials generated by neuronal activity [190]. The key to this technology lies in a specific defect center in diamonds known as the negatively charged nitrogen vacancy (NV) center [191]. The introduction of this defect into the diamond lattice, achieved through either nitrogen incorporation during the gas phase of CVD growth or via nitrogen ion implantation, has led to the creation of 2D magnetic imaging arrays. These arrays have proven instrumental in observing and reconstructing stray magnetic fields from current-carrying wires [192] and magnetic nanoparticles within living organisms [193]. The methodology for detecting neuronal activity through 2D MMA imaging relies on optically detected magnetic resonance (ODMR) of the NV magnetic sublevels. By employing wide-field microscopy, ODMR from diffraction-limited sensing volumes can be imaged with millisecond timing over extensive fields of view [194]. Although recent progress has expanded the field of view to mm^2, this expansion comes at the expense of spatial resolution [195]. The integration of magnetic sensors into the diamond substrate enables the use of total internal reflection microscopy (TIRF) to induce and report changes in the local magnetic field. This TIRF approach mitigates the photo-toxic effects of excitation light, ensuring a noninvasive methodology. Optical imaging not only facilitates swift and parallel signal acquisition but also holds the potential for integration with other optical techniques, such as optogenetics. This integration could establish a fully unified 2D solution for both optical stimulation and recording. The advantages of magnetic-based detection are notable, including the absence of local charge screening observed in electrical recording and the potential for higher spatial resolution due to the rapid decline of extracellular magnetic signals compared to complementary electrical signals [190]. Achieving proximity of NV magnetic sensors to the source of the magnetic field is crucial, and this can be accomplished by either directly cultivating the neuronal network on the diamond MMA or coating the homoepitaxial diamond with a thin adhesion layer to promote growth.

3. Conclusions and future outlook

Glassy carbon's high mechanical strength and chemical stability make it promising for neural interfaces. Its inert nature minimizes biological reactions and tissue irritation. Enhancing biocompatibility and optimizing electrical properties are future goals, with advances in surface modification and doping. Challenges include seamless integration with softer tissues and long-term in vivo stability, requiring innovative surface engineering and composite designs. On the other side, CNTs and CNFs offer exceptional electrical conductivity and mechanical strength, crucial for next-generation neural interfaces. Future research will focus on functionalization to enhance biocompatibility and reduce cytotoxicity. Scalable synthesis and precise alignment are critical. Challenges include dispersion control in biological environments and potential long-term health risks. Graphene's superior conductivity, mechanical properties, and flexibility make it promising for neural interfaces. Future goals include developing flexible, transparent, biocompatible graphene-based electrodes. Advances in fabrication and functionalization are expected. Challenges include large-scale production and managing biological interactions, such as inflammation. Synthetic diamonds, particularly doped with boron or nitrogen, offer appropriate conductivity and biocompatibility. They are stable for stimulating and recording neural activity. Future developments will optimize doping and scalable production. Challenges include high production costs and integrating diamonds with flexible substrates. Effective neural stimulation requires addressing low electrochemical capacitance through nanostructuring or heavy doping.

In general, the integration of carbon nanomaterials into neural interfaces faces several overarching challenges: (1) *Biocompatibility*: Ensuring that carbon nanomaterials do not induce inflammatory responses or cytotoxicity is critical. This requires extensive in vitro and *in vivo* testing to evaluate long-term interactions with neural tissues. (2) *Scalability*: The production of high-quality carbon nanomaterials at a scale sufficient for practical applications remains a significant hurdle. Advances in synthesis and processing techniques are needed to make these materials more accessible. (3) *Material stability*: The stability of carbon nanomaterials in the biological environment is crucial. They must retain their properties and functionality over extended periods *in vivo*. (4) *Integration with biological systems*: Achieving seamless integration with neural tissues involves addressing challenges related to the mechanical mismatch between the materials and soft biological tissues. Techniques to improve adhesion and reduce tissue response are essential. (5) *Ethical and safety concerns*: The long-term effects of carbon nanomaterials in medical applications must be carefully evaluated to address any potential health risks. Comprehensive safety assessments are required to ensure that these materials do not pose unforeseen risks to patients.

Conflict of interest

The authors declare no conflict of interest.

Author details

Shahab Ahmadi Seyedkhani
Sharif University of Technology, Tehran, Iran

*Address all correspondence to: ahmadiseyedkhani@gmail.com

References

[1] Voge CM, Stegemann JP. Carbon nanotubes in neural interfacing applications. Journal of Neural Engineering. 2011;**8**:11001

[2] Devi M, Vomero M, Fuhrer E, Castagnola E, Gueli C, Nimbalkar S, et al. Carbon-based neural electrodes: Promises and challenges. Journal of Neural Engineering. 2021;**18**:41007

[3] Seyedkhani SA, Mohammadpour R. In: Elnashar MMM, Karakuş S, editors. Perspective Chapter: Tissue-Electronics Interfaces. London, UK, Rijeka: IntechOpen; 2022. Ch. 6. DOI: 10.5772/intechopen.108129

[4] Kroto HW, Heath JR, O'Brien SC, Curl RF, Smalley RE. C60: Buckminsterfullerene. Nature. 1985;**318**:162-163. DOI: 10.1038/318162a0

[5] Chen J, Wei S, Xie H. A brief introduction of carbon nanotubes: History, synthesis, and properties. Journal of Physics Conference Series. 2021;**1948**:12184. DOI: 10.1088/1742-6596/1948/1/012184

[6] Geim AK, Novoselov KS. The rise of graphene. Nature Materials. 2007;**6**:183-191

[7] Ahmadi Seyedkhani S, Mohammadpour R. In: Elnashar PM, editor. Tissue-electronics Interfaces. London, UK, Rijeka: IntechOpen; 2022. Ch. 1. DOI: 10.5772/intechopen.108129

[8] Baldrighi M, Trusel M, Tonini R, Giordani S. Carbon nanomaterials interfacing with neurons: An in vivo perspective. Frontiers in Neuroscience. 2016;**10**:250

[9] Fraczek-Szczypta A. Carbon nanomaterials for nerve tissue stimulation and regeneration. Materials Science and Engineering: C. 2014;**34**:35-49

[10] Park D-W, Schendel AA, Mikael S, Brodnick SK, Richner TJ, Ness JP, et al. Graphene-based carbon-layered electrode array technology for neural imaging and optogenetic applications. Nature Communications. 2014;**5**:5258. DOI: 10.1038/ncomms6258

[11] Liu Y, Duan X. Carbon-based nanomaterials for neural electrode technology. Acta Physico-Chimica Sinica. 2020;**36**:95-107

[12] Li J, Cheng Y, Gu M, Yang Z, Zhan L, Du Z. Sensing and stimulation applications of carbon nanomaterials in implantable brain-computer interface. International Journal of Molecular Sciences. 2023;**24**:5182

[13] Sisubalan N, Shalini R, Ramya S, Sivamaruthi BS, Chaiyasut C. Recent advances in nanomaterials for neural applications: Opportunities and challenges. Nanomedicine. 2023;**18**:1979-1994

[14] Wang Y, Pham L, de Vasconcellos GPS, Madou M. Fabrication and characterization of micro PEM fuel cells using pyrolyzed carbon current collector plates. Journal of Power Sources. 2010;**195**:4796-4803

[15] Lin P-C, Park BY, Madou MJ. Development and characterization of a miniature PEM fuel cell stack with carbon bipolar plates. Journal of Power Sources. 2008;**176**:207-214

[16] Kassegne S, Moon K, Martín-Ramos P, Majzoub M, Őzturk G, Desai K, et al. Organic MEMS/

NEMS-based high-efficiency 3D ITO-less flexible photovoltaic cells. Journal of Micromechanics and Microengineering. 2012;**22**:115015

[17] Martinez-Duarte R, Renaud P, Madou MJ. A novel approach to dielectrophoresis using carbon electrodes. Electrophoresis. 2011;**32**:2385-2392

[18] Martinez-Duarte R, Gorkin RA III, Abi-Samra K, Madou MJ. The integration of 3D carbon-electrode dielectrophoresis on a CD-like centrifugal microfluidic platform. Lab on a Chip. 2010;**10**:1030-1043

[19] Hirabayashi M, Mehta B, Nguyen B, Kassegne S. DNA immobilization on high aspect ratio glassy carbon (GC-MEMS) microelectrodes for bionanoelectronics applications. Microsystem Technologies. 2015;**21**:2359-2365

[20] Vahidi NW, Hirabayashi M, Mehta B, Rayatparvar M, Wibowo D, Ramesh V, et al. Bionanoelectronics platform with DNA molecular wires attached to high aspect-ratio 3D metal microelectrodes. ECS Journal of Solid State Science and Technology. 2013;**3**:Q29

[21] Kassegne S, Vomero M, Gavuglio R, Hirabayashi M, Özyilmaz E, Nguyen S, et al. Electrical impedance, electrochemistry, mechanical stiffness, and hardness tunability in glassy carbon MEMS μECoG electrodes. Microelectronic Engineering. 2015;**133**:36-44

[22] Vomero M, Van Niekerk P, Nguyen V, Gong N, Hirabayashi M, Cinopri A, et al. A novel pattern transfer technique for mounting glassy carbon microelectrodes on polymeric flexible substrates. Journal of Micromechanics and Microengineering. 2016;**26**:25018

[23] VanDersarl JJ, Mercanzini A, Renaud P. Integration of 2D and 3D thin film glassy carbon electrode arrays for electrochemical dopamine sensing in flexible neuroelectronic implants. Advanced Functional Materials. 2015;**25**:78-84

[24] Vomero M, Castagnola E, Ciarpella F, Maggiolini E, Goshi N, Zucchini E, et al. Highly stable glassy carbon interfaces for long-term neural stimulation and low-noise recording of brain activity. Scientific Reports. 2017;**7**:40332. DOI: 10.1038/srep40332

[25] Vomero M, Zucchini E, Delfino E, Gueli C, Mondragon NC, Carli S, et al. Glassy carbon electrocorticography electrodes on ultra-thin and finger-like polyimide substrate: Performance evaluation based on different electrode diameters. Materials (Basel). 2018;**11**:2486

[26] Lin C-M, Lee Y-T, Yeh S-R, Fang W. Flexible carbon nanotubes electrode for neural recording. Biosensors & Bioelectronics. 2009;**24**:2791-2797

[27] Fernando KAS, Lin Y, Sun Y-P. High aqueous solubility of functionalized single-walled carbon nanotubes. Langmuir. 2004;**20**:4777-4778. DOI: 10.1021/la036217z

[28] Katz E, Willner I. Biomolecule-functionalized carbon nanotubes: Applications in nanobioelectronics. ChemPhysChem. 2004;**5**:1084-1104. DOI: 10.1002/cphc.200400193

[29] Kimura T, Ago H, Tobita M, Ohshima S, Kyotani M, Yumura M. Polymer composites of carbon nanotubes aligned by a magnetic field. Advanced Materials. 2002;**14**:1380-1383. DOI: 10.1002/1521-4095 (20021002)14:19<1380::AID-ADMA1380>3.0.CO;2-V

[30] Alexander JK, Fuss B, Colello RJ. Electric field-induced astrocyte alignment directs neurite outgrowth. Neuron Glia Biology. 2006;**2**:93-103. DOI: 10.1017/S1740925X0600010X

[31] Warheit DB, Laurence BR, Reed KL, Roach DH, Reynolds GAM, Webb TR. Comparative pulmonary toxicity assessment of single-wall carbon nanotubes in rats. Toxicological Sciences. 2004;**77**:117-125. DOI: 10.1093/toxsci/kfg228

[32] Poland CA, Duffin R, Kinloch I, Maynard A, Wallace WAH, Seaton A, et al. Carbon nanotubes introduced into the abdominal cavity of mice show asbestos-like pathogenicity in a pilot study. Nature Nanotechnology. 2008;**3**:423-428. DOI: 10.1038/nnano.2008.111

[33] Ding L, Stilwell J, Zhang T, Elboudwarej O, Jiang H, Selegue JP, et al. Molecular characterization of the cytotoxic mechanism of multiwall carbon nanotubes and Nano-onions on human skin fibroblast. Nano Letters. 2005;**5**:2448-2464. DOI: 10.1021/nl051748o

[34] Crouzier T, Nimmagadda A, Nollert MU, McFetridge PS. Modification of single walled carbon nanotube surface chemistry to improve aqueous solubility and enhance cellular interactions. Langmuir. 2008;**24**:13173-13181. DOI: 10.1021/la801999n

[35] Zhao Y-L, Stoddart JF. Noncovalent functionalization of single-walled carbon nanotubes. Accounts of Chemical Research. 2009;**42**:1161-1171. DOI: 10.1021/ar900056z

[36] Ren Y-J, Zhang H, Huang H, Wang X-M, Zhou Z-Y, Cui F-Z, et al. In vitro behavior of neural stem cells in response to different chemical functional groups. Biomaterials. 2009;**30**:1036-1044. DOI: 10.1016/j.biomaterials.2008.10.028

[37] Xie J, Willerth SM, Li X, Macewan MR, Rader A, Sakiyama-Elbert SE, et al. The differentiation of embryonic stem cells seeded on electrospun nanofibers into neural lineages. Biomaterials. 2009;**30**:354-362. DOI: 10.1016/j.biomaterials.2008.09.046

[38] Jan E, Kotov NA. Successful differentiation of mouse neural stem cells on layer-by-layer assembled single-walled carbon nanotube composite. Nano Letters. 2007;**7**:1123-1128. DOI: 10.1021/nl0620132

[39] Kam NWS, Jan E, Kotov NA. Electrical stimulation of neural stem cells mediated by humanized carbon nanotube composite made with extracellular matrix protein. Nano Letters. 2009;**9**:273-278. DOI: 10.1021/nl802859a

[40] Chao T-I, Xiang S, Chen C-S, Chin W-C, Nelson AJ, Wang C, et al. Carbon nanotubes promote neuron differentiation from human embryonic stem cells. Biochemical and Biophysical Research Communications. 2009;**384**:426-430. DOI: 10.1016/j.bbrc.2009.04.157

[41] Chao T-I, Xiang S, Lipstate JF, Wang C, Lu J. Poly(methacrylic acid)-grafted carbon nanotube scaffolds enhance differentiation of hESCs into neuronal cells. Advanced Materials. 2010;**22**:3542-3547. DOI: 10.1002/adma.201000262

[42] Sridharan I, Kim T, Wang R. Adapting collagen/CNT matrix in directing hESC differentiation. Biochemical and Biophysical Research Communications. 2009;**381**:508-512. DOI: 10.1016/j.bbrc.2009.02.072

[43] Yamada M, Tanemura K, Okada S, Iwanami A, Nakamura M, Mizuno H, et al. Electrical stimulation modulates fate determination of differentiating embryonic stem cells. Stem Cells. 2007;**25**:562-570. DOI: 10.1634/stemcells.2006-0011

[44] Gilmore JL, Yi X, Quan L, Kabanov AV. Novel nanomaterials for clinical neuroscience. Journal of Neuroimmune Pharmacology. 2008;**3**:83-94. DOI: 10.1007/s11481-007-9099-6

[45] Lacerda L, Bianco A, Prato M, Kostarelos K. Carbon nanotubes as nanomedicines: From toxicology to pharmacology. Advanced Drug Delivery Reviews. 2006;**58**:1460-1470. DOI: 10.1016/j.addr.2006.09.015

[46] Shvedova AA, Castranova V, Kisin ER, Schwegler-Berry D, Murray AR, Gandelsman VZ, et al. Exposure to carbon nanotube material: Assessment of nanotube cytotoxicity using human keratinocyte cells. Journal of Toxicology and Environmental Health. Part A. 2003;**66**:1909-1926. DOI: 10.1080/713853956

[47] Lam C-W, James JT, McCluskey R, Hunter RL. Pulmonary toxicity of single-wall carbon nanotubes in mice 7 and 90 days after intratracheal instillation. Toxicological Sciences. 2004;**77**:126-134. DOI: 10.1093/toxsci/kfg243

[48] Sato Y, Yokoyama A, Shibata K, Akimoto Y, Ogino S, Nodasaka Y, et al. Influence of length on cytotoxicity of multi-walled carbon nanotubes against human acute monocytic leukemia cell line THP-1 in vitro and subcutaneous tissue of rats in vivo. Molecular BioSystems. 2005;**1**:176-182. DOI: 10.1039/B502429C

[49] Sayes CM, Liang F, Hudson JL, Mendez J, Guo W, Beach JM, et al. Functionalization density dependence of single-walled carbon nanotubes cytotoxicity in vitro. Toxicology Letters. 2006;**161**:135-142. DOI: 10.1016/j.toxlet.2005.08.011

[50] Oberdörster E. Manufactured nanomaterials (fullerenes, C60) induce oxidative stress in the brain of juvenile largemouth bass. Environmental Health Perspectives. 2004;**112**:1058-1062. DOI: 10.1289/ehp.7021

[51] Smith CJ, Shaw BJ, Handy RD. Toxicity of single walled carbon nanotubes to rainbow trout, (Oncorhynchus mykiss): Respiratory toxicity, organ pathologies, and other physiological effects. Aquatic Toxicology. 2007;**82**:94-109. DOI: 10.1016/j.aquatox.2007.02.003

[52] Liopo AV, Stewart MP, Hudson J, Tour JM, Pappas TC. Biocompatibility of native and functionalized single-walled carbon nanotubes for neuronal interface. Journal of Nanoscience and Nanotechnology. 2006;**6**:1365-1374. DOI: 10.1166/jnn.2006.155

[53] Webster TJ, Waid MC, McKenzie JL, Price RL, Ejiofor JU. Nano-biotechnology: Carbon nanofibres as improved neural and orthopaedic implants. Nanotechnology. 2004;**15**:9

[54] Ni Y, Hu H, Malarkey EB, Zhao B, Montana V, Haddon RC, et al. Chemically functionalized water soluble single-walled carbon nanotubes modulate neurite outgrowth. Journal of Nanoscience and Nanotechnology. 2005;**5**:1707-1712

[55] Helland A, Wick P, Koehler A, Schmid K, Som C. Reviewing the environmental and human health knowledge base of carbon nanotubes. Ciência & Saúde Coletiva. 2008;**13**:441-452

[56] Hu H, Ni Y, Mandal SK, Montana V, Zhao B, Haddon RC, et al. Polyethyleneimine functionalized single-walled carbon nanotubes as a substrate for neuronal growth. The Journal of Physical Chemistry. B. 2005;**109**:4285-4289

[57] Kotov NA, Winter JO, Clements IP, Jan E, Timko BP, Campidelli S, et al. Nanomaterials for neural interfaces. Advanced Materials. 2009;**21**:3970-4004

[58] Huczko A, Lange H. Carbon nanotubes: Experimental evidence for a null risk of skin irritation and allergy. Fullerene Science and Technology. 2001;**9**:247-250

[59] Sharma CS, Sarkar S, Periyakaruppan A, Barr J, Wise K, Thomas R, et al. Single-walled carbon nanotubes induces oxidative stress in rat lung epithelial cells. Journal of Nanoscience and Nanotechnology. 2007;7:2466-2472

[60] Oberdörster G, Oberdörster E, Oberdörster J. Nanotoxicology: An emerging discipline evolving from studies of ultrafine particles. Environmental Health Perspectives. 2005;**113**:823-839

[61] Manna SK, Sarkar S, Barr J, Wise K, Barrera EV, Jejelowo O, et al. Single-walled carbon nanotube induces oxidative stress and activates nuclear transcription factor-κB in human keratinocytes. Nano Letters. 2005;**5**:1676-1684

[62] Pulskamp K, Diabaté S, Krug HF. Carbon nanotubes show no sign of acute toxicity but induce intracellular reactive oxygen species in dependence on contaminants. Toxicology Letters. 2007;**168**:58-74

[63] Cui D, Tian F, Ozkan CS, Wang M, Gao H. Effect of single wall carbon nanotubes on human HEK293 cells. Toxicology Letters. 2005;**155**:73-85

[64] Bottini M, Bruckner S, Nika K, Bottini N, Bellucci S, Magrini A, et al. Multi-walled carbon nanotubes induce T lymphocyte apoptosis. Toxicology Letters. 2006;**160**:121-126

[65] Wörle-Knirsch JM, Pulskamp K, Krug HF. Oops they did it again! Carbon nanotubes hoax scientists in viability assays. Nano Letters. 2006;**6**:1261-1268

[66] Monteiro-Riviere NA, Nemanich RJ, Inman AO, Wang YY, Riviere JE. Multi-walled carbon nanotube interactions with human epidermal keratinocytes. Toxicology Letters. 2005;**155**:377-384

[67] Pantarotto D, Briand J-P, Prato M, Bianco A. Translocation of bioactive peptides across cell membranes by carbon nanotubes. Chemical Communications. 2004:16-17

[68] Pantarotto D, Singh R, McCarthy D, Erhardt M, Briand J, Prato M, et al. Functionalized carbon nanotubes for plasmid DNA gene delivery. Angewandte Chemie International Edition. 2004;**43**:5242-5246

[69] Nimmagadda A, Thurston K, Nollert MU, McFetridge PS. Chemical modification of SWNT alters in vitro cell-SWNT interactions. Journal of Biomedical materials Research. Part A. Official Journal of: Society for Biomaterials, Japanese Society for Biomaterials, Australian Society for Biomaterials, and Korean Society for Biomaterials. 2006;**76**:614-625

[70] Tian F, Cui D, Schwarz H, Estrada GG, Kobayashi H. Cytotoxicity of single-wall carbon nanotubes on human fibroblasts. Toxicology in Vitro. 2006;**20**:1202-1212

[71] Magrez A, Kasas S, Salicio V, Pasquier N, Seo JW, Celio M, et al. Cellular toxicity of carbon-based nanomaterials. Nano Letters. 2006;**6**:1121-1125

[72] Kagan VE, Tyurina YY, Tyurin VA, Konduru NV, Potapovich AI, Osipov AN, et al. Direct and indirect effects of single walled carbon nanotubes on RAW 264.7 macrophages: Role of iron. Toxicology Letters. 2006;**165**:88-100

[73] Hu H, Ni Y, Montana V, Haddon RC, Parpura V. Chemically functionalized carbon nanotubes as substrates for neuronal growth. Nano Letters. 2004;**4**:507-511

[74] Pappas TC, Wickramanyake WMS, Jan E, Motamedi M, Brodwick M, Kotov NA. Nanoscale engineering of a cellular interface with semiconductor nanoparticle films for photoelectric stimulation of neurons. Nano Letters. 2007;**7**:513-519

[75] Mattson MP, Haddon RC, Rao AM. Molecular functionalization of carbon nanotubes and use as substrates for neuronal growth. Journal of Molecular Neuroscience. 2000;**14**:175-182

[76] Jia G, Wang H, Yan L, Wang X, Pei R, Yan T, et al. Cytotoxicity of carbon nanomaterials: Single-wall nanotube, multi-wall nanotube, and fullerene. Environmental Science & Technology. 2005;**39**:1378-1383

[77] Raja PMV, Connolley J, Ganesan GP, Ci L, Ajayan PM, Nalamasu O, et al. Impact of carbon nanotube exposure, dosage and aggregation on smooth muscle cells. Toxicology Letters. 2007;**169**:51-63

[78] Edgar D, Timpl R, Thoenen H. The heparin-binding domain of laminin is responsible for its effects on neurite outgrowth and neuronal survival. The EMBO Journal. 1984;**3**:1463-1468

[79] Dertinger SKW, Jiang X, Li Z, Murthy VN, Whitesides GM. Gradients of substrate-bound laminin orient axonal specification of neurons. Proceedings of the National Academy of Sciences. 2002;**99**:12542-12547

[80] Guimard NK, Gomez N, Schmidt CE. Conducting polymers in biomedical engineering. Progress in Polymer Science. 2007;**32**:876-921

[81] Wan M. A template-free method towards conducting polymer nanostructures. Advanced Materials. 2008;**20**:2926-2932. DOI: 10.1002/adma.200800466

[82] Ren ZF, Huang Z, Xu JW, Wang JH, Bush P, Siegal MP, et al. Synthesis of large arrays of well-aligned carbon nanotubes on glass. Science. 1998;**282**:1105-1107

[83] Wu Z, Chen Z, Du X, Logan JM, Sippel J, Nikolou M, et al. Transparent, conductive carbon nanotube films. Science. 2004;**305**:1273-1276

[84] Sano M, Kamino A, Okamura J, Shinkai S. Self-organization of PEO-g raft-single-walled carbon nanotubes in solutions and langmuir– Blodgett films. Langmuir. 2001;**17**:5125-5128

[85] Sun Y, Liu X, George MN, Park S, Gaihre B, Terzic A, et al. Enhanced nerve cell proliferation and differentiation on electrically conductive scaffolds embedded with graphene and carbon nanotubes. Journal of Biomedial Materials Research Part A. 2021;**109**:193-206

[86] Wang Z, Koratkar N, Ci L, Ajayan PM. Combined micro−/nanoscale surface roughness for enhanced hydrophobic stability in carbon nanotube arrays. Applied Physics Letters. 2007;**90**:1-3

[87] Galvan-Garcia P, Keefer EW, Yang F, Zhang M, Fang S, Zakhidov AA, et al. Robust cell migration and neuronal growth on pristine carbon nanotube sheets and yarns. Journal of Biomaterials Science Polymer Edition. 2007;**18**:1245-1261

[88] Zhou H, Gupta A, Zou T, Zhou J. Photoresist derived carbon for growth and differentiation of neuronal cells. International Journal of Molecular Sciences. 2007;**8**:884-893

[89] Sreekumar TV, Liu T, Kumar S, Ericson LM, Hauge RH, Smalley RE. Single-wall carbon nanotube films. Chemistry of Materials. 2003;**15**:175-178

[90] Thompson DM, Buettner HM. Schwann cell response to micropatterned laminin surfaces. Tissue Engineering. 2001;**7**:247-265

[91] Zhang X, Prasad S, Niyogi S, Morgan A, Ozkan M, Ozkan CS. Guided neurite growth on patterned carbon nanotubes. Sensors and Actuators B: Chemical. 2005;**106**:843-850

[92] Gabay T, Jakobs E, Ben-Jacob E, Hanein Y. Engineered self-organization of neural networks using carbon nanotube clusters. Physica A: Statistical Mechanics and its Applications. 2005;**350**:611-621

[93] Hung PJ, Lee PJ, Sabounchi P, Lin R, Lee LP. Continuous perfusion microfluidic cell culture array for high-throughput cell-based assays. Biotechnology and Bioengineering. 2005;**89**:1-8

[94] Gross GW, Rhoades BK, Azzazy HME, Wu M-C. The use of neuronal networks on multielectrode arrays as biosensors. Biosensors & Bioelectronics. 1995;**10**:553-567

[95] Chang JC, Brewer GJ, Wheeler BC. Modulation of neural network activity by patterning. Biosensors & Bioelectronics. 2001;**16**:527-533

[96] Nugent JM, Santhanam KSV, Rubio A, Ajayan PM. Fast electron transfer kinetics on multiwalled carbon nanotube microbundle electrodes. Nano Letters. 2001;**1**:87-91

[97] Gheith MK, Pappas TC, Liopo AV, Sinani VA, Shim BS, Motamedi M, et al. Stimulation of neural cells by lateral currents in conductive layer-by-layer films of single-walled carbon nanotubes. Advanced Materials. 2006;**18**:2975-2979

[98] MacDonald RA, Voge CM, Kariolis M, Stegemann JP. Carbon nanotubes increase the electrical conductivity of fibroblast-seeded collagen hydrogels. Acta Biomaterialia. 2008;**4**:1583-1592

[99] Malarkey EB, Fisher KA, Bekyarova E, Liu W, Haddon RC, Parpura V. Conductive single-walled carbon nanotube substrates modulate neuronal growth. Nano Letters. 2009;**9**:264-268

[100] Matsumoto K, Sato C, Naka Y, Kitazawa A, Whitby RLD, Shimizu N. Neurite outgrowths of neurons with neurotrophin-coated carbon nanotubes. Journal of Bioscience and Bioengineering. 2007;**103**:216-220

[101] Li J, Papadopoulos C, Xu JM, Moskovits M. Highly-ordered carbon nanotube arrays for electronics applications. Applied Physics Letters. 1999;**75**:367-369

[102] Meitl MA, Zhou Y, Gaur A, Jeon S, Usrey ML, Strano MS, et al. Solution casting and transfer printing single-walled carbon nanotube films. Nano Letters. 2004;**4**:1643-1647

[103] Vafaiee M, Mohammadpour R, Vossoughi M, Asadian E, Janahmadi M, Sasanpour P. Carbon nanotube modified microelectrode array for neural interface. Frontiers in Bioengineering and Biotechnology. 2021;**8**:582713

[104] Wei S, Jiang A, Sun H, Zhu J, Jia S, Liu X, et al. Shape-changing electrode array for minimally invasive large-scale intracranial brain activity mapping. Nature Communications. 2024;**15**:715

[105] Ruhunage C, Dhawan V, Nawarathne CP, Hoque A, Cui XT, Alvarez NT. Evaluation of polymer-coated carbon nanotube flexible microelectrodes for biomedical applications. Bioengineering. 2023;**10**:647

[106] Liu X, Xu Z, Fu X, Liu Y, Jia H, Yang Z, et al. Stable, long-term single-neuronal recording from the rat spinal cord with flexible carbon nanotube fiber electrodes. Journal of Neural Engineering. 2022;**19**:56024

[107] Sekitani T, Yokota T, Kuribara K, Kaltenbrunner M, Fukushima T, Inoue Y, et al. Ultraflexible organic amplifier with biocompatible gel electrodes. Nature Communications. 2016;7:11425

[108] Kostarelos K, Vincent M, Hebert C, Garrido JA. Graphene in the design and engineering of next-generation neural interfaces. Advanced Materials. 2017;**29**:1700909

[109] Ryu S, Kim B-S. Culture of neural cells and stem cells on graphene, tissue Eng. Regenerative Medicine. 2013;**10**:39-46

[110] Bouzid T, Sinitskii A, Lim JY. Graphene platform for neural regenerative medicine. Neural Regeneration Research. 2016;**11**:894-895

[111] Chen M, Qin X, Zeng G. Biodegradation of carbon nanotubes, graphene, and their derivatives. Trends in Biotechnology. 2017;**35**:836-846

[112] Park SY, Park J, Sim SH, Sung MG, Kim KS, Hong BH, et al. Enhanced differentiation of human neural stem cells into neurons on graphene. Advanced Materials. 2011;**23**:H263

[113] Li N, Zhang Q, Gao S, Song Q, Huang R, Wang L, et al. Three-dimensional graphene foam as a biocompatible and conductive scaffold for neural stem cells. Scientific Reports. 2013;**3**:1604

[114] Serrano MC, Patiño J, García-Rama C, Ferrer ML, Fierro JLG, Tamayo A, et al. 3D free-standing porous scaffolds made of graphene oxide as substrates for neural cell growth. Journal of Materials Chemistry B. 2014;**2**:5698-5706

[115] Yang W, Zhao Y, He X, Chen Y, Xu J, Li S, et al. Flexible conducting polymer/reduced graphene oxide films: Synthesis, characterization, and electrochemical performance. Nanoscale Research Letters. 2015;**10**:1-7

[116] Hu N, Zhang L, Yang C, Zhao J, Yang Z, Wei H, et al. Three-dimensional skeleton networks of graphene wrapped polyaniline nanofibers: An excellent structure for high-performance flexible solid-state supercapacitors. Scientific Reports. 2016;**6**:19777

[117] Apollo NV, Maturana MI, Tong W, Nayagam DAX, Shivdasani MN, Foroughi J, et al. Soft, flexible freestanding neural stimulation and recording electrodes fabricated from reduced graphene oxide. Advanced Functional Materials. 2015;**25**:3551-3559

[118] Lu Y, Lyu H, Richardson AG, Lucas TH, Kuzum D. Flexible neural electrode array based-on porous

graphene for cortical microstimulation and sensing. Scientific Reports. 2016;**6**:1-9

[119] Choi BG, Yang M, Hong WH, Choi JW, Huh YS. 3D macroporous graphene frameworks for supercapacitors with high energy and power densities. ACS Nano. 2012;**6**:4020-4028

[120] Aradilla D, Delaunay M, Sadki S, Gérard J-M, Bidan G. Vertically aligned graphene nanosheets on silicon using an ionic liquid electrolyte: Towards high performance on-chip micro-supercapacitors. Journal of Materials Chemistry A. 2015;**3**:19254-19262

[121] Fabbro A, Scaini D, León V, Vázquez E, Cellot G, Privitera G, et al. Graphene-based interfaces do not alter target nerve cells. ACS Nano. 2016;**10**:615-623

[122] Kuzum D, Takano H, Shim E, Reed JC, Juul H, Richardson AG, et al. Transparent and flexible low noise graphene electrodes for simultaneous electrophysiology and neuroimaging. Nature Communications. 2014;**5**:5259

[123] Blaschke BM, Tort-Colet N, Guimerà-Brunet A, Weinert J, Rousseau L, Heimann A, et al. Mapping brain activity with flexible graphene micro-transistors. 2D Materials. 2017;**4**:25040. DOI: 10.1088/2053-1583/aa5eff

[124] Kireev D, Shokoohimehr P, Ernst M, Montes VR, Srikantharajah K, Maybeck V, et al. Fabrication of ultrathin and flexible graphene-based devices for in vivo neuroprosthetics. MRS Advances. 2018;**3**:1621-1627

[125] Deng J, Yuk H, Wu J, Varela CE, Chen X, Roche ET, et al. Electrical bioadhesive interface for bioelectronics. Nature Materials. 2021;**20**:229-236

[126] Vincent M, De Lázaro I, Kostarelos K. Graphene materials as 2D non-viral gene transfer vector platforms. Gene Therapy. 2017;**24**:123-132

[127] Sun H, Zhang L, Xia W, Chen L, Xu Z, Zhang W. Fabrication of graphene oxide-modified chitosan for controlled release of dexamethasone phosphate. Applied Physics A: Materials Science & Processing. 2016;**122**:1-8

[128] Servant A, Leon V, Jasim D, Methven L, Limousin P, Fernandez-Pacheco EV, et al. Graphene-based electroresponsive scaffolds as polymeric implants for on-demand drug delivery. Advanced Healthcare Materials. 2014;**3**:1334-1343. DOI: 10.1002/adhm.201400016

[129] Tang M, Song Q, Li N, Jiang Z, Huang R, Cheng G. Enhancement of electrical signaling in neural networks on graphene films. Biomaterials. 2013;**34**:6402-6411

[130] Hess LH, Lyuleeva A, Blaschke BM, Sachsenhauser M, Seifert M, Garrido JA, et al. Graphene transistors with multifunctional polymer brushes for biosensing applications. ACS Applied Materials & Interfaces. 2014;**6**:9705-9710

[131] Venton BJ, Wightman RM. Psychoanalytical electrochemistry: Dopamine and behavior. American Chemical Society (Analytical Chemistry). 2003;**75**(19):414A-421A

[132] Choi J-H, Kim T-H, El-Said WA, Lee J-H, Yang L, Conley B, et al. In situ detection of neurotransmitters from stem cell-derived neural interface at the single-cell level via graphene-hybrid SERS nanobiosensing. Nano Letters. 2020;**20**:7670-7679

[133] Sakai K, Teshima TF, Goto T, Nakashima H, Yamaguchi M.

Self-folding graphene-based Interface for brain-like modular 3D tissue. Advanced Functional Materials. 2023;**33**:2301836

[134] May PW. Diamond thin films: A 21st-century material. Philosophical Transactions of the Royal Society of London. Series A: Mathematical, Physical and Engineering Sciences. 2000;**358**:473-495

[135] Konstanty J. Sintered diamond tools: Trends, challenges and prospects. Powder Metallurgy. 2013;**56**:184-188

[136] Kamo M, Sato Y, Matsumoto S, Setaka N. Diamond synthesis from gas phase in microwave plasma. Journal of Crystal Growth. 1983;**62**:642-644

[137] Lux B, Haubner R. Diamond as a wear-resistant coating. Philosophical Transactions of the Royal Society of London. Series A: Mathematical, Physical and Engineering Sciences. 1993;**342**:297-311

[138] Roy RK, Lee K. Biomedical applications of diamond-like carbon coatings: A review. Journal of Biomedical Materials Research. Part B, Applied biomaterials. An Official Journal of the: Society for Biomaterials; Japanese Society for Biomaterials; Australian Society for Biomaterials, and: Korean Society for Biomaterials. 2007;**83**:72-84

[139] Arnault J-C. Surface modifications of nanodiamonds and current issues for their biomedical applications. In: Nov. Asp. Diam. From Growth to Appl. Switzerland: Springer; 2014. pp. 85-122

[140] Mitura K, Wyre K, Zarzycki PK. Bioactive food packaging with nanodiamond particles manufactured by detonation and plasma-chemical methods. In: Food Packag. Amsterdam, The Netherlands: Elsevier; 2017. pp. 295-328

[141] Wilson WB. Evidence for hopping transport in boron-doped diamond. Physics Review. 1962;**127**:1549

[142] Ekimov EA, Sidorov VA, Bauer ED, Mel'Nik NN, Curro NJ, Thompson JD, et al. Superconductivity in diamond. Nature. 2004;**428**:542-545

[143] Amaral M, Gomes PS, Lopes MA, Santos JD, Silva RF, Fernandes MH. Nanocrystalline diamond as a coating for joint implants: Cytotoxicity and biocompatibility assessment. Journal of Nanomaterials. 2008;**2008**:9. DOI: 10.1155/2008/894352

[144] Amaral M, Dias AG, Gomes PS, Lopes MA, Silva RF, Santos JD, et al. Nanocrystalline diamond: In vitro biocompatibility assessment by MG63 and human bone marrow cells cultures. Journal of Biomedical Materials Research. Part A. An Official Journal of: Society for Biomaterials, Japanese Society for Biomaterials, Australian Society for Biomaterials, and Korean Society for Biomaterials. 2008;**87**:91-99

[145] Sommer AP, Zhu D, Gagsteiger F, Fecht H-J. Sperm performance better on diamond than on polystyrene. MRS Online Proceedings Library. 2013;**1511**:mrsf12-1511

[146] Bergonzo P, Bongrain A, Scorsone E, Bendali A, Rousseau L, Lissorgues G, et al. 3D shaped mechanically flexible diamond microelectrode arrays for eye implant applications: The MEDINAS project. IRBM. 2011;**32**:91-94

[147] Ganesan K, Garrett DJ, Ahnood A, Shivdasani MN, Tong W, Turnley AM, et al. An all-diamond, hermetic electrical feedthrough array for a retinal prosthesis. Biomaterials. 2014;**35**:908-915

[148] Nad S, Gu Y, Asmussen J. Growth strategies for large and high quality single

crystal diamond substrates. Diamond and Related Materials. 2015;**60**:26-34

[149] Bormashov VS, Tarelkin SA, Buga SG, Kuznetsov MS, Terentiev SA, Semenov AN, et al. Electrical properties of the high quality boron-doped synthetic single-crystal diamonds grown by the temperature gradient method. Diamond and Related Materials. 2013;**35**:19-23

[150] Butler JE, Mankelevich YA, Cheesman A, Ma J, Ashfold MNR. Understanding the chemical vapor deposition of diamond: Recent progress. Journal of Physics. Condensed Matter. 2009;**21**:364201

[151] Schwander M, Partes K. A review of diamond synthesis by CVD processes. Diamond and Related Materials. 2011;**20**:1287-1301

[152] Cogan SF. Neural stimulation and recording electrodes. Annual Review of Biomedical Engineering. 2008;**10**:275-309

[153] Al-Riyami S, Ohmagari S, Yoshitake T. X-ray photoemission spectroscopy of nitrogen-doped UNCD/aC: H films prepared by pulsed laser deposition. Diamond and Related Materials. 2010;**19**:510-513

[154] Garrett DJ, Ganesan K, Stacey A, Fox K, Meffin H, Prawer S. Ultra-nanocrystalline diamond electrodes: Optimization towards neural stimulation applications. Journal of Neural Engineering. 2011;**9**:16002

[155] Lu L, Jones MW, Wu RLC. Diamond-like carbon as biological compatible material for cell culture and medical application. Bio-medical Materials and Engineering. 1993;**3**:223-228

[156] Garrett DJ, Tong W, Simpson DA, Meffin H. Diamond for neural interfacing: A review. Carbon N. Y. 2016;**102**:437-454

[157] Specht CG, Williams OA, Jackman RB, Schoepfer R. Ordered growth of neurons on diamond. Biomaterials. 2004;**25**:4073-4078

[158] Cui FZ, Li DJ. A review of investigations on biocompatibility of diamond-like carbon and carbon nitride films. Surface and Coatings Technology. 2000;**131**:481-487

[159] Rezek B, Ukraintsev E, Kromka A. bOptimizing atomic force microscopy for characterization of diamond-protein interfaces. Nanoscale Research Letters. 2011;**6**:1-10

[160] Ariano P, Baldelli P, Carbone E, Gilardino A, Lo Giudice A, Lovisolo D, et al. Cellular adhesion and neuronal excitability on functionalised diamond surfaces. Diamond and Related Materials. 2005;**14**:669-674

[161] Kelly S, Regan EM, Uney JB, Dick AD, McGeehan JP, T.B.B. Group, et al. Patterned growth of neuronal cells on modified diamond-like carbon substrates. Biomaterials. 2008;**29**:2573-2580

[162] Bendali A, Agnès C, Meffert S, Forster V, Bongrain A, Arnault J-C, et al. Distinctive glial and neuronal interfacing on nanocrystalline diamond. PLoS One. 2014;**9**:e92562

[163] Nistor PA, May PW, Tamagnini F, Randall AD, Caldwell MA. Long-term culture of pluripotent stem-cell-derived human neurons on diamond–A substrate for neurodegeneration research and therapy. Biomaterials. 2015;**61**:139-149

[164] Tong W, Fox K, Ganesan K, Turnley AM, Shimoni O, Tran PA, et al.

Fabrication of planarised conductively patterned diamond for bio-applications. Materials Science and Engineering: C. 2014;**43**:135-144

[165] Babchenko O, Romanyuk N, Jendelova P, Kromka A. Tailoring morphologies of diamond thin films for neural stem cells culturing. Physica Status Solidi. 2013;**250**:2717-2722

[166] Piret G, Hébert C, Mazellier J-P, Rousseau L, Scorsone E, Cottance M, et al. 3D-nanostructured boron-doped diamond for microelectrode array neural interfacing. Biomaterials. 2015;**53**:173-183

[167] Chen Y-C, Lee D-C, Hsiao C-Y, Chung Y-F, Chen H-C, Thomas JP, et al. The effect of ultra-nanocrystalline diamond films on the proliferation and differentiation of neural stem cells. Biomaterials. 2009;**30**:3428-3435

[168] Edgington RJ, Thalhammer A, Welch JO, Bongrain A, Bergonzo P, Scorsone E, et al. Patterned neuronal networks using nanodiamonds and the effect of varying nanodiamond properties on neuronal adhesion and outgrowth. Journal of Neural Engineering. 2013;**10**:56022

[169] Ojovan SM, McDonald M, Rabieh N, Shmuel N, Erez H, Nesladek M, et al. Nanocrystalline diamond surfaces for adhesion and growth of primary neurons, conflicting results and rational explanation. Frontiers in Neuroengineering. 2014;**7**:17

[170] Yang N, Hees J, Nebel CE. Diamond ultramicro- and nano-electrode arrays. In: Yang N, editor. Novel Aspects of Diamond. Topics in Applied Physics. Vol. 121. Cham: Springer; 2015. DOI: 10.1007/978-3-319-09834-0_9

[171] Huang X, O'Mahony AM, Compton RG. Microelectrode arrays for electrochemistry: Approaches to fabrication. Small. 2009;**5**:776-788

[172] Park J, Quaiserová-Mocko V, Pecková K, Galligan JJ, Fink GD, Swain GM. Fabrication, characterization, and application of a diamond microelectrode for electrochemical measurement of norepinephrine release from the sympathetic nervous system. Diamond and Related Materials. 2006;**15**:761-772

[173] Halpern JM, Cullins MJ, Chiel HJ, Martin HB. Chronic in vivo nerve electrical recordings of Aplysia californica using a boron-doped polycrystalline diamond electrode. Diamond and Related Materials. 2010;**19**:178-181

[174] Chan H-Y, Aslam DM, Wiler JA, Casey B. A novel diamond microprobe for neuro-chemical and-electrical recording in neural prosthesis. Journal of Microelectromechanical Systems. 2009;**18**:511-521

[175] Ganesan K, Stacey A, Meffin H, Lichter S, Greferath U, Fletcher EL, et al. Diamond penetrating electrode array for epi-retinal prosthesis. In: 2010 Annu. Int. Conf. IEEE Eng. Med. Biol. Vol. 2010. Buenos Aires, Argentina: IEEE; 2010. pp. 6757-6760

[176] Pagels M, Hall CE, Lawrence NS, Meredith A, Jones TGJ, Godfried HP, et al. All-diamond microelectrode array device. Analytical Chemistry. 2005;**77**:3705-3708

[177] Bonnauron M, Saada S, Mer C, Gesset C, Williams OA, Rousseau L, et al. Transparent diamond-on-glass micro-electrode arrays for ex-vivo neuronal study. Physica Status Solidi. 2008;**205**:2126-2129

[178] Bonnauron M, Saada S, Rousseau L, Lissorgues G, Mer C, Bergonzo P. High

aspect ratio diamond microelectrode array for neuronal activity measurements. Diamond and Related Materials. 2008;**17**:1399-1404

[179] Kiran R, Rousseau L, Lissorgues G, Scorsone E, Bongrain A, Yvert B, et al. Multichannel boron doped nanocrystalline diamond ultramicroelectrode arrays: Design, fabrication and characterization. Sensors. 2012;**12**:7669-7681

[180] Güell AG, Meadows KE, Unwin PR, Macpherson JV. Trace voltammetric detection of serotonin at carbon electrodes: Comparison of glassy carbon, boron doped diamond and carbon nanotube network electrodes. Physical Chemistry Chemical Physics. 2010;**12**:10108-10114

[181] Dankerl M, Hauf MV, Stutzmann M, Garrido JA. Diamond solution-gated field effect transistors: Properties and bioelectronic applications. Physica Status Solidi. 2012;**209**:1631-1642

[182] Natori S, Yoshimi K, Takahashi T, Kagohashi M, Oyama G, Shimo Y, et al. Subsecond reward-related dopamine release in the mouse dorsal striatum. Neuroscience Research. 2009;**63**:267-272

[183] Suzuki A, Ivandini TA, Yoshimi K, Fujishima A, Oyama G, Nakazato T, et al. Fabrication, characterization, and application of boron-doped diamond microelectrodes for in vivo dopamine detection. Analytical Chemistry. 2007;**79**:8608-8615

[184] Bennet KE, Tomshine JR, Min H-K, Manciu FS, Marsh MP, Paek SB, et al. A diamond-based electrode for detection of neurochemicals in the human brain. Frontiers in Human Neuroscience. 2016;**10**:102

[185] Yoshimi K, Naya Y, Mitani N, Kato T, Inoue M, Natori S, et al. Phasic reward responses in the monkey striatum as detected by voltammetry with diamond microelectrodes. Neuroscience Research. 2011;**71**:49-62

[186] Joye N, Schmid A, Leblebici Y. Electrical modeling of the cell–electrode interface for recording neural activity from high-density microelectrode arrays. Neurocomputing. 2009;**73**:250-259

[187] Strobel P, Riedel M, Ristein J, Ley L. Surface transfer doping of diamond. Nature. 2004;**430**:439-441

[188] Dankerl M, Eick S, Hofmann B, Hauf M, Ingebrandt S, Offenhäusser A, et al. Diamond transistor array for extracellular recording from electrogenic cells. Advanced Functional Materials. 2009;**19**:2915-2923

[189] Fan B, Rusinek CA, Thompson CH, Setien M, Guo Y, Rechenberg R, et al. Flexible, diamond-based microelectrodes fabricated using the diamond growth side for neural sensing. Microsystems & Nanoengineering. 2020;**6**:42

[190] Hall LT, Beart GCG, Thomas EA, Simpson DA, McGuinness LP, Cole JH, et al. High spatial and temporal resolution wide-field imaging of neuron activity using quantum NV-diamond. Scientific Reports. 2012;**2**:401

[191] Doherty MW, Manson NB, Delaney P, Jelezko F, Wrachtrup J, Hollenberg LCL. The nitrogen-vacancy colour Centre in diamond. Physics Reports. 2013;**528**:1-45

[192] Steinert S, Dolde F, Neumann P, Aird A, Naydenov B, Balasubramanian G, et al. High sensitivity magnetic imaging using an array of spins in diamond. The Review of Scientific Instruments. 2010;**81**:1-5

[193] Le Sage D, Arai K, Glenn DR, DeVience SJ, Pham LM, Rahn-Lee L, et al. Optical magnetic imaging of living cells. Nature. 2013;**496**:486-489

[194] Simpson DA, Tetienne J-P, McCoey JM, Ganesan K, Hall LT, Petrou S, et al. Magneto-optical imaging of thin magnetic films using spins in diamond. Scientific Reports. 2016;**6**:22797

[195] Glenn DR, Lee K, Park H, Weissleder R, Yacoby A, Lukin MD, et al. Single-cell magnetic imaging using a quantum diamond microscope. Nature Methods. 2015;**12**:736-738

Chapter 2

Synthesis, Physico-Chemical Methods, and Characterization of Nanomaterials

Ananda Kumar Chettupalli, Kavitha Marati, Himabindu Peddapalli, Mounika Kuchukuntla, Rajendra Kumar Jadi and Shaik Abdul Rahaman

Abstract

Nanomaterials are at the leading edge of the rapidly developing field of nanotechnology. Their unique size-dependent properties make these materials superior and indispensable in many areas of human activity. Recent technological advancements have demonstrated that the evolution of nanotechnology and nanoscience is a crucial factor. Nanomaterials have been attracting considerable attention owing to their excellent electrical, optical, magnetic, catalytic properties with potential applications in almost all branches of science and technology. This review chapter highlights the types of nanoparticles, their synthesis methods, and their characterization techniques. Nanomaterials exhibit excellent properties such as outstanding chemical and physical stability, lower density, and a large surface area. When you mix membranes with the special properties of nanomaterials, you get very good physical and chemical stability with high rejection rates for target compounds by the nanomaterial-based membrane. This work thoroughly reviews the synthesis, physico-chemical procedures, and characterization techniques used in the creation of nanomaterials. To examine a range of top-down and bottom-up synthetic techniques, emphasizing their effects on the size, shape, and functionality of nanomaterials. Examples of this technique include hydrothermal synthesis, sol-gel procedures, and chemical vapor deposition. The application of physico-chemical techniques includes thermal stability, particle size distribution, and surface area analysis, to understand the fundamental properties of nanomaterials. These methods include X-ray diffraction, spectroscopy, and scanning and transmission electron microscopy. This work aims to provide a thorough understanding of the intricate interactions between the synthesis, characterization, and applications of nanomaterials, paving the way for further advancements in nanotechnology.

Keywords: nanotechnology, sol-gel process, condensation, nanomaterials, physicochemical process, synthesis

1. Introduction

Any "material having any external dimension in the nanoscale or having internal structure or surface structure in the nanoscale" (1–100 micrometers) is considered to be part of nanotechnology. According to Ref. [1], nanomaterials contain nanostructures or nano-objects. The EU defines a nanomaterial as "a natural, incidental, or manufactured material with particles in an unbound state, aggregate, or agglomerate, where 50% or more of the particles in the number size distribution have one or more external dimensions of 1 to 100 nm." Resolving concerns about the environment, public health, safety, and competition may require lowering the number size distribution barrier from 50% to 1% to 50% [2]. Nanomaterials are materials with at least one unit between 1 and 100 nm [3, 4]. Researchers in microfabrication have advanced materials metrology and synthesis, which nanomaterial research utilizes to approach nanotechnology. Nanoscale materials frequently exhibit unique optical, electrical, or mechanical features [5, 6].

2. Types of nanomaterials

2.1 Synthesis methods

Nanotechnology involves controlled construction of nanoscale objects to create functioning devices for technological purposes. Science and engineering have advanced rapidly in nanomaterial fabrication to achieve new features not found in bulk materials [7, 8]. Particles smaller than 100 nm display intriguing characteristics, mostly resulting from two physical mechanisms. Materiality nanoparticles, which are particularly well-suited for a wide range of niche uses, are characterized by sensitive size-dependent effects such as optical and magnetic properties, as well as alterations to their thermal, mechanical, and chemical characteristics due to the quantization of electronic states [9, 10].

2.2 Bottom-up and top-down methods

Nanofabrication's bottom-up, or self-assembly, approach employs minute physical or chemical forces to construct larger structures from smaller ones. Bottom-up nanofabrication methods complement top-down methods as component size reduces. Biological processes use chemical forces to produce most of life's structures, inspiring bottom-up techniques. Scientists are attempting to replicate the natural process of atomic clustering, which enables the self-assembly of larger structures [11].

Top-down methods use "force" (usually mechanical or laser) to split bulk materials into nanoparticles. The mechanical process of "ball milling" nanoparticles from bulk materials is popular. Laser ablation, which uses femtosecond lasers to ablate a material, can also create nanoparticles [12]. Atom coalescence in liquids and the condensation of atomic vapors on surfaces produce bottom-up nanoparticles. Liquid-phase methods using inverse micelles have produced size-selected semiconductor, magnetic, and other nanoparticles. Self-assembling quantum dots have limited control over creation and structure. Growing thin layers of InGa as on GaAs yields isolated quantum dots due to compressive strain's repulsive forces. Growing multiple layer pairs regularizes the dot spacing. Carbon nanotubes self-assemble at the right temperature and chemical conditions [13].

DNA-assisted assembly may assemble heterogeneous hybrids. Biology self-assembles and self-organizes well in fluidic settings without electrochemical forces. Surface molecules could direct fluid attachments *via* DNA-like recognition. Complementary DNA polymers act as intelligent "adhesive tape," sticking only when paired. To locate attachment locations, electrical fields can be used, followed by electrodeposition and metallization. DNA-assisted approaches offer numerous advantages: It is possible to sequence, replicate, and use codes to distinguish complementary DNA strands, hybridize to form strong bonds to their sequence, and tag large amounts of DNA molecules. Researchers study the properties of self-assembling nanoscale molecules. DNA that adheres exclusively to the faces of compound semiconductor crystals self-assembles. Complementary sequences at the opposite end of DNA allow certain faces of microscopic semiconductor construction components to either adhere or repel. A molecule with thiol groups adhere to gold, while those with carboxyl groups attach to silica. Directional assembly moving, electrically or magnetically, in quasi-equilibrium, and precisely placing parts is becoming more prevalent (**Figure 1**).

Chemically synthesizing atomic or molecular species produces precursor particles. Both procedures work in gas, liquid, supercritical fluid, solid, or vacuum. Most producers want to manage particle size, shape, dispersion, particle composition and agglomeration [14]. To make nanoparticles from atoms, solution-based chemical methods such as sol-gel processing, CVD, plasma or flame spraying synthesis, laser pyrolysis, and atomic or molecular condensation can be used. To carry out these chemical reactions, "metal-organic" molecules must first be present. Low-processing temperature distinguishes sol-gel processing from other chemical processes. The sol-gel technique is economical and flexible. Spraying entails the introduction of reactants, such as gas, liquid air pollutants, or combinations, into a more powerful

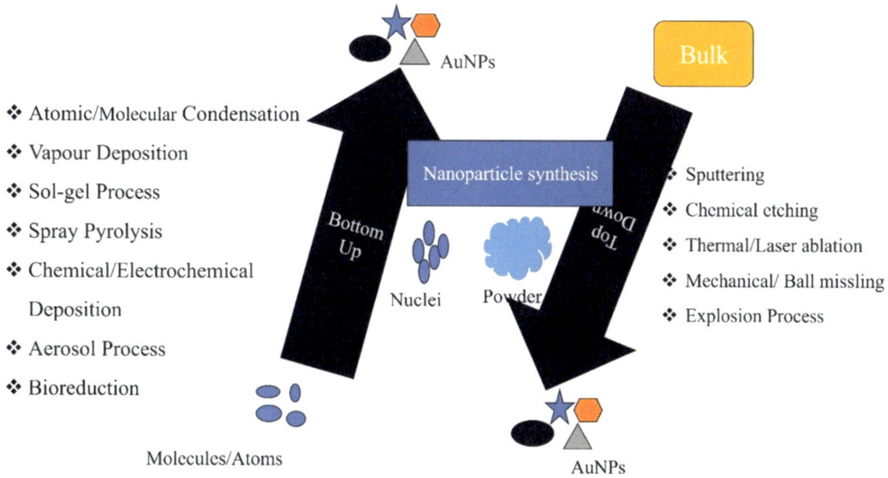

Figure 1.
Bottom-up and top-down AuNP synthesis using physical, chemical, and biological means.

flame, which can be achieved through plasma spraying or carbon dioxide lasers. Homogeneous nucleation and growth generate particles in a flame as reactants break down. Rapid cooling creates nanoparticles.

Solution-based chemical processes include hydro- or solvothermal syntheses, chemical vapor deposition (CVD), metal-organic decomposition (MOD), and sol-gel processing. Most chemical pathways require "metal-organic" precursors. Metal b-diketones and metal carboxylates are metal oxide precursors, although metal alkoxides are more flexible. Most elements contain them, and a cheap feedstock can synthesize some of them [15].

2.3 Sol-gel technology

The sol-gel method, a long-standing commercial method for synthesizing liquid-phase colloidal nanoparticles and coatings, has recently undergone improvements [16, 17]. The sol-gel process is useful for the synthesis of nanoparticles made of oxide and composite. Sol-gel processes for material preparation are versatile, have low-temperature, and have flexible rheology for shaping and embedding. They provide unique organic-inorganic material access. Alkoxides are the most popular oxide precursors because of their commercial availability and the M-OR bond's high liability. It enables processing-based on-site customization [18, 19].

2.4 Aerosol-based technique

Aerosol nanoparticle production is industry-favored. Aerosols are particles that are solid or liquid in the state of gas phase, ranging from molecules to 100 μm. Before their scientific and technological characteristics were known, people used industrial aerosols. Ti-tetrachloride oxidizes titanium pigment for paints and plastics, and flame pyrolyzes fumed silica and titania, and makes optical fibers [20, 21]. When chemical precursors are sprayed onto hot surfaces or heated air, they create particles and initiate pyrolysis. Oxford University invented the process of electro-spraying semiconductor-metal nanoparticle composites at room temperature. Aerosol micro-droplets containing Cd salts in hydrogen sulfide-rich atmospheres produced CdS nanoparticles [22–24].

2.5 Chemical vapor deposition (CVD)

CVD activates a substrate-gaseous precursor chemical process. Thermal CVD or plasma activation can occur. The key benefit is this technology's non-directivity. Plasma lowers process temperature significantly compared to thermal CVD. CVD routinely produces carbon nanotubes [25, 26].

2.6 Atomic or molecular condensation

This approach focuses on metal nanoparticles [27]. When a large mass of material is vacuum heated, it releases a jet of evaporated and fractured particles into a gas chamber, which can be either reactive or inert. Nanoparticles occur when metal atoms collide with gas molecules and cool quickly. Reactive gases like oxygen create metal oxide nanoparticles [28, 29].

2.7 Creating nanopowders of metals *via* gas-phase condensation

It is widely acknowledged that the gas-phase condensation hypothesis proposed in 1930 can be used to synthesize metal nanoparticles. Materials for gas-phase condensation include a heating element, a vacuum chamber, powder-collecting apparatus, metal to be nanopowdered, and vacuum equipment [30–32].

2.8 Determining the gas-phase condensation mechanism

This procedure utilized inert gases in a high-pressure environment, generating particles at a pressure sufficiently low to produce spherical particles [33]. Metal melts quickly on a hot element. The metal undergoes rapid heating at high melting and below the boiling point to achieve sufficient vapor pressure [34]. The pumps constantly add and remove gas from the chamber, taking the cooled metal from the source of heat. Gas cools metal vapor, forming nanoparticles. These tiny particles are liquid because they are too hot to be solid. In a controlled setting, liquid particles clash and coalesce to grow spherical and smooth. When chilled under control, liquid particles solidify and stop growing. A substance that resists agglomeration or other interactions coats these reactive nanoparticles [35] as shown in **Figure 2**.

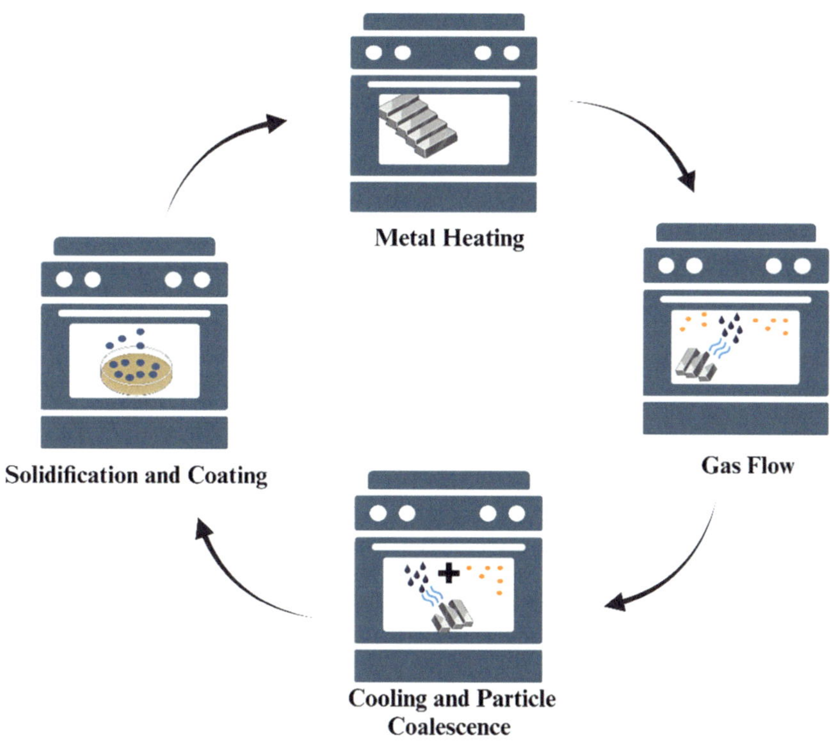

Figure 2.
Schematic representation of high-pressure inert gas method.

2.9 Supercritical fluid synthesis

Nanoparticle production using supercritical fluids is also effective. These methods exploit supercritical fluid properties to produce nanoparticles by rapidly expanding a supercritical solution. Continuous development is underway for the pilot-scale supercritical fluid approach [36, 37].

2.10 Spinning to make thin polymeric fibers

Using an electric field with a high voltage to spin-diluted polymer solutions is an innovative way to create thin polymer fibers. A polymer drop undergoes electrospinning at a voltage of hundreds of volts [38]. Tensile forces from an applied electric field, along with the jet's electrical charge, release a fine jet of polymer from the surface as the teardrop forms a Taylor cone at a typical voltage. This process produces polymer fiber bundles. A continuous web of fibers ranging from a few μm to no more than 100 nm can be formed by directing jets toward a grounded surface [39].

2.11 Making nanoparticles with the aid of templates

Use any substance with regular gaps or pores that are nano-sized to create nanoparticles. Some examples of molecules that can serve as templates include porous alumina, zeolites, dendrimers, diblock copolymers, and proteins. It is not necessary for the template to be three-dimensional. The formation of self-assembled monolayers on a flat surface or gas-liquid interface results in the creation of synthetic templates [40].

2.12 Self-assembly of nanoparticles

Chemical self-assembly can process nanoparticles of organic, biological, inorganic oxide, metal, and semiconductor materials [41]. These techniques employ selective molecular attachment to specific surfaces, biomolecular recognition, and ordering rules (e.g., DNA strands preferentially dock with complementary base pairs) to construct 1-D, 2-D, and 3-D self-assembled nanostructures. They also establish chemistry for molecular attachment to clusters and substrates, such as thiol (-SH) end groups. Like a jigsaw puzzle, molecular building blocks fit together perfectly without a driving force [42].

2.13 Recent advances

A significant portion of the previously mentioned technology has already undergone development, with some even securing commercial success. Biological and artificial materials and procedures for the development of novel systems and architectures are part of today's technology, along with self-healing and self-organizing biomimetic materials.

2.14 Using 3-D print innovation

The concept of additive manufacturing, sometimes known as 3-D printing, emerged in the late 1980s but has since gained immense popularity around the world [43–45]. Today, 3-D printing can be used to produce tissues, organs, meals, and garments. This process starts with a computer-aided design 3-D model of the intended building. Computer programs create thin layers based on designs. After receiving this file, the printer prints the 3-D structure [46]. Biological and metallic materials are both compatible with 3D printing. This method has produced pasta, candies, chocolates, shoes, clothes, components for industrial machines, orthopedic implants, and decorations [47].

2.15 A bio-based approach to nanofabrication

Use of template-assisted nanofabrication has persisted for quite some time. However, polymers, inorganic oxides, and metals made up the majority of the templates. One recent application of nanofabrication is the use of biomolecules as templates [48]. Layer by layer, viruses self-assemble and serve as templates for nanomaterial 3D films. DNA's manipulable chemistry and 3D shape make it a popular choice for bottom-up nanofabrication. One can directly manufacture DNA or use it as a template for nanomaterials [49].

2.16 DNA origami

DNA folding into 3-D structures, like origami, is a recent development. Short synthetic DNA sequences crosslink a lengthy DNA molecule to fold and "stitch" it [50]. Researchers have created complex structures such as rabbits and nano flasks from cubes and tetrahedrons. These structures also serve as nanofabrication molds for artificial materials, along with biocompounds [51].

2.17 Prospects for the future offered by molecular nanotechnology

These first strategies are just the beginning of the long-term goals of nanotechnology. Researchers use molecular nanotechnology to develop biomimetic materials capable of self-organization, self-healing, and self-replication. Industry uses synthetic and biological materials, structures, and systems to mimic biological processes. The last 10 years have seen tremendous progress in these domains. Animals and humans alike have tested a number of 3-D-printed implants. Tiny robots, capable of handling items too small for human hands, can administer medications. The field is advancing at a rapid pace, which means that some of these technologies might be ready for purchase shortly [52].

3. Top-down technique

Typically, the following steps are involved in the top-down strategy:

3.1 Condensation/evaporation

Making nanopowders requires heating and condensing metal vapor. This procedure utilizes the vaporization pressure of the metal, which is dependent on its chemical strength. Surface oxidation of Fe, Ni, Co, Cu, Pd, and Pt produces sufficient steam through radiative and inductive heating at temperatures of 1200 and 2000 degrees. Some of the materials undergo preparation in the lab. Oxygen (A1, C8, T1, Zr) and metals with reactivity (Mo, Hf, Ta, W) require a higher temperature. For instance, three thousand degrees of electric arc and bombardment heating, as well as fourteen thousand degrees of electric arc, serve as examples. A reactive environment (oxygen) would oxidize nanoparticles with metallic elements. A problem with this method is the loss of nanometric size control. The powders are obtained by rapidly chilling the expelled vapor, which forms many nanoparticles and restricts their growth and other processes. The process is used in the industrial manufacture of ceramic and metal nanopowders, producing millions of tons annually. Airborne nanopowders can catch fire and explode [53].

3.2 Laser pyrolysis

The nanopowder synthesis method is simple and effective. The process involves the interchange of reagents and carbon dioxide lasers. The energy current between the two raises reactive zone temperature quickly by increasing molecular vibration degrees. The nanoparticles produced by disconnecting the reactors from the flames extinguish themselves once the flames go out. Powers are placed for gathering. The following method makes 15–20 nanometer particle manufacturing at 100 g/h straightforward. Reaction flow is a benefit of this technology [54]. Additional benefits include high purity and excellent chemical and physical characteristics. This produced many nanopowders.

3.3 Ionic/electronic irradiation

Irradiations create unique nanoparticles. The approach relies on heavy ion facilitators like GANIL, which use local material changes in a 10 nm radius around the ion to produce energy. The approach enables the production of polymer membrane nanoparticles. The substance can undergo radiation exposure while still within the pore structure of a polymer membrane to form a "matrix" [55].

4. Bottom-up technique

Bottom-up integration uses nanoscale, physically as well as chemically, to incorporate primary components into larger structures. Because of the decrease in unit size, this technique replaces the top-down method. Biological systems inspire this technique, which uses chemical forces to build structures. Scientists are trying to mimic nature's self-integration into larger structures. White hydrogen and red oxygen combine to produce nanoparticles of a gold-palladium alloy by reacting carbon with acid and hydrogen peroxide, preventing decay. Atomic vapors and liquids have used condensation and coalescence to create nanoparticles [56].

5. Advanced chemical methods

5.1 Sol-gel technique

Nanomaterials from alkoxides or colloidal solutions. The process creates thin layers, crystallized nanopigments, and monoliths. They focus on polymerization reactions [57]. Sol-gel, metal-containing alkoxide, and pectin are examples. Sol-gel enables solution-grown chemicals. The sol becomes a three-sided solvent channel

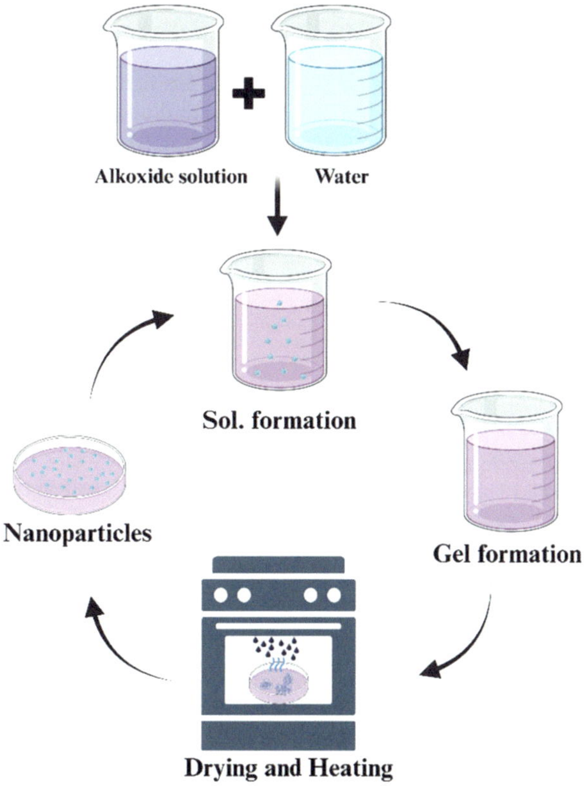

Figure 3.
Schematic illustration of sol-gel technology.

after species-liquid interactions. After freezing, the aerogel loses its liquid, and vaporization transforms sol-gels into distorted solids. This process aids in controlling the size and uniformity of the material [58]. They enable massive fiber production. The technique's principal drawback is that precursors are expensive, low-gain, low-weight, and harmful for living creatures as shown in **Figure 3**.

5.2 Highly pressurized fluid undergoing a chemical reaction

Supercritical solvent chemical conversions enhance precursor physical and chemical processes by elevating the temperature and pressure on tiny mediums. This control simplifies chemical reaction control. The approach divides inorganic elements, such as metals containing oxides as well as nitrate [59]. Two approaches to create metallic nanoparticles: the top-down technique, which involves slicing large pieces of material into nanostructures, and the bottom-up method, which involves incorporating specific molecules and atoms into broader nanostructures.

6. Physical methods

This is classified into two classes again. 1. Mechanical methods and 2. vapor methods.

6.1 Plasma method

Plasma is another nanoparticle-making process. Coils that generate heat at a certain frequency (RF) create plasma. The initial metal uses a pestle after evacuating the chamber. Encircling the chamber with high-voltage RF coils raises the metal's temperature above its evaporation peak. The coiled region generates a high-temperature plasma as helium (He) flows into the system. Metal vapor nucleates on helium gas atoms and dissipates to a cold collection rod, where oxygen gas passivates the nanoparticles [60].

6.2 Deposition of chemicals by vapor

Semiconductor fabrication typically uses a chemical reaction to deposit thin films of different materials. Any number of volatile precursors decompose on the substrate to create the target deposit. Substances in a vapor form can adsorb onto a heated substrate in a chemical vapor deposition (CVD) reactor. Adsorbed molecules react or break down to create crystals. CVD has three steps:

- A boundary layer carries reactants across the growing surface.
- The surface of the growing thing undergoes chemical reactions.
- The surface must be cleansed of byproducts generated by the gas-phase process. The gas phase is where homogeneous nucleation takes place, whereas a substrate is where heterogeneous nucleation takes place.

Gaseous phase chemical reaction of CVD can create ultrafine particles less than 1 millimeter. Controlling the process produces 10–100-nm nanoparticles [61, 62].

6.3 Inert-gas condensation

An inert-gas condensation (IGC) synthesizes nanostructured materials bottom-up in two phases. The process begins with the evaporation of the substance, followed by a rapidly regulated condensation that produces the desired particle size. An inert-gas condensation (IGC) chamber, backfilled with helium and evacuated to 10–8 mbar, evaporates, sputters, or abrades a material. Inert gas condensation is the easiest way to make metallic nanoparticles and nanocomposites. There are two steps. To produce the desired size of nanoparticle, the first step is to evaporate the inorganic or metal component and then to condense it rapidly under regulated conditions. Some methods of vaporizing the material include thermal vaporization, electric arc discharge, plasma heating, laser vaporization, and sputtering. When nanoparticles evaporate and collect on surfaces at low temperatures, they condense spontaneously. The procedure uses He, Xe, or Ar, or an inert gas environment. Atomic-level breakdown can be caused by both solid and gaseous sources. The size of the nanoparticles is determined by the pressure and the inert gas (He, Xe, or Ar) in the chamber. The catalyst that breaks down the metal nuclei determines whether condensation or chemical vapor deposition occurs [63]. In the chamber, atoms of material and gas collide, causing them to lose kinetic energy and eventually condense into tiny particles. A compaction device scrapes the condensed particles collected by a cold finger. Scraping and compaction under ultra-high vacuum conditions cleans particle surfaces and reduces trapped gases. Equiaxed crystallites result from inert gas condensation. The powder's size distribution is narrow, and its crystal size is just a few nanometers. The composition, rate of evaporation, sputtering or ablation, temperature, and pressure of the inert gas affect the size of the crystal [64].

The term "nanocrystalline materials" was first used in 1984 to describe materials with nanoscale grain sizes and interface volumes comparable to those of crystalline regions. The original process is used to produce nanostructured metals and ceramics after several modifications [65]. Early study focused on the production and characterization of pure metal nanoparticles, but IGC was also used to synthesize compounds, alloys, and particle combinations. In the beginning, oxides were studied by adding oxygen to the vapor phase [66]. Another way to produce metal nitride and carbides is to fill the chamber with gases of nitrogen or ammonia, or a carbonaceous atmosphere. Materials with intergranular contacts (IGC) nanostructures can be covalent substances (Si), ionic compounds (FeF_2, CaF_2, Fe_2O_3, and TiO_2), or metals (Cu, Fe, Ni, Pd, and W).

Inert gas condensation (IGC) synthesis involves volatile species evaporation and condensation under low pressure. The laser technique can produce monodispersed and loosely agglomerated nano powders in the lab, but not in commercial quantities. Commercial carbon black, TiO_2, and SiO_2 manufacturing has been accomplished using the flame technique. Experimental monoxide ceramics were made using plasma. The synthesis process produces highly agglomerated nanopowders. The IGC synthesis approach may solve nanoparticle aggregation in 1985. IGC generated experimental quantities of high-purity no-agglomerated nanopowders of diverse materials. However, this process cannot easily create many valuable ceramics and low-vapor-pressure materials, hence another way is needed [67, 68].

6.4 Arc discharge

Vasily V. Petrov initially characterized arc discharge as a "special fluid with electrical properties" in 1802. At the turn of the eighteenth century, Sir Humphry Davy demonstrated the arc by passing a current of electricity through two connecting carbon rods. He named this arc. Arc discharges using cathodes that contain graphite powder and metal catalysts such as cobalt, iron, or nickel result in the formation of single walled carbon nano tube (SWNT) deposits, as discovered by Sumio Iijima and Bethune in 1993 [69]. Arc discharge uses laser vaporization. An electrical arc vaporizes the substance directly, not through a laser pulse source. This eliminates the need for a separate cluster ionization stage, as ions make up 10% of the cluster. Usually, the target receives a 50-atom cluster. Nanoshell makes ultrafine Fe, Si, SiC, and Al2O3 powders by arc discharge heating. To achieve the desired particle size during evaporation, one must regulate gas pressure and arc current [70]. Two anodes and one cathode made of high-The arc discharge technique uses two high-purity graphite anodes and one cathode. The electrodes were turned into vapor by exposing them to a DC current (~100 A) in a 400-mbar helium atmosphere *via* two pieces of high-purity graphite that the cathode produces a carbon rod following an arc discharge. Although MWNTs are the most common products, technique not only produces MWNTs, but also manufactures SWNTs anodized or cathode with a metal catalyst such as Fe, Co, Ni, Y, or Mo type, plasma arc. The physical nanomaterial production process determines nanotube lengths, diameters, purity, and other parameters based on factors such as metal quantity, inert gas pressure, gas type, plasma arc, temperatures, current, and system geometry. A potential difference across two electrodes makes a gas conduct electricity, creating a plasma. Gas arc discharges lead to the formation of plasma. The most convenient way to generate hot plasma is arc discharge. It possesses substantial energy and achieves local thermal equilibrium [71].

Principle: Arc electrodes use thermionic electron emission to cool and evaporate electrode material. Heat causes charge carriers to flow out of a surface or potential energy barrier, a process known as thermoelectric emission. The work function, or binding potential, of the electrode material is less than the carrier thermal energy. The material, voltage, current, and inert gas pressure of the electrodes dictate the shape of the final product. One common method for producing carbon nanoparticles is anodic arc discharge using a consumable anode. Anodes made of graphite and cathodes made of metal make up a standard arc arrangement [72, 73]. Arc heating causes a significant amount of anode evaporation, which in turn deposits ablated material onto the cathode. A graphite anode anodic arc generates plasma mostly from carbon products evaporated in a buffer gas, often helium. Fullerene production calls for a buffer gas with a temperature range of 100 to 200 torr [74]. The production of MWNTs uses arcs ranges from 500 to 600 torr. Metal catalysts enable the production of SWNTs using atmospheric pressure carbon arcs. Magnetism and arc discharge can also create graphene flakes [75]. In all these synthesis applications, the arc typically controls itself with a few millimeters between electrodes and is current-limited.

Despite their prevalence in synthesis, little is known about transient anodic arcs caused by intense anode ablation. At a specific arc current, it was recently discovered that the ablation rate is almost constant at a crucial anode diameter for graphite anodes. When increasing the current density, ablation occurs more quickly at smaller anode diameters. The majority of synthesis arc methods employ ablation. Arcs rarely select nanotubes, but they do produce a large quantity of them [76]. How big and how flat a nanotube is depend on plasma conditions and how long it stays in the synthesis zone. Operating an arc can form a variety of materials, including carbon nanoparticles, nanotubes, and fullerenes. Actually, arc stability keeps synthesis processes going, like the birth and growth of nanoparticles, so that neutral and charged particles, fluxes, and heat fluxes can be made all the time. The results of this study demonstrate the instability of the conventional carbon arc method for synthesizing carbon nanotubes using a consumed anode. Because of instability, nanoparticle mixing decreases the selectivity of arc production.

A low current (5A) study was done on anode process arc instability problems in high-intensity lighting systems and plasma torches with atmospheric pressure. In sections 14–17, recent studies have linked the properties of the anode arc attachment, such as how the arc root moves and whether there are many, few, or diffuse attachments, to voltage and current oscillations. Theoretical models indicate that arc instability requires plasma column areas. Large gradients in gas, fluxes, arc plasma, and the anode surface trigger instability in 3D models. In 3D models, the high-anode ablation rates and the anode's gas and particle fluxes are often overlooked. Additional research is required to fully understand the role of medium currents (50–100 A) and brief breaks in the anodic arc during nanomaterial synthesis. In the appendix, other instability cases are discussed and a basic model is offered for arc instability [77].

6.5 RF-plasma method

Radio frequency (RF) induction thermal plasma technology is a novel method that can create composites, metal oxides, sophisticated materials, and nanoparticles [78].

The process produced Ag, FexOy, and composite nanoparticles. The RF plasma system has three main components: a high-frequency ultrasonic spray for aerosol formation, an RF induction system for thermal decomposition and electrostatic particle collection and filtration. In the studies, the ultrasonic atomizer took metal ion solutions ranging from 0.1 to 2 M and turned them into aerosol. It then used argon gas to move the aerosol into the plasma zone, to characterize the particles using SEM, TEM, and XRD. According to the characters, the particles are uniformly sized and have an average size of less than used for RF inductively coupled thermal plasma to generate aerosols, producing Ag, Ag/TiO, FexOy, and CuO/ZnO composite nanoparticles [79, 80].

Research revealed;

- By starting with a concentration of 2 M AgNO3, 30 nm Ag particles could be made.
- RF plasma systems are not suitable for the synthesis of composite Ag/TiO_2 particles.
- The precursor was likely exposed to quite high temperatures, which resulted in the generation of the distinct Ag and TiO_2 particles.
- A shift in particle shape from spherical to needle-like occurred as a result of increasing the solution concentration from 0.1 to 1 M, which in turn increased the average particle size. Agglomeration of parent particles produces secondary particles of CuO/ZnO with an average grain size of 30 nm, in contrast to the original particles' size of less than 10 nm [81].

6.6 Plasma arc technique

A new system is developed for pulse vacuum arc discharge nanoparticle deposition. Pulsed vacuum arc deposition allows easy metal ion generation to produce ultrathin films and nanoparticles. Other systems cannot accomplish film flatness and particle fineness like this technology. This technology has the ability to select nanoparticle diameters ranging from 1.5 to 6 nm by adjusting the condenser capacity. Any conductive target can produce plasma [82].

Target specific resistance: 0.01 Ωcm or less. This technique produces catalytic nanoparticles with higher activity than wet-processed ones. APD-S substrate sample size is 2 inches, while APD-P is a powder vessel with a stirring mechanism (internal dimensions: 95 × H 30 mm). Standard deposition numbers for APD-S and APD-P are the plasma arc method works by generating a high-temperature arc between electrodes, which vaporizes the target material. After chamber evacuation, a trigger initiates the discharge of the target rod arc. The target rod produces a highly ionized metal plasma, devoid of discharge gases, which deposits onto the substrate to form thin coatings and nanoparticles. The technology may select nanoparticle diameter between 1.5 and 6 nm by altering condenser capacity. If the mechanism is electrically conductive, it can playmate any material. Target specific resistance is 0.01-ohm cm or less. Changes in the environment easily produce oxide and nitride. Discharging graphite in H gas also produces ultra nanocrystalline diamond (UNCD). System-supported nanoparticles have active catalytic effects compared to the wet method. The APD-P model supports turning nanoparticles into powders. Nanoparticles adhere to 2-inch substrates with APD-S. The ones above 1, 3, and 4 are literature-based. With the arc-plasma method, Pt and Pd catalysts supported by CeO_2 were better at oxidizing CO than catalysts used in a wet process [83].

6.7 Ion sputtering

The produced TiOx/ZnO-containing silver nanoparticles in different fractions on silicon substrates use top-down ion beam sputtering (IBS) to characterize the nanoparticles using SEM and X-ray photoelectron spectroscopy. The crystal violet model analyte works well with these composites as SERS and SEIRAS substrates. When using ion beam sputtering, TiOx/AgNPs with 2 and 1 nm thicknesses produce the highest SEIRAS sensitivity. A substrate with 33% silver in it boosts SERS by a factor of 10,000. Under ultraviolet light, adsorbed crystal violet degrades [83]. Using TiO_2, Ag, and ZnO targets with a diameter of 4 inches, materials with a purity level of 99.999% were deposited. The article continues by describing the apparatus for ion beam sputtering deposition. Changing the ion beam energy and current, which are two deposition parameters of the sputtering ion sources, controls how fast the nanocomposite parts grow and how much metal is present. Their original intent was to deposit Ag nanoparticles onto ZnO at a rate of 0.03 nm/s. They coated 2 nm ZnO with 1 nm Ag nanoparticles. About 1 nm Ag NPs cover the 2 nm TiOx, to co-sputter the two initial targets to yield 33% Ag nanoparticle TiOx. For a 20-nm-top layer, adjust the deposition parameters of each target to get the Ag percentage. In order to find the quartz microbalance parameter for controlling the film thickness, measure the composite density (6.3 $g \cdot cm^{-3}$) prior to deposition [84]. Ion beam sputtering is used to deposit Ag nanoparticles and TiOx or ZnO NPs onto silicon chips and Si ATR waveguides for the fabrication of SERS and SEIRAS substrates. The manufacturing process evenly and quickly distributes the nanomaterials across the substrate. Nanocomposites of TiOx and Ag, with concentrations of 5.3% Ti and 18.3% Ag, enhance SERS signals. In SEIRAS studies, composites of TiOx and AgNPs that had an IBS thickness of less than 2 nanometers demonstrated the greatest signal amplification. The analyte's IR decreased over time when exposed to UV irradiation due to the photocatalytic activity of the substrate. There is need to clean and reuse the substrates. The oxidation stability of metal nanoparticles, however, is a determinant of substrate stability. The vibrational information that these substrates may provide to SERS and SEIRAS molecules is complementary. These substrates can employ any infrared or Raman-active analyte, with crystal violet serving as the model analyte. These substrates are utilized in both infrared reflection-absorption spectroscopy and attenuated total reflection IR. Results from SEIRAS show that substrates containing TiOx nanoparticles can degrade crystal violet when exposed to ultraviolet light [85].

6.8 Laser ablation

Laser ablation enables the creation of substitute quantum dots, core-shell nanoparticles, carbon nanotubes, and nanowires. A laser can evaporate gaseous species, converting them into nanoparticles. Quick vapor quenching is used to create highly pure nanoparticles, approximately 10 nm in size. You can create colloidal solutions or nanopowders by vaporizing or submerging a solid target with a laser. It synthesizes NPs quickly, in contrast to earlier approaches. There is no need for prolonged heating, complex chemical reactions, or multiple steps in the synthesis process. Furthermore, it is capable of producing polymeric, metallic, semiconducting, and complicated multi-element alloy NPs. Nanomaterial synthesis is environmentally friendly and safe for lab use because it does not involve any toxic, dangerous, or pyrophoric chemical precursors. Colloidal solutions and ultrapure nanoparticles produced by water do not include any counterions or reaction byproducts, making them ideal for use in *in vivo* biological and biochemical applications. To make NPs suitable for use in organic electronics, insulating capping ligands are not necessary [86]. After manufacture, one can functionalize the NPs by adding a ligand to the colloidal solution or by ablation in a solvent. This technique can directly produce NPs in a medium with practically any desired target materials and liquids. Some of the things that affect the shape, size, composition, and structure of the NPs made for each target material are the ablation laser parameters (wavelength, pulse width, pulse energy, and fluence) and the generation environment (vacuum, controlled gas atmosphere, or liquid). One method to fine-tune the size distribution during liquid NP production is to either alter the ablation period or irradiate the colloidal solution afterward. Laser ablation synthesis of NPs leads to colloidal solutions that have alloys or mixtures of NPs [87]. To create NPs through laser ablation, one must understand how a material responds to an intense laser beam, specifically one whose fluence, or energy per unit area, on the material's surface exceeds the material's ablation threshold. The electron-phonon coupling time constant of the material and the width of the laser pulse are the two most important parameters for laser ablation. The laser's energy propels the conduction band to the material's surface, carrying free electrons in metals and valence band electrons in semiconductors. Carriers transfer the energy to the lattice through electron-phonon interaction after a brief thermalization period of 100 fs. If the pulse width is greater than the electron-phonon coupling time constant, the laser beam will continue to flow energy into the material even after carriers have transmitted it into the lattice [88]. The laser beam heats the material's surface. The area immediately surrounding the laser beam's impact is prone to melting, heat-affected zones, recast layers, surface debris, mechanical cracks, and other defects. Vaporization and melting can occur at low fluences as well. A femtosecond (fs) laser limits its laser energy to the initial volume it irradiates because its pulse width is smaller than the electron-phonon coupling time constant. The ultrashort pulse width also stops the laser energy from diffusing into heat, and the surface enthalpy of the material is many orders of magnitude higher than its sublimation enthalpy. Sublimation with a low flux converts solids to vapors. When fluences exceed a particular value, thermal diffusion from the volume defined by the focused beam diameter and skin depth of the material occurs. This is because the heat penetration depth determines the energy penetration into the material, and the electron thermal diffusion length, not the optical absorption length of the laser radiation, primarily governs the ablation. The process of vaporization and subsequent melting mimic an ns pulse. The duration of a picosecond (ps) laser pulse ranges from nanoseconds (ns) to foci per second (fs) [89].

6.9 Ball milling

One of the most important industrial nanomaterial synthesis methods is ball milling, also known as mechanical alloying or attrition. This procedure uses top-down methods, such as molecular self-assembly and particle reduction. Frictional force from reactant surface collisions increases temperature, pressure, and internal energy in milling. In contrast to the other synthesis method, ultrafine grinding, also known as nanosizing, was used for particle sizes in the nanometer range. Metal, clay, and coal powder manufacture uses mechanical attrition to produce tones of nanoparticles every hour [90]. In ball milling, chemical reactions happen quickly. Modification occurs *via* breaking molecular bonds or modifying reagent reactivity. SWCNTs reduce in length and increase in surface area when ground in ball mills, which is necessary for different catalytic supports [91]. Recent studies have compared milled and unfilled CNT aniline absorption in fluid solution. Continuous alkyl chains, functionalized with aryl and alkyl groups and combined by a high-speed vibration mill, result in modified carbon surfaces that are well-soluble in organic solvents. Chemical and mechanical techniques inject ammonium bicarbonate containing amine and amide groups for *in situ* nanotube functionalization. A P-type to n-type semiconducting behavior conversion occurs. Ball milling at ambient settings lowers MWCNTs and broken end cleavage morphology, suggesting a distinct process. A mechanical-chemical process at room temperature with potassium hydroxide changed SWCNTs in the solid phase [92]. A simple mechanical-chemical technique placed metal nanoparticles on nanotubes in 10 min.

6.10 Molecular beam epitaxy

In order for molecular beam epitaxy (MBE) to function in ultra-high vacuum (UHV), one must heat a crystalline substrate and then react with molecular or atomic beams to grow crystals layer by layer. The term "molecular beam epitaxy" did not appear until 1970. Contrasted with the viscous fluid-like flows, molecular beams are the directional kinematic fluxes of atoms or molecules that do not involve collisions. The Greek roots of the English words "akin" and "taxis" (meaning "arrangement" or "order") form the "epitaxy" word. Epitaxy refers to the method of systematically building up successive crystalline layers that share the same crystal structure. Substrate and subsequent layer surface conditions are thus critical for epitaxial crystal growth. Ultra-high vacuum (UHV) and kinetic mechanisms such as adsorption, desorption, dissociation, migration, reaction, and incorporation are used to react atomic and molecular beams with a crystalline surface, which make molecular b These features allow the optimization of stoichiometry and epitaxy through *in situ* monitoring and control during substrate preparation and film growth in real time growth [65]. Even though molecular beam epitaxy (MBE) is a vacuum evaporation method, it uses atomic or molecular beams in a UHV environment to control the composition, growth conditions, purity, and monitoring in real time. The dominance of surface forces allows for the minimization of material and dopant interdiffusion at low substrate temperatures. So, MBE can make crystalline thin-film heterostructures that are very pure and have precise control over their edges, doping, and makeup, all within the range of a few nanometers, and with even growth along all sides and directions. The functional behavior of thin-film structures is through the use of MOCVD, its variants, the sub-nanometer structure, and doping control [93, 94].

6.11 Chemical vapor deposition method

A novel ensemble of graphitic carbon nanofibers and nanotubes is created. With or without a Ni catalyst, CVD produces carbon by using the holes in an alumina template membrane. You can generate tubules and fibrils of various materials, including carbon, metals, semiconductors, polymers, and composites, using a template membrane with cylindrical pores of uniform diameter. Fabrication techniques for these materials include electrochemical synthesis, *in situ* polymerization, electroless metal deposition, and sol-gel [95]. Although these methods are effective, other materials necessitate other synthesis procedures to produce fibrils and tubules. CVD produces films or particles by breaking down molecules in the gas phase into reactive species. CVD can deposit conductive, semiconducting, and insulating materials. The controlled synthesis of nanomaterials in porous hosts, including zeolite nanochannels, has lately been the focus of CVD. CVD controls the stoichiometry of films and deposits thin films of materials over nonuniform structures. They outperform thin-film TiS_2 electrodes in terms of capacity, resistance, and sensitivity to sluggish electron transport. Various applications and electrochemistry are driving the development of new carbon structures. Metal catalysts such as Ni, Fe, Fe-Cu, and Pt affect the size and shape of hydrocarbon-decomposed carbon nanostructures. Chemical vapor deposition, either with or without a nickel catalyst, is used for carbon synthesis in holes in alumina template membranes. Forming carbon nanofibers requires low temperatures. The annealing process uses a nickel catalyst to produce highly organized graphitic nanofibers. Researchers have tested metal catalysts such as Co and Fe [96].

6.12 Laser pyrolysis

This process makes use of a mixture of metallic composite precursors, which act as intermediate substances in the production of other materials, as they are not present in the final product. Heat produces nanoparticles from the solution by evaporating the solvent [97]. Here, the two phases of synthesis are combined by using gas to transport solution droplets instead of liquid. Most often, spray pyrolysis is used. Heat droplets chemically decompose into solid nanoparticles, while precursor solution particles in the air transform into atoms [98]. While the solute stays the same and the solvent decreases when the solution temperature rises, the solvent becomes more saturated.

A container is used to feed the precursor solution into the atomizer. The atomizer uses a heated solution to produce gaseous droplets. The equipment can transfer vaporizable droplets to different chambers and sections by injecting dry gas. The atomizer sends carrier gas droplets to the furnace. The temperatures of the various components within the furnace can vary widely [99]. How hot that area is. When the furnace turns off and the solvent evaporates completely, it separates from the carrier gas droplets. A furnace exhaust precipitator can collect nanoparticles and nanocomposites.

Nanocomposites, metallic, and metallic oxide nanoparticles are the end products. By adjusting the heating temperature of the furnace and the content of the precursor solution, one can enhance the physical properties of nanoparticles. The rate of liquid vaporization in the solution affects the shape, strength, and other properties of nanoparticles. This method produces nanoparticles that are hollow, dense, and foamy, or have a shell. Pyrolyzing materials yields nanoparticles and nanocomposite particles of several elements, including iron, titanium, bismuth oxide, cobalt oxide, and copper oxide [100].

Here are two of the most fundamental bottom-up approaches for creating metallic nanoparticles. Metal nanoparticles form through the condensation of inert gases. Solid metallic nanoparticles and nanocomposites of many metals include metallic oxides, by evaporating the solvent from the precursor solution and then creating aerosols from the mixture. Chemical vapor deposition or condensation refers to the process of depositing or condensing the composite in the precursor solution. Overheating the precursor solution during pyrolysis causes the solvent to evaporate, resulting in the formation of metallic nanoparticles. The precursor solution content and furnace heating temperature determine the characteristics of the nanoparticles. Always keep in mind that there is no magic bullet for creating metallic nanoparticles. Not all of these processes produce small metals; some produce metallic nanoparticles.

7. Chemical methods

7.1 Solvothermal synthesis

Solvothermal synthesis prepares metals, semiconductors, ceramics, and polymers. To optimize the synthesis process for precursor contact, using a solvent between 1 and 10,000 atm, and between 100 and 1000°C, water is the solvent of choice in hydrothermal synthesis. Hydrothermal synthesis takes place at temperatures lower than the supercritical water temperature (374°C). Nanocrystals, single crystals, bulk powders, and thin films are all byproducts of this procedure. The concentration of the chemical of interest, the degree of solvent supersaturation, and the applied kinetic control determine the crystal shape, which can be a sphere, rod, or wire. This approach can prepare compounds that are difficult to synthesize, as well as those that are thermodynamically stable and metastable. For last 10 years, nanocrystals have made up nearly all of the solvothermal synthesis literature. The latest developments are in a nanocrystalline solvothermal synthesis [101].

The "hydrothermal/solvothermal process" refers to the chemical reactions involving solvents in closed containers heated to almost critical temperatures and subjected to autogenous pressures. When water and a solvent are involved, it is referred as a "hydrothermal" process. It also details processes occurring in the presence of environmental factors. While "solvothermal process" refers to the use of organic solvents, this chapter focuses on hydrothermal processes that employ water.

It is common practice to use autoclaves, pressure containers, or high-pressure bombs to conduct hydrothermal or solvothermal reactions. Most hydrothermal/solvothermal reactors use metal autoclaves lined with Teflon or alloy, or a can, beaker, or tube made of Teflon, platinum, gold, or silver to protect the autoclave body from the very corrosive solvent that is placed under high temperatures and pressure. Bourdon gauges and stirring attachments can monitor pressure and concentration gradients in autoclaves. At experimental temperatures and pressures, the ideal hydrothermal/solvothermal autoclave has a long running time, is easy to install and disassemble, and is leakproof. Specialized autoclaves are required for the processing of glass, high-strength metals, and quartz. Material corrosion resistance, temperature, and pressure are the three most critical reactor selection criteria. Estimating and managing pressures below the autoclave material strength in sealed vessels is essential for safety reasons [102, 103]. Solvent methods allow for the synthesis of oxides, elements of groups III–V, II–IV, and VI, MOFs, and transitional metals. These approaches can produce organic NPs when organic compounds resist hydrolysis under extreme conditions. Stability concerns for organic compounds have slowed the advancement of organic NP synthesis techniques. Traditional wet chemistry can make organic NPs regardless of the use of surfactants. You can use a lot of different metal ions and organic ligands to connect MOF molecules, which makes hydrothermal and solvothermal MOF NP synthesis an interesting process to look into. In the next few decades, this study may produce some intriguing findings [104, 105].

7.2 Aerosol-based process

Nanoscale metal combinations yield multicomponent metal nanoparticles with diverse physical and chemical properties. Electrical properties make metals intriguing for catalysis, plasmonics, nanomedicine, and electronics. Researchers working on multicomponent metallic nanostructures favor high-entropy alloy nanoparticles. Nanostructures with five or more well-mixed metal components at equimolar ratios are high-entropy alloys [106]. High-entropy alloy nanoparticle production in the synthesis-structure study involves the production of Ni, Co, Cu, Fe, and Pt HEA nanoparticles. Five metal salts with equal amounts in DI water were turned into aerosol particles by a collision-type atomizer in a typical experiment. Zero-valent high-entropy alloys are made by nebulizing in a 10% H/90% Ar reducing gas. Silica dryers extracted an aqueous solvent from the gas mixture that nebulized aerosol droplets. Highly entropy alloy nanoparticles are formed by heating dried aerosol particles with five mixed metal salt species to approximately 1100°C, allowing for rapid thermal decomposition and hydrogen reduction. A membrane filter captured reactor particles. Depending on reactor geometry and flow rate, aerosol particles stay in the heating zone for 3 seconds. Ni, Co, Cu, Fe, and Pt HEA nanoparticles (~60 nm in diameter) show elemental distributions. Aerosol droplet-induced HEA nanoparticle formation is effective due to the equal distribution of Ni, Co, Cu, Fe, and Pt elements in the nanoparticle and their overlap in the combined EDS map. Bulk sample XRD reveals single-phase particles. EDS shows the particle's main phase is metallic with Ni, Co, Cu, Fe, Pt, and little oxygen. All five metals have an unfavorable 20% atomic proportion in the HEA nanoparticle, according to the quantitative EDS analysis. Nanoparticles comprise 19.16% Ni, 19.19% Co, 11.45% Cu, 24.59% Fe, and 25.60% Pt. Other metals are more abundant in nanoparticles than Cu. High-entropy alloy nanoparticles produced by carbothermal shock pCu have the highest metal vapor pressure, making this possible. In the previous study, they used normal synthesis methods like precursor compensation to balance the atomic ratios of volatile metals.

General approach for producing high-entropy alloy nanoparticles via aerosol-assisted techniques. To discuss the synthesis of HEA nanoparticles for Fe, Co, Ni, Cu, Pd, and Pt use heat and hydrogen gas to decrease metal salt precursors, as most metal elements have low standard potentials [107]. In our experiment, hydrogen is used to reduce metal nitrate precursors.

$$(NO3)2 + 2H2 \rightarrow M + 2NO2\uparrow + 2H2O\uparrow \qquad (1)$$

7.3 Sol-gel

Sol-gel has shown promise in controlling oxide bulk and surface characteristics. Sol-gel allows different precursors, which is a benefit. Recent research uses TEOS as a precursor. Sol-gel manufacturing involves precursor hydrolysis and resultant polycondensation [108]. The XRD, FTIR, SEM, TGA, and DTA analyzes nano-thiamine hydrochloride composite (NTHC) synthesized using a novel sol-gel method.

Sol-gel produces a thiamine hydrochloride nanostructure formed this sol by hydrolyzing TEOS using water, nitric acid, and ethanol. Components have a molar ratio of 1:10:2:1 dissolved TEOS in alcohol, while continuously stirring, and then added deionized water and acid drop by drop at 80°C. Aging the solution for 3 hours at 80°C under reflux yielded a transparent silica sol, for adding thiamine hydrochloride in varying amounts to the translucent and uniform silica sol. Following 45 minutes,

120 μL of Triton X-114 was added to the solution to prevent gel fracturing. The surfactant was added at a concentration of 0.2 mM, which was below its critical micelle concentration, and stirred it for 120 minutes at 100°C with creating a nano-thiamine hydrochloride structure with uniform diameters of 5–15 nm at 100°C using the sol-gel strategy. The analysis included powder X-ray diffraction, FTIR, SEM, TGA, and DTA. The nano-thiamine hydrochloride demonstrated high thermal stability. The sol-gel approach produces pure-phase nanomaterials with controlled size, morphology, and shape [109].

7.4 Molecular/atomic condensation

Metal nanoparticles typically use this method. The method involves vacuum-heating a bulk substance to evaporate and atomize materials for both inert and reactive gas chambers. Nanoparticles occur when metal atoms collide with gas molecules and cool quickly. Use of reactive gas like oxygen produces metal oxide nanoparticles [110].

7.5 Spray drying

Spray drying produces micron-sized products from nanoparticles, nanocapsules, and other nanometer-sized components. Thus, the spray drying technique may link nanomaterial production to existing procedures that manufacture end products from micron-sized materials [111]. Spray drying converts liquid into micron-sized droplets and dries them with hot gas in one step. A dry and insulated cylindrical container with a conical base is used. Air dispensers on the chamber roof supply the hot drying gas. There are various ways to atomize. Although there are more types of nozzles, the three most common are centrifugal, pressure, and air blast. The time required for the first droplet to dry is between 1 and 5 seconds. Liquid evaporation energy consumption cools the drying gas to temperatures ranging from 50 to 120°C, depending on the product and application. Products can have their temperatures lowered below the gas outlet temperature by the use of evaporative cooling, which reduces to use cyclones or bag filters to separate the drying gas from the dried product. From the dried product, rapid, low-temperature spray drying maintains the potential properties of nanocomponents. Spray drying re-disperses granules by suspending them first.

The four main areas of instrumentation in a spray dryer are atomization, drying, collecting, and waste filtration zones. Atomization zone nozzles disperse liquid input into the air as minute droplets for compressed gas drying. The air intake, heater, drying chamber, and aspirator work together to filter and heat the air to the appropriate temperature to dry the sprayed droplets. Air temperature sensors for the drying chamber's entrance and exit are located in this area. A separator gathers particles in the collecting zone. To prevent particles from escaping and to heat the air prior to the next drying cycle, the waste filtration zone absorbs both the undesirable particles and the damp air. Based on the design of the atomizer and the size of the droplets, the drying chamber can have a diameter-to-length ratio of 1 to 5, or even higher. Spray drying transforms nanotitanium dioxide suspensions into non-dusty products. The goal was to create granules for coatings that employ titanium's anatase phase's photocatalytic properties. A product that could flow freely and had an average granule size of less than 50 μm was required for the coating equipment to be handled easily. In order to determine the optimal conditions for product manufacturing, researchers have examined several spray drying setups and settings [112].

7.6 Cryo-chemical synthesis

Cryo-chemical synthesis yields materials and nanomaterials using low-temperature chemical processes. Low-temperature chemical synthesis in aqueous solutions is most common. The cation of the synthetic material is kept in a stoichiometric ratio in a liquid, suspension, or chemical coprecipitate. This is then quickly frozen and freeze-dried in a vacuum. Thermal decomposition synthesizes the oxide powder, which has a crystallite size ranging from 40 to 300 nm [113]. Most commonly used are cryo-chemical synthesis processes using aqueous solutions. Quickly freeze the synthesized material cations, freeze-dry them in a vacuum, and then thermally break them down in solutions, suspensions, or *via* chemical co-precipitation of the original solution components. The crystallite sizes range from 40 to 300 nm, and the levels of agglomeration change depend on whether the parts are frozen in a solution, a suspension, or sludge. Method to remove ice is by using polar organic solvents for low-temperature extraction or by coprecipitating frozen components in a precipitant solution. Cryo-impregnation is recommended for porous supports, where substrates soaked in a catalytically active solution freeze and then freeze-dry. Low temperatures can greatly improve substrate micro-component dispersion uniformity. Cryo-chemical processes include low-temperature mechanochemical techniques like cryo-milling, and according to certain writer can use inert gases or organic compounds to deposit metal vapors on very cold surfaces, creating 1–10-nm metal nanoparticles.

Cryo-chemical methods have been used in recent studies to improve the production of salts of nanocrystalline metal oxides (such as NiO, FeO, and CeO) and NaNO. According to the study, mixing sols and solutions with liquid nitrogen (−196°C) can make the specific surface area 10 times bigger and the particle dispersion 3 to 8 times bigger. The proposed technique not only decreases nanoparticle aggregation, but also yields crystallites of NiO, Fe_2O_3, and CeO_2 with diameters smaller than 10 nm. Freezing in liquid nitrogen can cut the size of $NaNO_3$ salt crystals to less than 50 nm, which is a big improvement over current methods that use an average diameter of 300–1000 nm.

8. Biological methods

8.1 Nanoparticle synthesis by plant extracts

Recent decades have seen a dramatic increase in research into metal nanoparticles, thanks to their unusual physical, chemical, and biological properties [114]. Because of its significance and intrigue, nanoparticle synthesis using chemical and physical methods has been the subject of much study. One can create gold nanoparticles using methods such as ion sputtering, chemical reduction, sol-gel, hydrothermal, and reverse micelle. To make inorganic nanoparticles in other ways [115, 116], you can use biodegradable polymers such as chitosan and plant, fungal, sugar, bacterial, and reductant/stabilizer extracts. Plant extract nanoparticle synthesis is easy, one-step, cost-effective, environmentally friendly, and reproducible [117, 118]. Plant extracts are easier to use to make nanoparticles than whole plant extracts and tissue. Plant extract-mediated synthesis is gaining popularity. Plant extracts can reduce and stabilize nanoparticle synthesis. The source of the plant extract influences the properties of the nanoparticles. Extracts have varying concentrations and combinations of organic reducing agents [119]. Plant extract-mediated bio-reduction usually entails combining the extract with a metal salt solution in water. At room temperature, it takes minutes to complete the reaction. Bio-reduction is complicated due to the quantity of chemicals involved. Researchers are interested in the processes of plant metal ion absorption and bio-reduction, as plants and their metabolites are more conducive to the formation of nanoparticles. Researchers manufacture AgNPs from leaves such as neem, black tea employed Ocimum sanctum (Tulsi) leaf extracts, and quercetin (flavonoid) to produce silver nanoparticles (AgNPs) from cationic silver using green chemistry. This study investigated the role of Tulsi biomolecules in the production of AgNPs. Different types of XRD, TEM, and optical spectroscopy (absorption, photoluminescence (PL), PL-lifetime, and Fourier transform infrared) were used to look at the size, shape, morphology, and stability of AgNPs [120, 121].

Azadirachta indica, *Aloe vera* leaf extract, etc., produce gold nanoparticles [122]. Green synthesis of Pd and Pt nanoparticles has been done using *Cinnamomum camphora*, *Gardenia jasminoides*, *Pinus resinosa*, *Anogeissus latifolia*, *Glycine max*, *Ocimum sanctum*, *Curcuma longa*, *Musa paradisiaca*, *Cinnamon zeylanicum*, *Pulicaria glutinosa*, and *Doispyros kaki*. *Aloe vera* flower extract reduces aqueous copper ions to nanoparticles [123]. A 578-nm peak at the UV-visible spectrometer confirmed the presence of 40-nm Cu nanoparticles [124]. The leaves of Meliaceae species *Azadirachta indica*, flower, and *Vitex negundo* are the best for ZnO synthesis [125].

8.2 Nanoparticle synthesis by bacteria

Researchers have synthesized inorganic nanoparticles mostly using prokaryotes. Bacteria are ideal for INP production due to their abundance and adaptability [126]. In addition to its abundance, it has low production costs, a rapid growth rate, cheap culture, and simple control over oxygenation, temperature, pH, and incubation time. Bacteria are ideal for nanoparticle manufacturing because they remove heavy metal ions. Some bacteria can employ specialized defensive mechanisms to counteract stressors such as heavy metal ions or metal poisoning. High amounts of metal ions can promote the growth of *Pseudomonas stutzeri* and *Pseudomonas aeruginosa*. Brock and Gustafson [127] found that when cultured in sulfur, thiobacillus ferro-oxidants, thio-oxidants, and *Sulfolobus acidocaldarius* were capable of converting ferric ion to ferrous ions. Low-pH *T. thiooxidans* achieved the aerobic reduction of ferrous. The rapid bacterial reoxidation in oxygen obstructed aerobic bio-reduction, preventing *T. thiooxidans* from oxidizing the produced ferrous iron [128]. Some bacteria, including *Enterobacter cloacae*, *Desulfovibrio desulfuricans*, and *Rhodospirillum rubrum*, change selenite into selenium. Other bacteria, such as Geobacter metalloproteins (formerly strain GS-15) and *Shewanella putrefaciens*, directly lower Tc (VII), and *Escherichia coli* K12, make tellurium (Te). Various bacterial species, such as *Escherichia coli*, *P. aeruginosa*, *B. subtilis*, and *B. cereus*, have the ability to extract Ag+, Cd2+, Cu2+, and La3+ from solutions. Bacteria bound large amounts of metal cations together. Bacteria with intrinsic magnetic properties can synthesize magnetite nanoparticles [129].

To carry out the synthesis using Rhod pseudomonas capsulata, *Bacillus megaterium* D01, *Delftia acidovorans*, *Escherichia coli* DH5FC, and *Plectonema boryanum* UTEX 485, the generated AuNPs remained consistent for several weeks and had a size of 1.9 ± 0.8 nm. In a different study, Li et al. made antibacterial AuNPs using *Deinococcus radiodurans*, a bacterium that is resistant to radiation and oxidants (page 161). Several experiments have shown that *Bacillus licheniformis* and other species of Bacillus can create AgNPs intracellularly. Industrial use of these techniques was not feasible due to their slowness. Aspergillus Niger YESM 1 produces magnetite nanoparticles by decomposing ferric chloride and ferrous sulfate into ferrous oxide and ferrous sulfate, respectively.

To make iron nanoparticles (18 nm in diameter) and Fe_0O_4 nanoparticles (50 nm in diameter), the particles were heated very quickly in EtOH at 300°C and 850 psi for 1 hour. There is evidence that bacteria can produce luminous quantum dot nanoparticles based on semiconductors such as ZnS, CdS, and PbS. These nanoparticles find uses in cell labeling, imaging, and fluorescent biomarkers. After being grown in 1 mM $CdSO_4$ for 72 hours, *Rhodpseudomonas palustris* was able to make CdS QDs [130]. Microbes created nanoparticles of copper, nickel, zinc, palladium, titanium, and cobalt. *Lactobacillus* sp. generated spherical Ti NPs of 40–60 nm at normal temperature.

8.3 Nanoparticles synthesis by fungi

Fungal NP synthesis is the cheapest, greenest, and least energy-intensive technique. Synthesizing NPs of a certain size using fungus is a huge advance. To alter this trait, researchers have improved several situations. TEM and SEM analyzers are the most frequent methods for detecting NP sizes since they are simple and easy to analyze. The broad-spectrum antibacterial activity of silver nanoparticles makes them stand out. Microorganisms' cell walls and membranes may allow these nanoparticles to enter. They harm cells, produce reactive oxygen species, and disrupt signal transduction.

Fungal-based NP production reduces metal ions with internal and extracellular enzymes and biomolecules to biomineralize. Trichoderma species like *T. reesei* are better for mass AgNP generation than harmful fungi. Industrially adapted *T. reesei* has not caused any harm. AuNPs from *Verticillium luteoalbum* and AgNPs from the Penicillium strain are present.

8.4 Nanoparticle synthesis by yeast

Everywhere you look, you can find fungi, which are eukaryotic decomposers. Only 70,000 of the estimated 1.5 million species of fungi on Earth have received positive identification. To need to conduct more research on fungi in nanobiotechnology, the bioaccumulation of metals and the fact that fungi are tolerable make them attractive candidates for the natural production of metallic nanoparticles. The production of nanoparticles, such as the fermentation of thin solid substrates, benefits from the ease of scaling. Fungi effectively secrete extracellular enzymes, thereby enabling large-scale enzyme synthesis possible. Through biomimetic mineralization and reducing enzymes intracellularly or extracellularly, fungi can create metal nanoparticles/meso- and nanostructures. Verticillium sp., a filamentous fungus, synthesizes silver NPs extracellularly in one of the most studied biosynthesis studies. NP synthesis commonly involves the filamentous fungus *Fusarium oxysporum*.

The *R. oryzae*-made nanogold-bio combination demonstrated strong antibacterial action against *B. subtilis*, *P. aeruginosa*, *E. coli*, *Salmonella* sp., *S. aureus*, and the yeasts *S. cerevisiae* and *C. albicans*. *Verticillium* and *Fusarium oxysporum*, among other fungi, can produce metal nanoparticles inside or outside of cells by combining hydrous ions with metals such as AuCl or Ag. *S. cerevisiae*-produced silver nanoparticles inhibited fluconazole-susceptible and fluconazole-resistant *Candida albicans*.

8.5 Nanoparticle synthesis by biological particles

Nanoparticles can be biosynthesized from viruses, proteins, peptides, and enzymes. Cowpea chlorotic mottle and mosaic viruses mineralize inorganic materials.

Tobacco mosaic virus mineralizes sulfide and crystalline nanowires generated in nanoshell have found peptides that nucleate nanocrystals on M13 bacteriophage and in combinatorial screens.

9. Conclusion

The synthesis and characterization of nanomaterials require a variety of methods, showcasing their complexity and adaptability. Nanomaterial qualities and applications depend on the synthesis method—physical, chemical, or biological. Nanotechnology advances through nanomaterial creation and characterization. Physical and chemical processes are precise, while biological methods are sustainable and unique. The desired nanomaterial properties and applications determine the synthesis procedure. Validating nanomaterial qualities and verifying their appropriateness for varied applications require effective characterization methods. Advances in synthesis and characterization will let us create sophisticated nanomaterials with customized characteristics for many applications.

Acknowledgements

The author acknowledges (or authors acknowledge) the use of Grammarly premium account (AI tool) for language polishing of the manuscript.

Author details

Ananda Kumar Chettupalli[1*], Kavitha Marati[2], Himabindu Peddapalli[3], Mounika Kuchukuntla[3], Rajendra Kumar Jadi[3] and Shaik Abdul Rahaman[1]

1 Department of Pharmacy, School of Medical and Allied Sciences, Galgotias University, Greater Noida, Uttar Pradesh, India

2 Department of Pharmacology, Raja Bahadur Venkata Rama Reddy (RBVRR) Women's College of Pharmacy, Barkatpura, Hyderabad, Telangana, India

3 Department of Pharmaceutical Sciences, Center for Nanomedicine, School of Pharmacy, Anurag University, Hyderabad, Telangana, India

*Address all correspondence to: anandphd88@gmail.com

References

[1] Adams FC, Barbante C. Nanoscience, nanotechnology and spectrometry. Spectrochimica Acta Part B: Atomic Spectroscopy. 2013;**86**:3-13. DOI: 10.1016/j.sab.2013.04.008

[2] Rauscher H, Rasmussen K, Sokull-Klüttgen B. Regulatory aspects of nanomaterials in the EU. Chemie Ingenieur Technik. 2017;**89**:224-231. DOI: 10.1002/cite.201600076

[3] Algarni TS, Al-Mohaimeed AM, Al-Odayni A-B, Abduh NAY. Activated carbon/ZnFe2O4 nanocomposite adsorbent for efficient removal of crystal violet cationic dye from aqueous solutions. Nanomaterials. 2022;**12**:3224. DOI: 10.3390/nano12183224

[4] Sapavatu SN, Chinthala R, Jadi RK. An overview on pharmacokinetics of polymeric nanoparticles intended for oral delivery. Journal of Young Pharmacists. 2020;**12**:187-189. DOI: 10.5530/jyp.2020.12.57

[5] Liu S, Yu B, Wang S, Shen Y, Cong H. Preparation, surface functionalization and application of Fe3O4 magnetic nanoparticles. Advances in Colloid and Interface Science. 2020;**281**:102165. DOI: 10.1016/j.cis.2020.102165

[6] Lavernia EJ, Han BQ, Schoenung JM. Cryomilled nanostructured materials: Processing and properties. Materials Science and Engineering A. 2008;**493**: 207-214. DOI: 10.1016/j.msea.2007.06.099

[7] Popok VN, Kylián O. Gas-phase synthesis of functional nanomaterials. Applied Nano. 2020;**1**:25-58. DOI: 10.3390/applnano1010004

[8] Liu P, Qin R, Fu G, Zheng N. Surface coordination chemistry of metal nanomaterials. Journal of the American Chemical Society. 2017;**139**:2122-2131. DOI: 10.1021/jacs.6b10978

[9] Mai H, Mutlu R, Tawk C, Alici G, Sencadas V. Ultra-stretchable MWCNT–Ecoflex piezoresistive sensors for human motion detection applications. Composites Science and Technology. 2019;**173**:118-124. DOI: 10.1016/j.compscitech.2019.02.001

[10] Biswas A, Bayer IS, Biris AS, Wang T, Dervishi E, Faupel F. Advances in top–down and bottom–up surface nanofabrication: Techniques, applications & future prospects. Advances in Colloid and Interface Science. 2012;**170**:2-27. DOI: 10.1016/j.cis.2011.11.001

[11] Cai Z, Li Z, Ravaine S, He M, Song Y, Yin Y, et al. From colloidal particles to photonic crystals: Advances in self-assembly and their emerging applications. Chemical Society Reviews. 2021;**50**:5898-5951. DOI: 10.1039/D0CS00706D

[12] Busseron E, Ruff Y, Moulin E, Giuseppone N. Supramolecular self-assemblies as functional nanomaterials. Nanoscale. 2013;**5**:7098. DOI: 10.1039/c3nr02176a

[13] Duan H, Wang D, Li Y. Green chemistry for nanoparticle synthesis. Chemical Society Reviews. 2015;**44**: 5778-5792. DOI: 10.1039/C4CS00363B

[14] Ciriminna R, Fidalgo A, Pandarus V, Béland F, Ilharco LM, Pagliaro M. The sol–gel route to advanced silica-based materials and recent applications. Chemical Reviews. 2013;**113**:6592-6620. DOI: 10.1021/cr300399c

[15] Manawi YM, Ihsanullah SA, Al-Ansari T, Atieh MA. A review of carbon nanomaterials' synthesis via the chemical vapor deposition (CVD) method. Materials. 2018;**11**:822. DOI: 10.3390/ma11050822

[16] Barrino F. Hybrid organic–inorganic materials prepared by sol–gel and sol–gel-coating method for biomedical use: Study and synthetic review of synthesis and properties. Coatings. 2024;**14**:425. DOI: 10.3390/coatings14040425

[17] Singh LP, Bhattacharyya SK, Kumar R, Mishra G, Sharma U, Singh G, et al. Sol-gel processing of silica nanoparticles and their applications. Advances in Colloid and Interface Science. 2014;**214**:17-37. DOI: 10.1016/j.cis.2014.10.007

[18] Danks AE, Hall SR, Schnepp Z. The evolution of 'sol–gel' chemistry as a technique for materials synthesis. Materials Horizons. 2016;**3**:91-112. DOI: 10.1039/C5MH00260E

[19] Mackenzie JD, Bescher EP. Chemical routes in the synthesis of nanomaterials using the sol–gel process. Accounts of Chemical Research. 2007;**40**:810-818. DOI: 10.1021/ar7000149

[20] Zhang J, Wu L, Chan H-K, Watanabe W. Formation, characterization, and fate of inhaled drug nanoparticles. Advanced Drug Delivery Reviews. 2011;**63**:441-455. DOI: 10.1016/j.addr.2010.11.002

[21] Swanson J, Kittelson D, Pui D, Watts W. Alternatives to the gravimetric method for quantification of diesel particulate matter near the lower level of detection. Journal of the Air & Waste Management Association. 2010;**60**:1177-1191. DOI: 10.3155/1047-3289.60.10.1177

[22] Gyulasaryan H, Kuzanyan A, Manukyan A, Mukasyan AS. Combustion synthesis of magnetic nanomaterials for biomedical applications. Nanomaterials. 2023;**13**:1902. DOI: 10.3390/nano13131902

[23] Gautam M, Kim JO, Yong CS. Fabrication of aerosol-based nanoparticles and their applications in biomedical fields. Journal of Pharmaceutical Investigation. 2021;**51**:361-375. DOI: 10.1007/s40005-021-00523-1

[24] Chen Y-C, Lin W-H, Tseng W-S, Chen C-C, Georger R, Chen C-D, et al. Direct growth of mm-size twisted bilayer graphene by plasma-enhanced chemical vapor deposition. Carbone New York. 2020;**156**:212-224. DOI: 10.1016/j.carbon.2019.09.052

[25] Droes SR, Kodas TT, Hampden-Smith MJ. Plasma-enhanced chemical vapor deposition (PECVD). In: Carbide, Nitride and Boride Materials Synthesis and Processing. Dordrecht: Springer Netherlands; 1997. pp. 579-603. DOI: 10.1007/978-94-009-0071-4_23

[26] Yan Y, Miao J, Yang Z, Xiao F-X, Bin YH, Liu B, et al. Carbon nanotube catalysts: Recent advances in synthesis, characterization and applications. Chemical Society Reviews. 2015;**44**:3295-3346. DOI: 10.1039/C4CS00492B

[27] Harish V, Ansari MM, Tewari D, Gaur M, Yadav AB, García-Betancourt M-L, et al. Nanoparticle and nanostructure synthesis and controlled growth methods. Nanomaterials. 2022;**12**:3226. DOI: 10.3390/nano12183226

[28] Mbarek A, Moussa G, Leblond CJ. Pharmaceutical applications of molecular tweezers, clefts and clips. Molecules. 2019;**24**:1803. DOI: 10.3390/molecules24091803

[29] Vasić K, Knez Ž, Leitgeb M. Multifunctional iron oxide nanoparticles as promising magnetic biomaterials

in drug delivery: A review. Journal of Functional Biomaterials. 2024;**15**:227. DOI: 10.3390/jfb15080227

[30] Whitehead CB, Özkar S, Finke RG. LaMer's 1950 model of particle formation: A review and critical analysis of its classical nucleation and fluctuation theory basis, of competing models and mechanisms for phase-changes and particle formation, and then of its application to silver halide, semiconductor, metal, and metal-oxide nanoparticles. Materials Advances. 2021;**2**:186-235. DOI: 10.1039/D0MA00439A

[31] Moisala A, Nasibulin AG, Kauppinen EI. The role of metal nanoparticles in the catalytic production of single-walled carbon nanotubes—A review. Journal of Physics: Condensed Matter. 2003;**15**:S3011-S3035. DOI: 10.1088/0953-8984/15/42/003

[32] Grammatikopoulos P, Steinhauer S, Vernieres J, Singh V, Sowwan M. Nanoparticle design by gas-phase synthesis. Advances in Physics: X. 2016;**1**:81-100. DOI: 10.1080/23746149.2016.1142829

[33] Zhang H, Zhou H, Wang Y, Li S, Biswas P. Mini review on gas-phase synthesis for energy nanomaterials. Energy & Fuels. 2021;**35**:63-85. DOI: 10.1021/acs.energyfuels.0c03264

[34] Grammatikopoulos P, Bouloumis T, Steinhauer S. Gas-phase synthesis of nanoparticles: Current application challenges and instrumentation development responses. Physical Chemistry Chemical Physics. 2023;**25**:897-912. DOI: 10.1039/D2CP04068A

[35] Moiz SA, Alahmadi ANM, Aljohani AJ. Design of silicon nanowire array for PEDOT:PSS-silicon nanowire-based hybrid solar cell. Energies (Basel). 2020;**13**:3797. DOI: 10.3390/en13153797

[36] Tighe CJ, Cabrera RQ, Gruar RI, Darr JA. Scale up production of nanoparticles: Continuous supercritical water synthesis of Ce–Zn oxides. Industrial and Engineering Chemistry Research. 2013;**52**:5522-5528. DOI: 10.1021/ie3025642

[37] Franco P, De Marco I. Nanoparticles and nanocrystals by supercritical CO2-assisted techniques for pharmaceutical applications: A review. Applied Sciences. 2021;**11**:1476. DOI: 10.3390/app11041476

[38] Bhardwaj N, Kundu SC. Electrospinning: A fascinating fiber fabrication technique. Biotechnology Advances. 2010;**28**:325-347. DOI: 10.1016/j.biotechadv.2010.01.004

[39] Teo W-E, Inai R, Ramakrishna S. Technological advances in electrospinning of nanofibers. Science and Technology of Advanced Materials. 2011;**12**:013002. DOI: 10.1088/1468-6996/12/1/013002

[40] Pérez-Page M, Yu E, Li J, Rahman M, Dryden DM, Vidu R, et al. Template-based syntheses for shape controlled nanostructures. Advances in Colloid and Interface Science. 2016;**234**:51-79. DOI: 10.1016/j.cis.2016.04.001

[41] Ilievski F, Mirica KA, Ellerbee AK, Whitesides GM. Templated self-assembly in three dimensions using magnetic levitation. Soft Matter. 2011;**7**:9113. DOI: 10.1039/c1sm05962a

[42] Thorkelsson K, Bai P, Xu T. Self-assembly and applications of anisotropic nanomaterials: A review. Nano Today. 2015;**10**:48-66. DOI: 10.1016/j.nantod.2014.12.005

[43] Tamir TS, Xiong G, Shen Z, Leng J, Fang Q, Yang Y, et al. 3D printing in materials manufacturing industry: A realm of industry 4.0. Heliyon. 2023;**9**:e19689. DOI: 10.1016/j.heliyon.2023.e19689

[44] Goh GD, Yap YL, Tan HKJ, Sing SL, Goh GL, Yeong WY. Process–structure–properties in polymer additive manufacturing via material extrusion: A review. Critical Reviews in Solid State and Materials Sciences. 2020;**45**:113-133. DOI: 10.1080/10408436.2018.1549977

[45] Vaezi M, Drescher P, Seitz H. Beamless metal additive manufacturing. Materials. 2020;**13**:922. DOI: 10.3390/ma13040922

[46] Liu J, Sun L, Xu W, Wang Q, Yu S, Sun J. Current advances and future perspectives of 3D printing natural-derived biopolymers. Carbohydrate Polymers. 2019;**207**:297-316. DOI: 10.1016/j.carbpol.2018.11.077

[47] Walker M, Humphries S. 3D printing: Applications in evolution and ecology. Ecology and Evolution. 2019;**9**:4289-4301. DOI: 10.1002/ece3.5050

[48] Barthelat F, Yin Z, Buehler MJ. Structure and mechanics of interfaces in biological materials. Nature Reviews Materials. 2016;**1**:16007. DOI: 10.1038/natrevmats.2016.7

[49] Fu D, Reif J. 3D DNA nanostructures: The nanoscale architect. Applied Sciences. 2021;**11**:2624. DOI: 10.3390/app11062624

[50] Tong S, Zhao W, Zhao D, Zhang W, Zhang Z. Biomaterials-mediated tumor infarction therapy. Frontiers in Bioengineering and Biotechnology. 2022;**10**. DOI: 10.3389/fbioe.2022.916926

[51] Seeman NC, Sleiman HF. DNA nanotechnology. Nature Reviews Materials. 2017;**3**:17068. DOI: 10.1038/natrevmats.2017.68

[52] Park H-J, Hong H, Thangam R, Song M-G, Kim J-E, Jo E-H, et al. Static and dynamic biomaterial engineering for cell modulation. Nanomaterials. 2022;**12**:1377. DOI: 10.3390/nano12081377

[53] Zeng H, Du X, Singh SC, Kulinich SA, Yang S, He J, et al. Nanomaterials via laser ablation/irradiation in liquid: A review. Advanced Functional Materials. 2012;**22**:1333-1353. DOI: 10.1002/adfm.201102295

[54] Laurent S, Forge D, Port M, Roch A, Robic C, Vander Elst L, et al. Magnetic iron oxide nanoparticles: Synthesis, stabilization, vectorization, physicochemical characterizations, and biological applications. Chemical Reviews. 2008;**108**:2064-2110. DOI: 10.1021/cr068445e

[55] Grancharova T, Zagorchev P, Pilicheva B. Iron oxide nanoparticles: Parameters for optimized photoconversion efficiency in synergistic cancer treatment. Journal of Functional Biomaterials. 2024;**15**:207. DOI: 10.3390/jfb15080207

[56] Chen Y-J, Groves B, Muscat RA, Seelig G. DNA nanotechnology from the test tube to the cell. Nature Nanotechnology. 2015;**10**:748-760. DOI: 10.1038/nnano.2015.195

[57] Livage J, Henry M, Sanchez C. Sol-gel chemistry of transition metal oxides. Progress in Solid State Chemistry. 1988;**18**:259-341. DOI: 10.1016/0079-6786(88)90005-2

[58] Ismail WNW. Sol–gel technology for innovative fabric finishing—A review. Journal of Sol-Gel Science and Technology. 2016;**78**:698-707. DOI: 10.1007/s10971-016-4027-y

[59] Ruiz-Jorge F, Portela JR, Sánchez-Oneto J, Martínez de la Ossa EJ. Synthesis of micro- and nanoparticles in sub- and supercritical water: From the laboratory to larger scales. Applied Sciences. 2020;**10**:5508. DOI: 10.3390/app10165508

[60] Primc G, Brenčič K, Mozetič M, Gorjanc M. Recent advances in the plasma-assisted synthesis of zinc

oxide nanoparticles. Nanomaterials. 2021;**11**:1191. DOI: 10.3390/nano11051191

[61] Chen F, Zhang D, Yang Q, Yong J, Du G, Si J, et al. Bioinspired wetting surface via laser microfabrication. ACS Applied Materials & Interfaces. 2013;**5**:6777-6792. DOI: 10.1021/am401677z

[62] Hachem K, Ansari MJ, Saleh RO, Kzar HH, Al-Gazally ME, Altimari US, et al. Methods of chemical synthesis in the synthesis of nanomaterial and nanoparticles by the chemical deposition method: A review. BioNanoScience. 2022;**12**:1032-1057. DOI: 10.1007/s12668-022-00996-w

[63] Kruis FE, Fissan H, Peled A. Synthesis of nanoparticles in the gas phase for electronic, optical and magnetic applications—A review. Journal of Aerosol Science. 1998;**29**:511-535. DOI: 10.1016/S0021-8502(97)10032-5

[64] Saha A, Mishra P, Biswas G, Bhakta S. Greening the pathways: A comprehensive review of sustainable synthesis strategies for silica nanoparticles and their diverse applications. RSC Advances. 2024;**14**:11197-11216. DOI: 10.1039/D4RA01047G

[65] Cushing BL, Kolesnichenko VL, O'Connor CJ. Recent advances in the liquid-phase syntheses of inorganic nanoparticles. Chemical Reviews. 2004;**104**:3893-3946. DOI: 10.1021/cr030027b

[66] Choy K. Vapor processing of nanostructured materials. In: Handbook of Nanostructured Materials and Nanotechnology. Elsevier; 2000. pp. 533-577. DOI: 10.1016/B978-012513760-7/50015-0

[67] Alshammari BH, Lashin MMA, Mahmood MA, Al-Mubaddel FS, Ilyas N, Rahman N, et al. Organic and inorganic nanomaterials: Fabrication, properties and applications. RSC Advances. 2023;**13**:13735-13785. DOI: 10.1039/D3RA01421E

[68] Dhand C, Dwivedi N, Loh XJ, Jie Ying AN, Verma NK, Beuerman RW, et al. Methods and strategies for the synthesis of diverse nanoparticles and their applications: A comprehensive overview. RSC Advances. 2015;**5**:105003-105037. DOI: 10.1039/C5RA19388E

[69] Das R, Shahnavaz Z, Ali ME, Islam MM, Abd Hamid SB. Can we optimize arc discharge and laser ablation for well-controlled carbon nanotube synthesis? Nanoscale Research Letters. 2016;**11**:510. DOI: 10.1186/s11671-016-1730-0

[70] Schultrich B. Structure and characterization of vacuum arc deposited carbon films—A critical overview. Coatings. 2022;**12**:109. DOI: 10.3390/coatings12020109

[71] Ashkarran AA. Metal and metal oxide nanostructures prepared by electrical arc discharge method in liquids. Journal of Cluster Science. 2011;**22**:233-266. DOI: 10.1007/s10876-011-0376-4

[72] Chaudhary KT, Rizvi ZH, Bhatti KA, Ali J, Yupapin PP. Multiwalled carbon nanotube synthesis using arc discharge with hydrocarbon as feedstock. Journal of Nanomaterials. 2013;**2013**. DOI: 10.1155/2013/105145

[73] Sinha S, Sanfo K, Dallas P, Kumar S, Porfyrakis K. Process parameter optimisation for endohedral metallofullerene synthesis via the arc-discharge method. Inorganics (Basel). 2024;**12**:38. DOI: 10.3390/inorganics12020038

[74] Szabó A, Perri C, Csató A, Giordano G, Vuono D, Nagy JB. Synthesis methods of carbon nanotubes and related materials. Materials. 2010;**3**:3092-3140. DOI: 10.3390/ma3053092

[75] Corbella C, Portal S, Kundrapu MN, Keidar M. Nanosynthesis by atmospheric arc discharges excited with pulsed-DC power: A review. Nanotechnology. 2022;**33**:342001. DOI: 10.1088/1361-6528/ac6bad

[76] de Jonge N, Bonard J. Carbon nanotube electron sources and applications. Philosophical Transactions of the Royal Society of London Series A: Mathematical, Physical and Engineering Sciences. 2004;**362**:2239-2266. DOI: 10.1098/rsta.2004.1438

[77] Kaushik NK, Kaushik N, Linh NN, Ghimire B, Pengkit A, Sornsakdanuphap J, et al. Plasma and nanomaterials: Fabrication and biomedical applications. Nanomaterials. 2019;**9**:98. DOI: 10.3390/nano9010098

[78] Jang HJ, Jung EY, Parsons T, Tae H-S, Park C-S. A review of plasma synthesis methods for polymer films and nanoparticles under atmospheric pressure conditions. Polymers (Basel). 2021;**13**:2267. DOI: 10.3390/polym13142267

[79] Szabó D, Schlabach S. Microwave plasma synthesis of materials—From physics and chemistry to nanoparticles: A materials scientist's viewpoint. Inorganics (Basel). 2014;**2**:468-507. DOI: 10.3390/inorganics2030468

[80] Linkov P, Artemyev M, Efimov AE, Nabiev I. Comparative advantages and limitations of the basic metrology methods applied to the characterization of nanomaterials. Nanoscale. 2013;**5**:8781. DOI: 10.1039/c3nr02372a

[81] Pechyen C, Tangnorawich B, Toommee S, Marks R, Parcharoen Y. Green synthesis of metal nanoparticles, characterization, and biosensing applications. Sensors International. 2024;**5**:100287. DOI: 10.1016/j.sintl.2024.100287

[82] Huang X, Song J, Yung BC, Huang X, Xiong Y, Chen X. Ratiometric optical nanoprobes enable accurate molecular detection and imaging. Chemical Society Reviews. 2018;**47**:2873-2920. DOI: 10.1039/C7CS00612H

[83] Rodriguez-Otero A, Vargas V, Galarneau A, Castillo J, Christensen JH, Bouyssiere B. Sustainable harnessing of SiO2 nanoparticles from rice husks: A review of the best synthesis and applications. PRO. 2023;**11**:3373. DOI: 10.3390/pr11123373

[84] Dufour T. From basics to Frontiers: A comprehensive review of plasma-modified and plasma-synthesized polymer films. Polymers (Basel). 2023;**15**:3607. DOI: 10.3390/polym15173607

[85] Ning J, Lei Y, Hu H, Gai C. A comprehensive review of surface acoustic wave-enabled acoustic droplet ejection technology and its applications. Micromachines (Basel). 2023;**14**:1543. DOI: 10.3390/mi14081543

[86] Ali A, Shah T, Ullah R, Zhou P, Guo M, Ovais M, et al. Review on recent progress in magnetic nanoparticles: Synthesis, characterization, and diverse applications. Frontiers in Chemistry. 2021;**9**. DOI: 10.3389/fchem.2021.629054

[87] Pan M, Yang J, Liu K, Yin Z, Ma T, Liu S, et al. Noble metal nanostructured materials for chemical and biosensing systems. Nanomaterials. 2020;**10**:209. DOI: 10.3390/nano10020209

[88] Shirk MD, Molian PA. A review of ultrashort pulsed laser ablation of materials. Journal of Laser Applications. 1998;**10**:18-28. DOI: 10.2351/1.521827

[89] Galbács G, Kéri A, Kohut A, Veres M, Zs G. Nanoparticles in analytical laser and plasma spectroscopy – A review of recent developments in methodology and applications. Journal of Analytical

Atomic Spectrometry. 2021;**36**:1826-1872. DOI: 10.1039/D1JA00149C

[90] Du F-H, Wang K-X, Chen J-S. Strategies to succeed in improving the lithium-ion storage properties of silicon nanomaterials. Journal of Materials Chemistry A. 2016;**4**:32-50. DOI: 10.1039/C5TA06962A

[91] Lima F de A, Medeiros GB, Chagas PAM, Aguiar ML, Guerra VG. Aerosol nanoparticle control by electrostatic precipitation and filtration processes—A review. Powders. 2023;**2**:259-298. DOI: 10.3390/powders2020017

[92] Abu-Dief AM, Abdel-Fatah SM. Development and functionalization of magnetic nanoparticles as powerful and green catalysts for organic synthesis. Beni-Suef University Journal of Basic and Applied Sciences. 2018;**7**:55-67. DOI: 10.1016/j.bjbas.2017.05.008

[93] Chen X, Ma X, He K, Jia J, Xue Q. Molecular beam epitaxial growth of topological insulators. Advanced Materials. 2011;**23**:1162-1165. DOI: 10.1002/adma.201003855

[94] Nakayama Y, Tsuruta R, Koganezawa T. 'Molecular beam epitaxy' on organic semiconductor single crystals: Characterization of well-defined molecular interfaces by synchrotron radiation X-ray diffraction techniques. Materials. 2022;**15**:7119. DOI: 10.3390/ma15207119

[95] Vafaeva KM, Zegait R. Carbon nanotubes: Revolutionizing construction materials for a sustainable future: A review. Research on Engineering Structures and Materials. 2023. DOI: 10.17515/resm2023.42ma0818rv

[96] Rajabathar JR, Periyasami G, Alanazi AM, Govindasamy M, Arunachalam P. Review on carbon nanotube varieties for healthcare application: Effect of preparation methods and mechanism insight. PRO. 2020;**8**:1654. DOI: 10.3390/pr8121654

[97] Devi M, Rawat S, Sharma S. A comprehensive review of the pyrolysis process: From carbon nanomaterial synthesis to waste treatment. Oxford Open Materials Science. 2020;**1**. DOI: 10.1093/oxfmat/itab014

[98] Jeyaraj M, Gurunathan S, Qasim M, Kang M-H, Kim J-H. A comprehensive review on the synthesis, characterization, and biomedical application of platinum nanoparticles. Nanomaterials. 2019;**9**:1719. DOI: 10.3390/nano9121719

[99] Yang L, Wei J, Ma Z, Song P, Ma J, Zhao Y, et al. The fabrication of micro/nano structures by laser machining. Nanomaterials. 2019;**9**:1789. DOI: 10.3390/nano9121789

[100] Khan Y, Sadia H, Ali Shah SZ, Khan MN, Shah AA, Ullah N, et al. Classification, synthetic, and characterization approaches to nanoparticles, and their applications in various fields of nanotechnology: A review. Catalysts. 2022;**12**:1386. DOI: 10.3390/catal12111386

[101] Balapure A, Ray Dutta J, Ganesan R. Recent advances in semiconductor heterojunctions: A detailed review of the fundamentals of photocatalysis, charge transfer mechanism and materials. RSC Applied Interfaces. 2024;**1**:43-69. DOI: 10.1039/D3LF00126A

[102] Ndlwana L, Raleie N, Dimpe KM, Ogutu HF, Oseghe EO, Motsa MM, et al. Sustainable hydrothermal and solvothermal synthesis of advanced carbon materials in multidimensional applications: A review. Materials. 2021;**14**:5094. DOI: 10.3390/ma14175094

[103] Kumar P, Saini M, Dehiya BS, Sindhu A, Kumar V, Kumar R, et al. Comprehensive survey on

nanobiomaterials for bone tissue engineering applications. Nanomaterials. 2020;**10**:2019. DOI: 10.3390/nano10102019

[104] Dutta V, Verma R, Gopalkrishnan C, Yuan M-H, Batoo KM, Jayavel R, et al. Bio-inspired synthesis of carbon-based nanomaterials and their potential environmental applications: A state-of-the-art review. Inorganics (Basel). 2022;**10**:169. DOI: 10.3390/inorganics10100169

[105] Liu L, Wang S, Jiang G, Zhang B, Yang J, Wang J, et al. Solvothermal synthesis of zirconia nanomaterials: Latest developments and future. Ceramics International. 2022;**48**:32649-32676. DOI: 10.1016/j.ceramint.2022.07.290

[106] Kharisov BI, Rasika Dias HV, Kharissova OV, Manuel Jiménez-Pérez V, Olvera Pérez B, Muñoz FB. Iron-containing nanomaterials: Synthesis, properties, and environmental applications. RSC Advances. 2012;**2**:9325. DOI: 10.1039/c2ra20812a

[107] Akinwekomi AD, Akhtar F. Bibliometric mapping of literature on high-entropy/multicomponent alloys and systematic review of emerging applications. Entropy. 2022;**24**:329. DOI: 10.3390/e24030329

[108] Bokov D, Turki Jalil A, Chupradit S, Suksatan W, Javed Ansari M, Shewael IH, et al. Nanomaterial by sol-gel method: Synthesis and application. Advances in Materials Science and Engineering. 2021;**2021**. DOI: 10.1155/2021/5102014

[109] Garrido MD, Benítez M, Ros-Lis JV, Amorós P. A travel though the atrane route, a versatile tool for the materials soft-synthesis: A twenty-five years perspective. Nano Select. 2024;**5**. DOI: 10.1002/nano.202300169

[110] Zhang R, Khalizov A, Wang L, Hu M, Xu W. Nucleation and growth of nanoparticles in the atmosphere. Chemical Reviews. 2012;**112**:1957-2011. DOI: 10.1021/cr2001756

[111] Lintingre E, Lequeux F, Talini L, Tsapis N. Control of particle morphology in the spray drying of colloidal suspensions. Soft Matter. 2016;**12**:7435-7444. DOI: 10.1039/C6SM01314G

[112] Cal K, Sollohub K. Spray drying technique. I: Hardware and process parameters. Journal of Pharmaceutical Sciences. 2010;**99**:575-586. DOI: 10.1002/jps.21886

[113] An L-Y, Dai Z, Di B, Xu L-L. Advances in cryochemistry: Mechanisms, reactions and applications. Molecules. 2021;**26**:750. DOI: 10.3390/molecules26030750

[114] Abbas R, Luo J, Qi X, Naz A, Khan IA, Liu H, et al. Silver nanoparticles: Synthesis, structure, properties and applications. Nanomaterials. 2024;**14**:1425. DOI: 10.3390/nano14171425

[115] Yadi M, Mostafavi E, Saleh B, Davaran S, Aliyeva I, Khalilov R, et al. Current developments in green synthesis of metallic nanoparticles using plant extracts: A review. Artificial Cells, Nanomedicine, and Biotechnology. 2018;**46**:336-343. DOI: 10.1080/21691401.2018.1492931

[116] Jaison JP, Balasubramanian B, Gangwar J, James N, Pappuswamy M, Anand AV, et al. Green synthesis of bioinspired nanoparticles mediated from plant extracts of Asteraceae family for potential biological applications. Antibiotics. 2023;**12**:543. DOI: 10.3390/antibiotics12030543

[117] Weerasinghe UA, Wu T, Chee PL, Yew PYM, Lee HK, Loh XJ, et al. Deep eutectic solvents towards green polymeric materials. Green Chemistry. 2024;**26**:8497-8527. DOI: 10.1039/D4GC00532E

[118] Rani N, Singh P, Kumar S, Kumar P, Bhankar V, Kumar K. Plant-mediated synthesis of nanoparticles and their applications: A review. Materials Research Bulletin. 2023;**163**:112233. DOI: 10.1016/j.materresbull.2023.112233

[119] Carrapiço A, Martins MR, Caldeira AT, Mirão J, Dias L. Biosynthesis of metal and metal oxide nanoparticles using microbial cultures: Mechanisms, antimicrobial activity and applications to cultural heritage. Microorganisms. 2023;**11**:378. DOI: 10.3390/microorganisms11020378

[120] Barhoum A, García-Betancourt ML, Jeevanandam J, Hussien EA, Mekkawy SA, Mostafa M, et al. Review on natural, incidental, bioinspired, and engineered nanomaterials: History, definitions, classifications, synthesis, properties, market, toxicities, risks, and regulations. Nanomaterials. 2022;**12**:177. DOI: 10.3390/nano12020177

[121] Aljarba NH, Imtiaz S, Anwar N, Alanazi IS, Alkahtani S. Anticancer and microbial activities of gold nanoparticles: A mechanistic review. Journal of King Saud University, Science. 2022;**34**:101907. DOI: 10.1016/j.jksus.2022.101907

[122] Naganthran A, Verasoundarapandian G, Khalid FE, Masarudin MJ, Zulkharnain A, Nawawi NM, et al. Synthesis, characterization and biomedical application of silver nanoparticles. Materials. 2022;**15**:427. DOI: 10.3390/ma15020427

[123] Dadkhah M, Tulliani J-M. Green synthesis of metal oxides semiconductors for gas sensing applications. Sensors. 2022;**22**:4669. DOI: 10.3390/s22134669

[124] Goel M, Sharma A, Sharma B. Recent advances in biogenic silver nanoparticles for their biomedical applications. Sustainable Chemistry. 2023;**4**:61-94. DOI: 10.3390/suschem4010007

[125] Zambonino MC, Quizhpe EM, Jaramillo FE, Rahman A, Santiago Vispo N, Jeffryes C, et al. Green synthesis of selenium and tellurium nanoparticles: Current trends, biological properties and biomedical applications. International Journal of Molecular Sciences. 2021;**22**:989. DOI: 10.3390/ijms22030989

[126] Alghuthaymi MA, Abd-Elsalam KA, AboDalam HM, Ahmed FK, Ravichandran M, Kalia A, et al. Trichoderma: An eco-friendly source of nanomaterials for sustainable agroecosystems. Journal of Fungi. 2022;**8**:367. DOI: 10.3390/jof8040367

[127] Kamanina OA, Saverina EA, Rybochkin PV, Arlyapov VA, Vereshchagin AN, Ananikov VP. Preparation of hybrid sol-gel materials based on living cells of microorganisms and their application in nanotechnology. Nanomaterials. 2022;**12**:1086. DOI: 10.3390/nano12071086

[128] Cano-Prieto C, Undabarrena A, de Carvalho AC, Keasling JD, Cruz-Morales P. Triumphs and challenges of natural product discovery in the postgenomic era. Annual Review of Biochemistry. 2024;**93**:411-445. DOI: 10.1146/annurev-biochem-032620-104731

[129] Boroumand Moghaddam A, Namvar F, Moniri M, Tahir P Md, Azizi S, Mohamad R. Nanoparticles biosynthesized by fungi and yeast: A review of their preparation, properties, and medical applications. Molecules. 2015;**20**:16540-16565. DOI: 10.3390/molecules200916540

[130] Sharma RK, Dey G, Banerjee P, Maity JP, Lu C-M, Siddique JA, et al. New aspects of lipopeptide-incorporated nanoparticle synthesis and recent advancements in biomedical and environmental sciences: A review. Journal of Materials Chemistry B. 2023;**11**:10-32. DOI: 10.1039/D2TB01564A

Section 2

State-of-the-Art Characterization

Chapter 3

Green Synthesis–Mediated Nanoparticles: Characterization, Antimicrobial Activity and Genomics Analysis

Akamu J. Ewunkem, Niore Johnson, A'lyiha F. Beard, Zahirah J. Williams, Ilunga Tshimanga, Brittany Justice, Dinesh K. Singh and Jeffery Meixner

Abstract

Infectious diseases are the leading cause of health loss globally and have long been highlighted as a global public health priority. Antibiotics have revolutionized the treatment of infectious diseases by increasing life expectancy and reducing mortality. However, the misuse and overuse of antibiotics have contributed to antibiotic resistance. Antibiotic resistance is a threat to the achievement of antibiotic era and a significant public health issue. Consequentially, there is an urgent hunt for new treatments to help fight infections before antibiotics become obsolete. Nanoparticles have been a significant breakthrough in recent years and have emerged as an alternative to antibiotics. Therefore, there is an increased amount of research devoted to the synthesis of nanoparticles. Nowadays the synthesis of nanoparticles employs chemical and green synthesis methods. This chapter describes the green synthesis route for the synthesis of silver nanoparticles using easily available honeybee (*Apis mellifera*) wings aiming to control *Staphylococcus aureus* and to explore whole genome sequencing in this strain.

Keywords: nanoparticles, green synthesis, infections, bacteria, honeybee, genomics, mutations, *Staphylococcus aureus*

1. Introduction

Many human infections are caused by either viruses or bacteria. Bacterial infections (ranging from mildly inconvenient to deadly infections) have long been highlighted as a global public priority and recognized as the second cause of death globally [1, 2]. More than 7 million people die each year as a direct result of bacterial infections. *Staphylococcus aureus* is responsible for about 30% of the death among the bacteria studied [3]. Common infections with *S. aureus* include bacteremia, infective endocarditis osteoarticular, skin and pulmonary infections [4, 5]. An integral part of the treatment of these infections is the administration of antibiotics that target etiologic agents.

Antibiotics are a critical public health tool (since the discovery of penicillin in 1928) in the fight against bacterial infections. The consumption rate of antibiotics has increased by 46% in the last 20 years and has revolutionized public health [6]. The use of antibiotics has increased survival chances of bacterial infections by 10–90% and life expectancy by 5–10 years increasing population longevity [7]. Antibiotics work by: interference with cell wall synthesis; inhibition of protein synthesis; interference with nucleic acid synthesis; inhibition of a metabolic pathway; inhibition of membrane function and inhibition of ATP Synthase. However, the effectiveness of antibiotics is now in jeopardy because bacteria have developed resistance [8, 9]. Mechanisms of bacteria resistance can be divided into four main categories: limiting uptake of a drug; modifying a drug target; inactivating a drug and active drug efflux [10, 11]. Nanotechnology has received considerable attention and is being employed across various fields to overcome antibacterial resistance.

Nanotechnology is the manipulation of matter at nanoscale level to produce structures, devices and materials. A class of materials that range in size from 1 to 100 nm is known as nanoparticle [12–14]. Nanoparticles have gained the status of a new and excellent antimicrobial agent that overcomes antibiotic-resistance mechanisms due to their unique physio-chemical properties. The unique physio-chemical properties of nanoparticles execute multiple bactericidal/bacteriostatic pathways to achieve antimicrobial activity by multiple means including disruption of cell membrane integrity, disruption of the electron transport chain, inhibition of the citric acid cycle, catalytic killing, cell division arrest, prolonged ionic killing and nanoparticle aggregation mediated cell trapping [15]. Hence nanoparticles have significant applications in sectors such as pharmaceutical, biochemical, food, agriculture and the environment where scientists have exploited various synthetic methods [16–19].

Nanoparticles can be synthesized using chemical and biological methods [20, 21]. The chemical- based production of nanoparticles is carried out under extremely harsh conditions and it is highly expensive and environmentally harmful [22]. In contrast, the biological or green synthesis procedures that use organisms are simple, cost effective and environmentally friendly [23]. Green synthesis of nanoparticles using fungi, plants, bacteria and insects have been reported by several researchers [13, 14, 21, 24–28]. Thus, the present study demonstrates green synthesis of silver nanoparticles using honeybee (*Apis mellifera*) wings and its antimicrobial potential against *Staphylococcus aureus*. The study also evaluates the nature of genomics changes associated with the green synthetic silver nanoparticles from honeybee wings.

2. Materials and methods

2.1 Green synthesis and characterization of silver nanoparticles from honey-bee wings

Dead honeybees' wings were incised and washed with distilled water to remove particulate matter. Each set of wings was weighed (0.1 g) and added to distilled water in separated containers and incubated at 90°C for 60 min. The samples were filtered and centrifuged at 8000 rpm for 10 min. The supernatants were collected, and the pH adjusted to neutral pH. Afterward, 1.0 ml of the supernatant was added

0.1 M of silver nitrate. The reaction was then incubated at 28°C ± 1°C for 30 minutes under static conditions for the synthesis of silver nanoparticles via color change. The wavelength of maximum absorbance was scanned through UV–Vis spectra at the range of 200–1000 nm at a resolution of 1 nm at room temperature using a UV–Visible spectrophotometer (Genesys 10 UV Thermoelectron Corporation, UK). Morphology and size of the biosynthesized silver nanoparticles were determined by JEOL's new Field Emission Scanning Electron Microscope, the JSM-IT800 at the Joint School of Nanosciences and Nanotechnology, North Carolina USA.

2.2 High performance liquid chromatography (HPLC) system

HPLC was carried out at Chemistry Laboratory in the Department of Chemistry Winston Salem State University using Agilent 1100 series with a C18 column (4.6 mm × 250 mm i.d., 5 μm) at 35°C to identify and quantify compounds present in honeybee wing extracts. The mobile phase consisted of HPLC water (A) and acetonitrile (B) at a flow rate of 1 ml/min. The multi-wavelength detector was monitored at 280 nm.

2.3 Antimicrobial activity

The antibacterial activity of the biologically synthesized silver nanoparticles was determined, using broth microdilution method. The test organism *Staphylococcus aureus* ATCC 25923 was grown in nutrient broth (NB) medium in 50-ml Erlenmeyer flasks and placed in a shaking incubator overnight at 115 rpm at 37°C and the turbidity was fixed around 1×10^6 CFUs/ml. Ten microliters of test bacterial (1×10^6 CFUs/ml) suspension was added into 96-well plates. Then, an equal volume of biologically synthesized silver nanoparticles with concentrations (0–10 μM) was added, and finally, the plates were incubated in a 37°C shaking incubator for 24 h at 115 rpm. Bacterial growth was assessed by measuring turbidity at 600 nm for hours 0 and 24 h, using a 96-well plate format Glomax multi plate reader (Promega, USA). Structural and morphological changes of the treated *S. aureus* were examined by JEOL's new Field Emission Scanning Electron [21].

2.4 Genomic analysis

Following 24 h of incubation, DNA was extracted from the control and treated cells (cells exposed to the biosynthesized silver nanoparticles) using DNeasy 96 PowerSoil Pro QIAcube HT Kit as per manufacturer instructions. Samples were prepared for whole genome sequencing using the Illumina DNA Prep tagmentation kit and IDT For Illumina Unique Dual Indexes. Sequencing was performed on the Illumina NextSeq2000 platform using a 300-cycle flow cell kit to produce 2 × 150 bp paired reads. 1–2% PhiX control was spiked into the run to support optimal base calling. Read demultiplexing, read trimming, and run analytics were performed using DRAGEN v4.2.7, an on-board analysis software on the NextSeq2000.

3. Results

3.1 Biosynthesis of silver nanoparticles from worker honeybee wings

During the biosynthesis of silver nano-particle's reaction, the distinct color change of aqueous silver nitrate within 45 minutes of incubation with honeybee wing extracts confirmed the synthesis of silver nanoparticles (**Figure 1**) due to excitation of surface plasmon vibrations in silver nanoparticles.

3.2 Characterizations of silver nanoparticles from honeybee wings

The synthesized silver nanoparticles were further confirmed with the UV–Vis characteristic absorption band of silver nanoparticles at a wavelength of 400 nm (**Figure 2**).

3.3 Scanning electron microscopy (SEM)

The Scanning Electron Microscope (SEM) analysis finally confirmed the synthesis of silver nanoparticle. The silver nanoparticle possesses a variety of morphological characteristics like sphericity and cuboid (**Figure 3**).

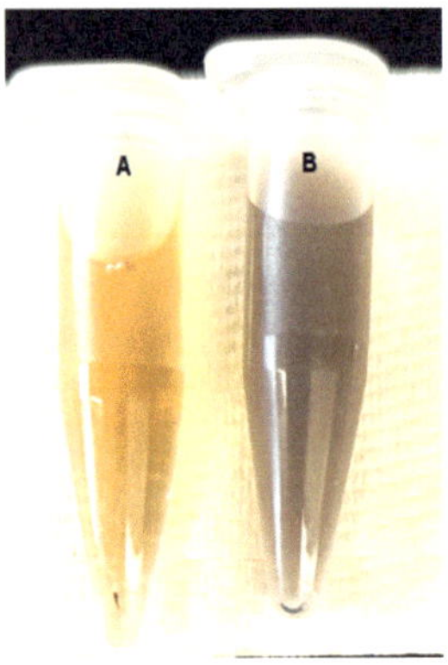

Figure 1.
(A) Honeybee wing extract with no silver nitrate. (B) Honeybee wings extract with added silver nitrate.

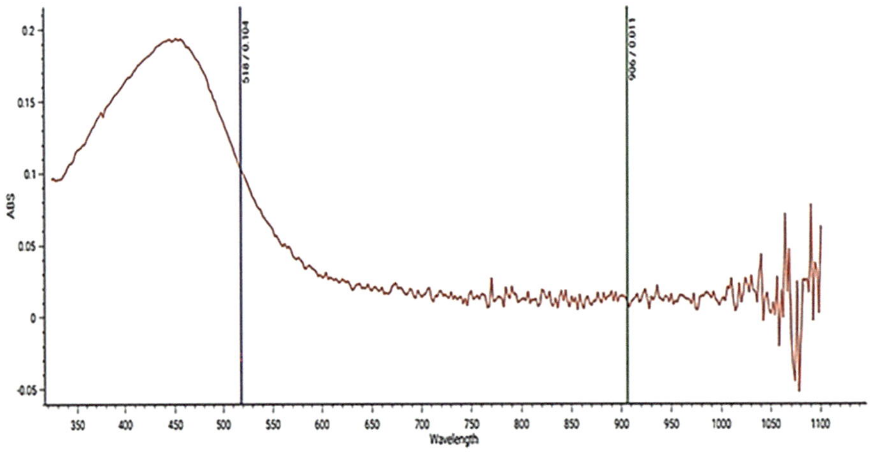

Figure 2.
UV–visible spectra of green synthesized silver nanoparticles from honeybee wings.

DOI: http://dx.doi.org/10.5772/intechopen.1006905

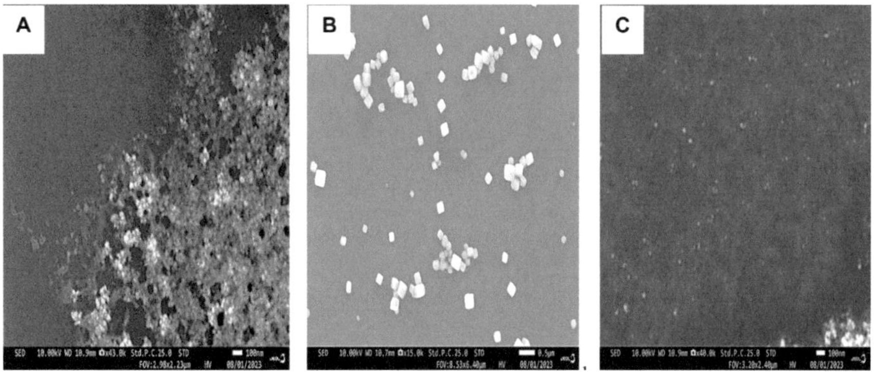

Figure 3.
SEM micrographs of silver nanoparticle synthesized from honeybee wings extracts (A) biosynthesized silver nanoparticles; (B) cube-like nanoparticle; (C) spherical shape silver nanoparticles.

3.4 HPLC

HPLC was used to identify and quantify the possible components of the honeybee extract responsible for the reduction of silver ions. The analysis revealed four major components (**Figure 4**).

3.5 Antimicrobial activity of biosynthesized silver nanoparticles from honeybee wings

The antibacterial activity of silver nanoparticles from honeybee wings was determined against *Staphylococcus aureus*. The results revealed that biosynthesized silver nanoparticles from honeybee wings extract significantly ($P < 0.05$) suppressed the growth of *S. aureus* (**Figure 5**). To further understand *S. aureus* responses

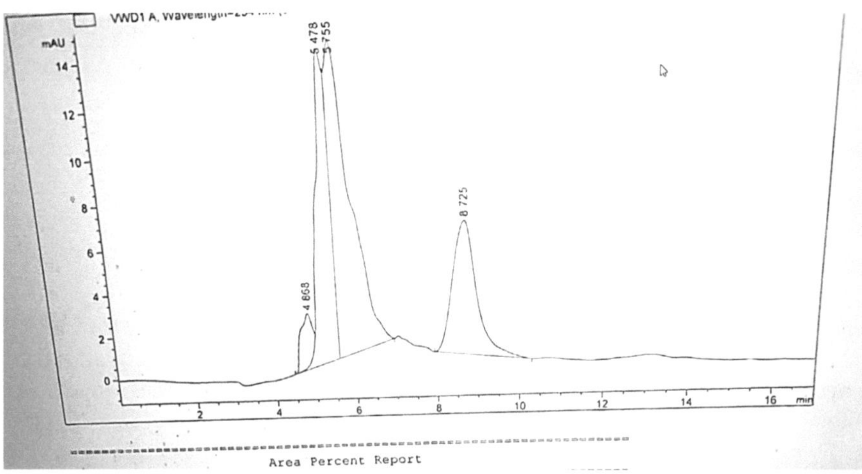

Figure 4.
HPLC profile of water extract derived from honeybee wing.

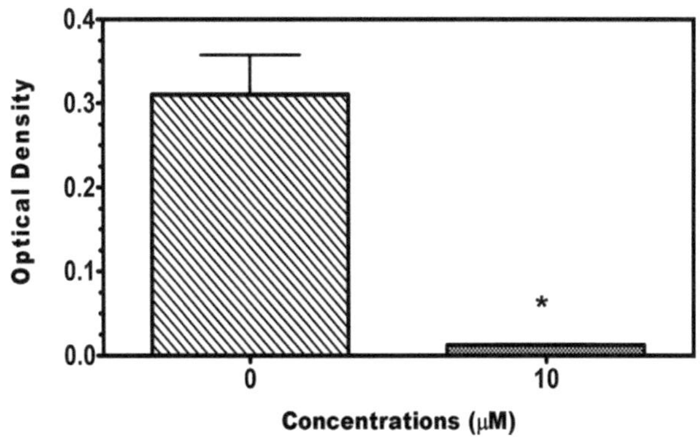

Figure 5.
Antibacterial potential of biosynthesized silver nanoparticles from honeybee wing extracts against S. aureus.

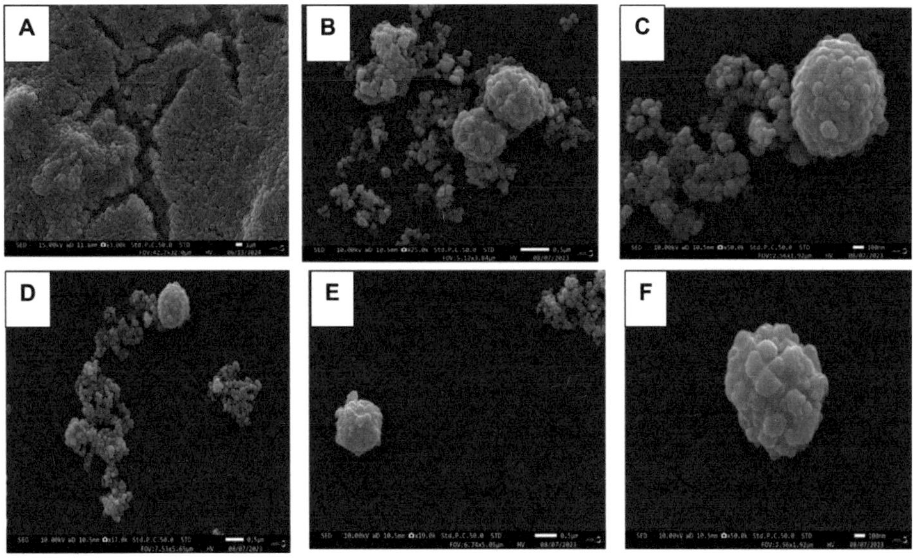

Figure 6.
SEM micrograph of biosynthesized silver nanoparticles untreated S. aureus *cells (A) and the treated* S. aureus *cells (B,C,D,E,F).*

to the biosynthesized silver nanoparticles stress, the bacterial morphology and nanoparticle-cell interactions were examined with SEM. The SEM images of bacterial cells treated or untreated with the biosynthesized silver NPS are shown in **Figure 6**. The untreated or controlled cells were smooth and intact surface clearly visible while the silver nanoparticles aggregated on the surfaces of the treated bacterial cells (**Figure 6A** and **D**). The tested bacterial cells strains captured different amounts of the biosynthesized silver nanoparticles (**Figure 6E** and **F**).

4. Genomics results

The genomic variants found in the control and exposed cells are listed in **Tables 1** and **2**. The following hard sweeps were observed in both the control and

DOI: http://dx.doi.org/10.5772/intechopen.1006905

Position	Annotations	Genes	Products
1,886,098	K4N (AAA → AAC)	*KQ76_RS09235*	DUF1433 domain-containing protein
2,389,192	R188P (CGA → CCA)	*hssR*	DNA-binding heme response regulator HssR
2,574,726	G69A (GGC → GCC) ‡	*KQ76_RS13020*	alpha/beta hydrolase
986,858	Q59K (CAA → AAA)	*KQ76_RS04770*	ATP-binding protein
2,665,478	E162V (GAA → GTA)	*KQ76_RS13475*	glutathione peroxidase
2,574,727	G69R (GGC → CGC) ‡	*KQ76_RS13020*	alpha/beta hydrolase
2,389,188	E187Q (GAA → CAA)	*hssR*	DNA-binding heme response regulator HssR
2,776,116	H117Q (CAT→CAA)	*mnmG*	tRNA uridine-5-carboxymethylaminomethyl(34) synthesis enzyme MnmG
411,301	M76I (ATG → ATC)	*KQ76_RS01915*	superantigen-like protein SSL10
2,123,947	Y65N (TAC → AAC)	*KQ76_RS10630*	heavy metal-associated domain-containing protein
453,752	pseudogene (4457/4500 nt	*gltB*	glutamate synthase large subunit
2,233,551	T39S (ACT→AGT)	*KQ76_RS11175*	BCCT family transporter
460,550	intergenic (+70/−69)	*KQ76_RS02145/KQ76_RS02150*	hypothetical protein/N-acetyltransferase
1,213,003	E184D (GAG→GAC)	*ylqF*	ribosome biogenesis GTPase YlqF
986,867	H62N (CAT→AAT)	*KQ76_RS04770*	ATP-binding protein
2,746,925	A172G (GCT → GGT)	*KQ76_RS13825*	ECF-type riboflavin transporter substrate-binding protein
2,252,747	intergenic (−54/+157)	*KQ76_RS11280/KQ76_RS11285*	M23 family metallopeptidase/HAD-IIB family hydrolase
2,190,680	T500S (ACG → TCG)	*KQ76_RS10985*	BglG family transcription antiterminator
2,541,135	Y291* (TAC → TAG)	*gntK*	gluconokinase
2,766,496	I591I (ATA → ATT)	*vraE*	peptide resistance ABC transporter permease subunit VraE

Table 1.
Position and description of selective sweeps in control S. aureus.

Position	Annotations	Genes	Products
1,886,098	K4N (AAA → AAC)	*KQ76_RS09235*	DUF1433 domain-containing protein
2,574,726	G69A (GGC → GCC) ‡	*KQ76_RS13020*	alpha/beta hydrolase
453,752	pseudogene (4457/4500 nt)	*ltB*	glutamate synthase large subunit
2,574,727	G69R (GGC → CGC) ‡	*KQ76_RS13020*	alpha/beta hydrolase
2,389,192	R188P (CGA → CCA)	*hssR*	DNA-binding heme response regulator HssR
2,190,680	T500S (ACG → TCG)	*KQ76_RS10985*	BglG family transcription antiterminator
2,389,188	E187Q (GAA → CAA)	*hssR*	DNA-binding heme response regulator HssR
287,661	V42E (GTA → GAA)	*KQ76_RS01230*	TIGR01741 family protein
986,858	Q59K (CAA → AAA)	*KQ76_RS04770*	ATP-binding protein
2,574,728	E68D (GAG→GAC)	*KQ76_RS13020*	alpha/beta hydrolase
2,665,478	E162V (GAA → GTA)	*KQ76_RS13475*	glutathione peroxidase
2,252,747	intergenic (−54/+157)	*KQ76_RS11280/KQ76_RS11285*	M23 family metallopeptidase/HAD-IIB family hydrolase
2,236,063	R52P (CGT → CCT)	*KQ76_RS11185*	NADP-dependent oxidoreductase
986,867	H62N (CAT→AAT)	*KQ76_RS04770*	ATP-binding protein
1,890,405	S137T (AGT → ACT)	*KQ76_RS09255*	hypothetical protein
2,746,925	A172G (GCT → GGT)	*KQ76_RS13825*	ECF-type riboflavin transporter substrate-binding protein
2,776,116	H117Q (CAT→CAA)	*mnmG*	tRNA uridine-5-carboxymethylaminomethyl(34) synthesis enzyme MnmG
1,027,271	A87P (GCA → CCA)	*purS*	phosphoribosylformylglycinamidine synthase subunit PurS
1,284,341	G97A (GGC → GCC)	*KQ76_RS06205*	alpha/beta hydrolase
2,564,194	A17P (GCA → CCA)	*KQ76_RS12955*	D-lactate dehydrogenase

Table 2.
Position and description of selective sweeps in S. aureus *exposed to green synthesized silver nanoparticles from honeybee wings.*

treated cells at 24 hr. incubation: DUF1433 domain containing protein (*KQ76_RS09235*), DNA binding heme response regulator HssR (*hssR*), alpha/beta hydrolase (*KQ76_RS13020*), ATP binding protein (*KQ76_RS04770*), glutathione peroxidase (*KQ76_RS13475*), ribosome biogenesis GTPase YlqF (*ylqF*), ECF type riboflavin transporter substrate binding protein (*KQ76_RS13825*) (**Tables 1** and **2**). Only the control cells displayed mutations in gluconokinase (*gntK*) and peptide resistance ABC transporter permease subunit VraE (*vraE*) (**Table 1**). The most notable mutations observed in the treated cellsbut not control cells was phosphoribosylformylglycinamidine synthase subunit PurS (*purS*) (**Table 2**).

5. Discussion

A rise in antimicrobial resistance is causing the spread of drug-resistant infections posing a serious threat to global public health. Developing technology in the field of antimicrobials can make a significant contribution to overcoming antimicrobial resistance [29]. Over the past few years, nanoparticles have been developed to prevail against antimicrobial resistance by acting as transporter for natural antibiotics and antimicrobials [30]. Therefore, scientists, industries and companies have developed methods to synthesize nanoparticles for their use in antimicrobial therapeutic applications to combat bacterial pathogens.

Nanoparticles synthesis can be subdivided into two main categories: conventional methods and green methods [31]. The conventional methods of nanoparticles synthesis are reproducible, reliable and efficiently producing high yield of nanoparticles. However, the conventional method of nanoparticles synthesis is not environmentally friendly and costly [32]. To overcome these problems, a large amount of time and effort have been devoted to synthesizing nanoparticles using green synthesis.

Green synthesis approach explores organisms in the development of nanoparticles having advantages regarding biocompatibility, cost effectiveness and reduced environmental toxicity [33]. This study delved into the biosynthesis and characterization of silver nanoparticles derived from the extract of honeybee wings and studied its antimicrobial potential against *S. aureus,* a common human pathogen that causes a wide range of clinical infections. The study also examined the genomics changes associated with the antibacterial activity of the biosynthesized silver nanoparticles.

During the biosynthesis of silver nanoparticles, the observed change in color from light brown to dark brown reaction solution primarily indicates the recovery of silver ions and the formation of silver nanoparticles [24]. The change occurred within 15 minutes of incubation is due to the rapid excitation of surface plasmon vibrations in metal nanoparticles and quantum confinement [34]. The sensible color change in the reaction mixture, from light brown to dark brown in this study corroborates with the recent report of Jakinala et al., [27] who asserted that the color change in of the reaction mixture to dark brown after incubation suggests the formation of *M. mao* silver nanoparticles. It is worth noting that the color change from light yellow to dark brown was due to the reduction of silver ions by compounds present in honeybee wing extracts (**Figure 2**). Insect wing extracts contain several compounds such as aliphatic hydrocarbons, nonadecanoic acid, ethyl 9-octadecenoate, heptacosanoic acid, 1- [2-Deoxy-beta. -d-erythro-pentofuranosyl], hexadecanoic acid, 3-Nonyn-2-ol, 8-acetoxy-6-benzenesulfonyl-2-th, ethanone and 1,5-Cyclooctadiene which act as reducing agents to produce nanoparticles from metal salts [27, 35, 36].

During the process of silver nanoparticles formation, the silver ions form a complex with compounds in the wing extracts resulting in oxidation of the compounds and a reduction in silver ions to a zero -oxidation state [37]. To confirm this, the biosynthesized silver nanoparticles were monitored by measuring the UV–Vis absorbance with spectrum ranging 350–900 nm from for the reaction medium after diluting with silver nitrate solution. A peak was observed at 430 nm which showed a stable range of silver nanoparticles formation. Our results are in agreement with previous studies [38–41] that suggested that UV–vis spectra of silver nanoparticles give a sharp peak at 400–500 nm after incubation which is a characteristic surface plasmon resonance peak of biosynthesized silver nanoparticles. The exhibited absorption peak at a given wavelength in the UV–vis region is also a characteristic of the morphology of nanoparticles [42].

The morphology of the green-synthesized silver nanoparticles examined using SEM were found to be in polymorphic shapes; some of them were spherical, cuboid irregularly granulated, and highly aggregated with size distribution in the range of 10–30 nm as shown in **Figure 3**. These results are in accordance with the previous results that found biosynthesized silver nanoparticles with size range between 20 and 40 nm [13, 14, 43, 44]. The uniquely small size of the biosynthesized silver nanoparticles gives them a larger surface-to-volume ratio resulting in greater interaction with bacterial cells enhancing antibacterial activity, hence reversing bacterial resistance [45].

Bacterial resistance is based on three fundamental mechanisms: enzymatic degradation of drugs; alteration of bacterial biomolecules that are drug targets and changes in bacterial cell membrane permeability to antibiotics [46, 47]. Silver nanoparticles are known to successfully eschew the occurrence of bacterial resistance by several bactericidal strategies such as the production of reactive oxygen species (ROS), destruction of intracellular elements and by binding and disrupting cell wall and bacterial membrane causing leakage of cytoplasmic components [48, 49]. In the present study these biosynthesized silver nanoparticles from honeybee wing extracts showed a prominent antibacterial activity against a deadly Gram-positive bacterial pathogen, *S. aureus* by adhering the cell wall and cytoplasmic membrane due to electrostatic attraction [50]. The adhered biosynthesized silver nanoparticles damage the cell wall and membrane of bacterial cells, causing morphological changes paving way for the continuous release of silver ions into the cells interacting with the respiratory enzymes, generating ROS and interrupting adenosine triphosphate production [51, 52].

Metal nanoparticles can change metabolic pathways of pathogens which is an advantage in terms of destroying infectious agents. In this study we used whole genomics sequencing to investigate further the antibacterial mechanisms of green synthesized silver nanoparticles from honeybee wings against *S. aureus.* Even though cells exposed to the biosynthesized silver nanoparticles displayed a statistically significant reduction in the growth, we did not find sweeps of variants associated with oxidative stress induction, metal ion release, or non-oxidative mechanisms [53, 54].

Mutations were observed in genes that play major roles in metabolism, homeostasis, nutrient uptake and defense mechanisms suggesting the presence of antimicrobial resistance genes in *S. aureus* allowing it to express resistance phenotypes. *S. aureus* is known to resist a plethora of antimicrobial agents and stress through a variety of mechanisms, including, enzymatic inactivation, target alteration, trapping, efflux pumps, limiting uptake and synthesis of nucleic acid [55–57]. In our present study, only cells treated with biosynthesized silver nanoparticles acquired mutation in phosphoribosylformylglycinamidine synthase subunit PurS (*purS*), a highly conserved enzyme for de novo purine synthesis [58]. Purine biosynthesis is a key part of nucleotide metabolism and plays a role in bacterial pathogenesis, antibiotic resistance and bacterial growth [59].

Mutations in *Purs* may have modulated the formation of bacterial persisters. Persisters are generally slow growing cells that survive exposure to antimicrobial agents or stresses but are still susceptible to drugs and stresses upon exiting dormancy [60]. Persisters are genetically identical to the rest of the population of cells but exhibit a phenotype that allows them to be drug-tolerant [61]. Persister cells were first described in *S. aureus* in 1944 where the action of an antibiotic was not effective against 1% of *S. aureus* cells [62]. These observations indicate that persister cells are the major culprit accountable for recalcitrance of chronic infections because they can remain viable and repopulate when the concentration of antibiotics decreases and or in the absence of antibiotics [63].

However, the control or untreated cells did not display mutations in *Purs,* genomics analysis illustrated mutations polymorphisms in superantigen like protein SSL10 (*KQ76_RS01915*), heavy metal associated domain containing protein (*KQ76_RS10630*), BCCT family transporter (*KQ76_RS11175*), hypothetical protein/N acetyltransferase (*KQ76_RS02145/KQ76_RS02150*) and gluconokinase (*gntK*) and peptide resistance ABC transporter permease subunit VraE (*vraE*). These genes have been shown to be associated with *S. aureus* growth, colonization and an increase in resistance against the host defense mechanisms of the body [64–66]. For example, in the case of superantigen like protein SSL10 (*KQ76_RS01915*) which plays a role in the formation of amyloid structures that enhance biofilm formation; heavy metal associated domain containing protein (*KQ76_RS10630*) transports and detoxifies heavy metals. Furthermore, *KQ76_RS10630* is involved in microbial adjustment to high osmolarity surroundings via uptake of solutes; *KQ76_RS11175* ensures cellular hydration at high osmolarity; *KQ76_RS02145/KQ76_RS02150* plays a key role in bacterial virulence and *vraE* plays a vital role in bacterial pathogenesis and virulence.

The controls and treatments shared the following sweeps: DUF1433 domain containing protein (*KQ76_RS09235*) a protein of unknown function; alpha/beta hydrolase (*KQ76_RS13020*) plays key roles in metabolism; glutamate synthase large subunit (*gltB*) plays an essential role in the metabolism of nitrogen; DNA binding heme response regulator HssR (*hssR*) is involved in intracellular heme homeostasis and tempering of staphylococcal virulence; BglG family transcription antiterminator (*KQ76_RS10985*) regulates carbohydrate; ATP binding protein (*KQ76_RS04770*) translocates a wide variety of substrates across the membrane; glutathione peroxidase (*KQ76_RS13475*) protects cells from oxidative stress and hydrogen peroxide; M23 family metallopeptidase/HAD IIB family hydrolase (*KQ76_RS11280/KQ76_RS11285*) is used by bacteria to lyse cell walls of other bacteria; ECF type riboflavin transporter substrate binding protein (*KQ76_RS13825*) mediates the uptake of micronutrients; tRNA uridine 5 carboxymethylaminomethyl (34) synthesis enzyme MnmG (*mnmG*) is crucial is important for protein synthesis and alpha/beta hydrolase (KQ76_RS06205) plays a role in growth and development.

Considering the surge in infectious diseases and antibiotic resistance, biosynthetically synthesized silver nanoparticles from honeybee wing extracts can be potentially used as antibacterial agent against pathogens. Thus, biosynthesized silver nanoparticles from honeybee wing extracts can be used in antibacterial coatings for implantable devices and medicinal materials to prevent infection and to treat disease.

6. Conclusion

Bacterial infections are highlighted as one of the top global public health threats. About 7 million deaths are associated with bacterial infections caused by at least 30 bacterial pathogens. *Staphylococcus aureus* is one of the leading pathogenic bacteria linked to the estimated burden of bacterial infections. *S. aureus* is a Gram-positive bacterium that causes a range of human infections including skin infection, pneumonia, endocarditis, and osteomyelitis. Reducing bacterial infections is an urgent global health priority.

The use of antibiotics is used to protect the health care systems against the threat of bacterial infections. However, extensive and injudicious use of antibiotics are the main drivers in the development of resistant pathogens contributing to millions death annually [67]. This problem urgently requires alternative to antibiotics that are novel, trailblazing and less prone to bacterial resistance [68]. Today nanoparticles are developed (to replace antibiotics) to reduce and combat bacterial resistance [30]. Nanoparticles synthesis is typically performed by chemical and green synthesis approaches. In this study, the green synthesis approach was used to synthesize silver nanoparticles using honeybee wing extracts. The biosynthesized silver nanoparticles were characterized using UV–Vis spectroscopy and SEM. All these characterizations confirmed the synthesis of silver nanoparticles with average size less than 100 nm and absorbance peak at 430 nm. Moreover, these biosynthesized silver nanoparticles from honeybee wing extracts had potential antibacterial activity against *S. aureus* by binding to the cell wall resulting in mutations associated with purine biosynthesis.

Author details

Akamu J. Ewunkem[1*], Niore Johnson[1], A'lyiha F. Beard[1], Zahirah J. Williams[2], Ilunga Tshimanga[1], Brittany Justice[1], Dinesh K. Singh[1] and Jeffery Meixner[3]

1 Department of Biological Sciences, Winston Salem State University, Winston Salem, North Carolina, USA

2 Division of Nursing, Winston Salem State University, Winston Salem, North Carolina, USA

3 Department of Clinical and Laboratory Sciences, Winston Salem State University, Winston Salem, North Carolina, USA

*Address all correspondence to: judeakamu@gmail.com

References

[1] Ikuta KS, Swetschinski LR, Aguilar GR, Sharara F, Mestrovic T, Gray AP, et al. Global mortality associated with 33 bacterial pathogens in 2019: A systematic analysis for the global burden of disease study 2019. The Lancet. 2022;**400**(10369):2221-2248

[2] Chen X, Gallagher F, Sellmyer MA, Ordonez AA, Kjaer A, Ohliger M, et al. Visualizing bacterial infections with novel targeted molecular imaging approaches. The Journal of Infectious Diseases. 2023;**228**(Suppl. 4): S249-S258

[3] Zhang C, Fu X, Liu Y, Zhao H, Wang G. Burden of infectious diseases and bacterial antimicrobial resistance in China: A systematic analysis for the global burden of disease study 2019. The Lancet Regional Health–Western Pacific. 2024;**43**

[4] Bergenman O, Nilson B, Rasmussen M. Risk of infective endocarditis and complicated infection in *Staphylococcus aureus* bacteremia–A retrospective cohort study on the role of bacteriuria. European Journal of Clinical Microbiology & Infectious Diseases. 2024;**43**:1419-1426

[5] Jamilu I, Lema S. Antibacterial activity of vitamin c and sweet orange juice (Citrus sinensis) extract against *Staphylococcus aureus* isolated from wound infection among patient attending Bodinga General Hospital, Sokoto, Nigeria. Asian Journal of Biotechnology and Genetic Engineering. 2024;**7**(1):121-127

[6] Tang KWK, Millar BC, Moore JE. Antimicrobial resistance (AMR). British Journal of Biomedical Science. 2023;**80**:11387

[7] Rossolini GM, Arena F, Pecile P, Pollini S. Update on the antibiotic resistance crisis. Current Opinion in Pharmacology. 2014;**18**:56-60

[8] Halawa EM, Fadel M, Al-Rabia MW, Behairy A, Nouh NA, Abdo M, et al. Antibiotic action and resistance: Updated review of mechanisms, spread, influencing factors, and alternative approaches for combating resistance. Frontiers in Pharmacology. 2024;**14**:1305294

[9] Saikia S, Chetia P. Antibiotics: From mechanism of action to resistance and beyond. Indian Journal of Microbiology. 2024;**64**:821-845

[10] Munita JM, Arias CA. Mechanisms of antibiotic resistance. Microbiol Spectrum. 2016;**4**(2):VMBF-0016-2015

[11] Peterson E, Kaur P. Antibiotic resistance mechanisms in bacteria: Relationships between resistance determinants of antibiotic producers, environmental bacteria, and clinical pathogens. Frontiers in Microbiology. 2018;**9**:2928

[12] Prakash A, Praveen RP, Daisy PA, Abeena PB. A review on nanoparticles. International Journal of Pharmaceutical Sciences Review and Research. 2020;**64**:64-68

[13] Ewunkem AJ, Williams ZJ, Johnson NS, Brittany JL, Maselugbo A, Nowlin K. Exploring the "carpenter" as a substrate for green synthesis: Biosynthesis and antimicrobial potential. Gene & Protein in Disease. 2023;**2**(4):2155

[14] Ewunkem AJ, Williams D, Ariyon B, Tshimanga I, Justice B, Singh DK. Rapid green synthesis of silver nanoparticles by

Reishi and their antibacterial activity and mechanisms. Journal of Biomaterials and Nanobiotechnology. 2024;**15**(3):51-63

[15] Summer M, Ali S, Tahir HM, Abaidullah R, Fiaz U, Mumtaz S, et al. Mode of action of biogenic silver, zinc, copper, titanium and cobalt nanoparticles against antibiotics resistant pathogens. Journal of Inorganic and Organometallic Polymers and Materials. 2024;**34**(4):1417-1451

[16] Al-Khayri JM, Rashmi R, Surya Ulhas R, Sudheer WN, Banadka A, Nagella P, et al. The role of nanoparticles in response of plants to abiotic stress at physiological, biochemical, and molecular levels. Plants. 2023;**12**(2):292

[17] Jiang M, Althomali RH, Ansari SA, Saleh EAM, Gupta J, Kambarov KD, et al. Advances in preparation, biomedical, and pharmaceutical applications of chitosan-based gold, silver, and magnetic nanoparticles: A review. International Journal of Biological Macromolecules. 1 Nov 2023;**251**:126390

[18] Wasilewska A, Bielicka M, Klekotka U, Kalska-Szostko B. Nanoparticle applications in food–A review. Food & Function. 2023;**14**(6):2544-2567

[19] Mgadi K, Ndaba B, Roopnarain A, Rama H, Adeleke R. Nanoparticle applications in agriculture: Overview and response of plant-associated microorganisms. Frontiers in Microbiology. 2024;**15**:1354440

[20] Ijaz I, Gilani E, Nazir A, Bukhari A. Detail review on chemical, physical and green synthesis, classification, characterizations and applications of nanoparticles. Green Chemistry Letters and Reviews. 2020;**13**(3):223-245

[21] Ewunkem AJ, Tshimanga I, Justice B, Meixner J. Synthesis of silver nanoparticles from honeybees and its antibacterial potential. Open Journal of Medical Microbiology. 2024;**14**(1):77-92

[22] Vijayaram S, Razafindralambo H, Sun YZ, Vasantharaj S, Ghafarifarsani H, Hoseinifar SH, et al. Applications of green synthesized metal nanoparticles—A review. Biological Trace Element Research. 2024;**202**(1):360-386

[23] Suppiah DD, Julkapli NM, Sagadevan S, Johan MR. Eco-friendly green synthesis approach and evaluation of environmental and biological applications of iron oxide nanoparticles. Inorganic Chemistry Communications. 2023;**152**:110700

[24] Khatami M, Iravani S, Varma RS, Mosazade F, Darroudi M, Borhani F. Cockroach wings-promoted safe and greener synthesis of silver nanoparticles and their insecticidal activity. Bioprocess and Biosystems Engineering. 2019;**42**:2007-2014

[25] Thi TUD, Nguyen TT, Thi YD, Thi KHT, Phan BT, Pham KN. Green synthesis of ZnO nanoparticles using orange fruit peel extract for antibacterial activities. RSC Advances. 2020;**10**(40):23899-23907

[26] Hano C, Abbasi BH. Plant-based green synthesis of nanoparticles: Production, characterization and applications. Biomolecules. 2021;**12**(1):31

[27] Jakinala P, Lingampally N, Hameeda B, Sayyed RZ, Khan MY, Elsayed EA, et al. Silver nanoparticles from insect wing extract: Biosynthesis and evaluation for antioxidant and antimicrobial potential. PLoS One. 2021;**16**(3):e0241729

[28] Alsaiari NS, Alzahrani FM, Amari A, Osman H, Harharah HN, Elboughdiri N, et al. Plant and microbial approaches

as green methods for the synthesis of nanomaterials: Synthesis, applications, and future perspectives. Molecules. 2023;**28**(1):463

[29] Yılmaz GE, Göktürk I, Ovezova M, Yılmaz F, Kılıç S, Denizli A. Antimicrobial nanomaterials: A review. Hygiene. 2023;**3**(3):269-290

[30] Hetta HF, Ramadan YN, Al-Harbi AI, Ahmed EA, Battah B, Abd Ellah NH, et al. Nanotechnology as a promising approach to combat multidrug resistant bacteria: A comprehensive review and future perspectives. Biomedicine. 2023;**11**:Article 413. Molecules, 28, 637

[31] Deepak P, Amutha V, Kamaraj C, Balasubramani G, Aiswarya D, Perumal P. Chemical and green synthesis of nanoparticles and their efficacy on cancer cells. In: Green Synthesis, Characterization and Applications of Nanoparticles. Amsterdam, Netherlands: Elsevier; 2019. pp. 369-387

[32] Vishwanath R, Negi B. Conventional and green methods of synthesis of silver nanoparticles and their antimicrobial properties. Current Research in Green and Sustainable Chemistry. 2021;**4**:100205

[33] Nguyen NPU, Dang NT, Doan L, Nguyen TTH. Synthesis of silver nanoparticles: From conventional to 'modern'methods—A review. PRO. 2023;**11**(9):2617

[34] Tesfaye M, Gonfa Y, Tadesse G, Temesgen T, Periyasamy S. Green synthesis of silver nanoparticles using Vernonia amygdalina plant extract and its antimicrobial activities. Heliyon. Jun 2023;**9**(6):e17356

[35] Ivanova EP, Nguyen SH, Webb HK, Hasan J, Truong VK, Lamb RN, et al. Molecular organization of the nanoscale surface structures of the dragonfly *Hemianax papuensis* wing epicuticle. PLoS One. 2013;**8**(7):e67893

[36] Bhuvaneswari T, Thiyagarajan M, Geetha N, Venkatachalam P. Bioactive compound loaded stable silver nanoparticle synthesis from microwave irradiated aqueous extracellular leaf extracts of *Naringi crenulata* and its wound healing activity in experimental rat model. Acta Tropica. 2014;**135**:55-61

[37] More PR, Pandit S, Filippis AD, Franci G, Mijakovic I, Galdiero M. Silver nanoparticles: Bactericidal and mechanistic approach against drug resistant pathogens. Microorganisms. 2023;**11**(2):369

[38] Bhainsa KC, D'souza SF. Extracellular biosynthesis of silver nanoparticles using the fungus Aspergillus fumigatus. Colloids and Surfaces B: Biointerfaces. 2006;**47**(2):160-164

[39] Meva FEA, Segnou ML, Ebongue CO, Ntoumba AA, Kedi PBE, Deli V, et al. Spectroscopic synthetic optimizations monitoring silver nanoparticles formation from *Megaphrynium macrostachyum* leaf extract. Revista Brasileira de Farmacognosia. 2016;**26**:640-646

[40] Sur UK, Ankamwar B, Karmakar S, Halder A, Das P. Green synthesis of silver nanoparticles using the plant extract of Shikakai and Reetha. Materials Today: Proceedings. 2018;**5**(1):2321-2329

[41] Amr M, Abu-Hussien SH, Ismail R, Aboubakr A, Wael R, Yasser M, et al. Utilization of biosynthesized silver nanoparticles from *Agaricus bisporus* extract for food safety application: Synthesis, characterization, antimicrobial efficacy, and toxicological

assessment. Scientific Reports. 2023;**13**(1):15048

[42] Siddiqi KS, Husen A, Rao RA. A review on biosynthesis of silver nanoparticles and their biocidal properties. Journal of Nanobiotechnology. 2018;**16**:1-28

[43] Raj S, Mali SC, Trivedi R. Green synthesis and characterization of silver nanoparticles using *Enicostemma axillare* (Lam.) leaf extract. Biochemical and Biophysical Research Communications. 2018;**503**(4):2814-2819

[44] Asif M, Yasmin R, Asif R, Ambreen A, Mustafa M, Umbreen S. Green synthesis of silver nanoparticles (AgNPs), structural characterization, and their antibacterial potential. Dose-Response. 2022;**20**(2):15593258221088709

[45] Sayed FAZ, Eissa NG, Shen Y, Hunstad DA, Wooley KL, Elsabahy M. Morphologic design of nanostructures for enhanced antimicrobial activity. Journal of Nanobiotechnology. 2022;**20**(1):536

[46] Yin IX, Zhang J, Zhao IS, Mei ML, Li Q, Chu CH. The antibacterial mechanism of silver nanoparticles and its application in dentistry. International Journal of Nanomedicine. 2020;**15**:2555-2562

[47] Xin X, Qi C, Xu L, Gao Q, Liu X. Green synthesis of silver nanoparticles and their antibacterial effects. Frontiers in Chemical Engineering. 2022;**4**:941240

[48] Gupta A, Mumtaz S, Li CH, Hussain I, Rotello VM. Combatting antibiotic-resistant bacteria using nanomaterials. Chemical Society Reviews. 2019;**48**(2):415-427

[49] Wang S, Zhang Y, Chen X, Mourdikoudis S, Fan S, Li H, et al. Disentangling the "tip-effects" enhanced antibacterial mechanism of Ag nanoparticles. Dalton Transactions. 2024;**53**(29):12281-12290

[50] Jin J, Wu Y, Liang L, Wei Y, Zheng X, Chen Y. Altering sliver nanoparticles-induced inhibition to bacterial denitrification via visible light by regulating silver transformation and adaptive mechanism under anaerobic conditions. Chemical Engineering Journal. 2023;**452**:139268

[51] Das B, Dash SK, Mandal D, Ghosh T, Chattopadhyay S, Tripathy S, et al. Green synthesized silver nanoparticles destroy multidrug resistant bacteria via reactive oxygen species mediated membrane damage. Arabian Journal of Chemistry. 2017;**10**:862-876

[52] Xu GY, Zhao IS, Lung CY, Yin IX, Lo EC, Chu CH. Silver compounds for caries management. International Dental Journal. Apr 2023;**74**(2):179-186

[53] Dakal TC, Kumar A, Majumdar RS, Yadav V. Mechanistic basis of antimicrobial actions of silver nanoparticles. Frontiers in Microbiology. 2016;**7**:1831

[54] Wu K, Li H, Cui X, Feng R, Chen W, Jiang Y, et al. Mutagenesis and resistance development of bacteria challenged by silver nanoparticles. Antimicrobial Agents and Chemotherapy. 2022;**66**(10):e00628-e00622

[55] Akpaka PE, Roberts R, Monecke S. Molecular characterization of antimicrobial resistance genes against Staphylococcus aureus isolates from Trinidad and Tobago. Journal of Infection and Public Health. 2017;**10**(3):316-323

[56] Pantosti A, Sanchini A, Monaco M. Mechanisms of antibiotic resistance

in *Staphylococcus aureus*. Future Microbiology. 2007;**2**(3):323-334

[57] Foster TJ. Antibiotic resistance in Staphylococcus aureus. Current status and future prospects. FEMS Microbiology Reviews. 2017;**41**(3):430-449

[58] Lu A, Disoma C, Zhou Y, Chen Z, Zhang L, Shen Y, et al. Protein interactome of the deamidase phosphoribosylformylglycinamidine synthetase (PFAS) by LC-MS/MS. Biochemical and Biophysical Research Communications. 2019;**513**(3):746-752

[59] Li L, Abdelhady W, Donegan NP, Seidl K, Cheung A, Zhou YF, et al. Role of purine biosynthesis in persistent methicillin-resistant *Staphylococcus aureus* infection. The Journal of Infectious Diseases. 2018;**218**(9):1367-1377

[60] Niu H, Gu J, Zhang Y. Bacterial persisters: Molecular mechanisms and therapeutic development. Signal Transduction and Targeted Therapy. 2024;**9**(1):174

[61] Yee R, Cui P, Shi W, Feng J, Zhang Y. Genetic screen reveals the role of purine metabolism in Staphylococcus aureus persistence to rifampicin. Antibiotics. 2015;**4**(4):627-642

[62] Wood TK, Knabel SJ, Kwan BW. Bacterial persister cell formation and dormancy. Applied and Environmental Microbiology. 2013;**79**(23):7116-7121

[63] Lewis K. Persister cells. Annual Review of Microbiology. 2010;**64**(1):357-372

[64] Sochocka M, Boratyński J. Osmoregulation--An important parameter of bacterial growth. Postepy Higieny i Medycyny Doswiadczalnej (Online). 2011;**65**:714-724

[65] Vestby LK, Grønseth T, Simm R, Nesse LL. Bacterial biofilm and its role in the pathogenesis of disease. Antibiotics. 2020;**9**(2):59

[66] Soni S, Mukherjee R, De UK, Bharti D, Paul BR, Gandhar JS, et al. In-vitro synergistic interactions of baicalein in combination with β-lactam antibiotics against methicillin-resistant *Staphylococcus aureus* Exploratory Animal and Medical Research. (Online) 2024;**14**(1). ISSN 2277-470X (Print), ISSN 2319-247X

[67] Llor C, Bjerrum L. Antimicrobial resistance: Risk associated with antibiotic overuse and initiatives to reduce the problem. Therapeutic Advances in Drug Safety. 2014;**5**(6):229-241

[68] Alaoui Mdarhri H, Benmessaoud R, Yacoubi H, Seffar L, Guennouni Assimi H, Hamam M, et al. Alternatives therapeutic approaches to conventional antibiotics: Advantages, limitations and potential application in medicine. Antibiotics. 2022;**11**(12):1826

Chapter 4

Recent Advances in Mechano-Bactericidal Nanostructured Surfaces for Biomedical Implants and Frequently Touched Surface Applications

Henry Agbe and Samuel Kwofie

Abstract

Microbial contamination of medical devices and infection prone areas in hygiene critical environments is one of the main causes of healthcare-associated infections. Healthcare-associated infection is the fourth leading cause of death after cancer, heart disease and stroke. Besides antibiotic, one potential strategy for breaking the chain of transmission of infections is *via* engineering novel antimicrobial surfaces. However, conventional antimicrobial coatings are not ideal for practical application due to coating instability, possible toxicity and antibacterial resistance problems. Contrarily, mechano-bactericidal nanostructured surfaces are biocompatible and kill bacteria *via* mechanical stress (induced by physical contact), without releasing toxic-agents. In other words, they are eco-sustainable and hold great potential for fabricating the next-generation antimicrobial surface technology. This review illuminates on recent advances in antimicrobial surface technology, with particular focus on mechano-bactericidal nanostructured surfaces, the killing mechanisms, clinical applications, challenges and future prospects in the field of mechano-bactericidal nanostructured antimicrobial surface technology.

Keywords: mechano-bactericidal, nanostructured surface, biomedical devices, healthcare-associated infections, implants, frequently touched surfaces

1. Introduction

Antimicrobial coatings for biomedical implant devices (such as catheters, vascular stents, contact lenses, prostheses, medical instruments and respiratory machines) and high-touch surface applications (healthcare facility, long-term care homes, outpatient surgical centres and other hygiene-critical environment) have gained much attention in recent times due to their ability to reduce the prevalence of healthcare-associated infections (HCAI) [1, 2]. HCAI is a serious public health problem [3], and the fourth leading causes of death after cancer, heart disease and stroke [4]. In the USA, about 2 million patients contract HCAI annually, leading to about 100,000 fatalities, and a financial loss of $4.5 billion [5]. Furthermore, up to 7% of patients in developed countries and 10% in developing countries are affected each year by at least one of HCAI [6]. Also, HCAI problem is compounded by the fact that most causative pathogens have become resistant to most conventional antibiotics (a condition, leading to the rise of super bugs and global antimicrobial resistance problem) [7], thus necessitating the need for novel antimicrobial solutions. This is particularly important, given that 99% of pathogens exist as community of bacteria called biofilms (in vegetative or spores stage) on either implant devices (within 24 hours of implantation) and frequently touched surfaces (such as door knobs, push plates and hand rails) for days, weeks and months [1, 2, 8, 9]. Additionally, bacteria in biofilms require ~10–1000 order of magnitude higher minimum inhibitory concentration (MIC) of antimicrobial agent than their free-floating planktonic forms [10]. As a result, biofilms provide a safe haven for transmission of pathogenic HCAIs, occurring in a form of: central line-associated bloodstream infection (CLABSI), surgical site infections (SSI), catheter-associated urinary tract infections (CAUTI) and frequently touched surfaces-mediated infections [11].

Besides administering conventional antibacterial, antifungal and antiviral therapies, one potential strategy for breaking the chain of HCAI transmission is to install or retrofit existing surfaces with inherently antimicrobial surface to curb attachment of pathogens (such as *Staphylococcus aureus, Pseudomonas aeruginosa and Escherichia coli*) and their subsequent biofilm formation. While conventional leachable antimicrobial coatings (Ag, Cu, ZnO etc.,) provide excellent antimicrobial properties against biofilm formation and post-operative implant-associated bacterial infections, they pose risks for patients' health due to uncontrolled release of antimicrobial agents, possible detachment from substrates, cytotoxicity in *in vitro* and *in vivo* studies [12], and tendency for antimicrobial resistance problems [13]. On the other hand, mechano-bactericidal nanostructured surfaces are biocompatible and environmentally friendly, and inactivate microorganisms without releasing toxic-agents. Thus, they appear most favourable and ideal candidates for fabricating the next-generation antimicrobial surface technology.

Note that the first report of antimicrobial activity of mechano-bactericidal nanostructured surfaces was made in 2012 by Ivanova et al. [14]. In their work, the mechanical stresses, induced by the physical contact between the micro-nanoscale patterns of the *Psaltoda claripennis* cicada wings (the mechano-bactericidal nanostructured surfaces) and the cell envelope of the *Pseudomonas aeruginosa* (*P. aeruginosa*) (gram-ve) bacterium, was reported to be responsible for the death of the *P. aeruginosa* under 30 minutes of contact. Since this pioneering work, our research group and others [11, 13, 15] have demonstrated the antimicrobial properties of mechano-bactericidal nanostructured surfaces on fomites and other abiotic surfaces such as plastic, metallic and ceramics. Although antibacterial mechanism of mechano-bactericidal surface was attributed to mechanical stress, the exact mechanism and their real-world application are still scientific challenge. Furthermore, problems such as long-term stability and mechanical durability of nanostructured surfaces under various environmental conditions, scalability, biocompatibility, multi-functional potency, lack of standards and regulations have hindered efforts at rapid full-scale application of mechano-bactericidal nanostructured antimicrobial surface technology.

This review highlights recent advances in the antimicrobial surface technology, with particular focus on mechano-bactericidal nanostructured surfaces, the killing mechanisms of these surfaces, their clinical applications and future prospects. Challenges and limitations in their development and implementation are also discussed.

2. Antimicrobial surface technology

Antimicrobial surfaces are fabricated from build-up thin films of antibacterial agents on substratum surfaces [16]. Antimicrobial surfaces can inhibit the extent of attachment and presumably the proliferation of microorganisms. They are generally classified into two main categories (**Figure 1**): (a) *Anti-biofouling surfaces* that repel initial attachment of microbial organisms *via* the use of steric hindrance, electrostatic repulsion or the presence of low surface energy molecules; and (b) *bactericidal surfaces* that kill microorganisms by the release of toxin biocides or mechanical stresses, induced by physical contacts on mechano-bactericidal nanostructured surfaces [18]. It should be mentioned that one strategy for achieving steric repulsive anti-biofouling surface is through polyethylene glycol (PEG) coating. Similarly, electrostatic repulsive surfaces are fabricated *via antimicrobial peptides* (AMP) coating. On the other hand, superhydrophobic surfaces have been fabricated in the laboratory by creating micro-nano topographical patterns, followed by low-surface energy molecule coatings. As noted, *Bactericidal surfaces* disrupt cell membrane to affect bacterial death [16] and can be grouped into those that inactivate bacteria by releasing biocides such as metallic silver and copper or inactivate microorganisms through physical contact of mechano-bactericidal nanostructured surface with microbial cell membrane. Overall, antimicrobial surfaces are fabricated by chemical, biological and physical surface modification strategies [8].

Chemical and biological strategies may include surface functionalization, polymerization, plasma-assisted and surface treatment; in contrast, mechano-bactericidal nanostructured surface constitutes to example of the physical modification process

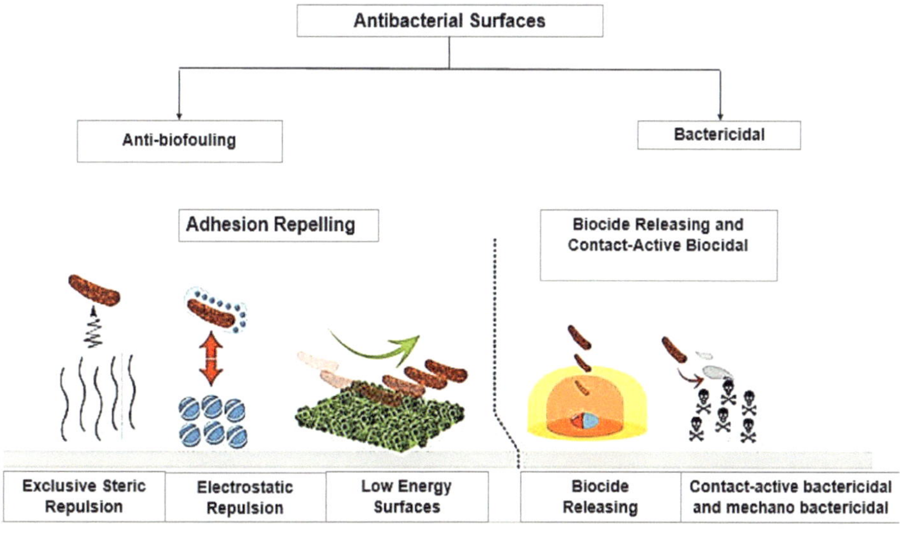

Figure 1.
Classification of antibacterial surfaces (Adapted from Ref. [17]).

[19]. However, for the purpose of this chapter, attention is focused mainly on chemical and physical modification strategies, which lead to hydrophilic, superhydrophobic and mechano-bactericidal nanostructured surfaces.

2.1 Fabrication of hydrophilic antimicrobial surfaces

These surfaces are fabricated *via* physical, chemical and biological processes, and exhibit water contact angle between 10° and 90°. They are generally useful for orthopaedic protheses and medical implants device, where biomolecules adhesion and in particular protein conditioning and implant-tissue integration are required [20]. Hydrophilic antimicrobial surfaces are generally fabricated by immobilizing antimicrobial agents such as: (a) metallic Ag and Cu; (b) inorganic metal oxides of TiO_2, ZnO; and (c) organic compounds such as quaternary ammonium compounds (**Table 1**). These are typically immobilized on metallic, ceramic and polymeric substrates. Of particular interest are silver and copper antimicrobial agents, and therefore discussed further herein.

2.1.1 Silver-based antimicrobial coating

Silver-based antimicrobial coating involves the immobilization of silver nanoparticles in matrix, mostly polymeric, as coating. The antimicrobial properties of silver, along with copper, have been known throughout history. For example, the ancient Chaldean used silver as cutleries as early as 4000 B.C, before the discovery of microbes [26]. Furthermore, the Greeks and the Romans stored water in silver vessels for their antimicrobial properties. Again, the Mediterranean and Asiatic cultures used silver storage containers and silver foil as preventive measure against liquid spoilage and wound infections, respectively [27]. Additionally, the Americans travelling west in the 1880s left silver coins in their water barrels to inhibit the growth of microorganisms such as bacteria and algae [28]. From the foregoing, it is reasonable to conclude that antimicrobial properties and their broad-spectrum effectiveness against variety of pathogens (such as bacteria, fungus, yeast, virus and protozoa) have been well known though out history. In fact, due to its antimicrobial effectiveness, silver was used in many applications, and by the twentieth century, it had become the main antimicrobial agent. However, its use was discouraged as a result of: (a) the discovery of penicillin by Sir Alexander Fleming in 1928, the advent of antibiotics, and (b) the high costs associated with its production [29]. It should be mentioned that prior to the twenty-first century, silver was utilized in many domestic products such as needles, vessels, plates, cutleries and silver fillings due to its superior antimicrobial properties (**Figure 2**) [30].

Substrates	Representative antibacterial material	Bacterial type	Ref.
Glass	Tetraethoxysilane (TEOS) and Q_4N^+ $Si(OR)_3^-$	*S. aureus* and *E. coli*	[21]
Glassware	CuS-NPs	*S. aureus* and *E. coli*	[22]
5 L carbon steel	TiO_2/ZnO ceramic coatings	*E. coli*	[23]
316 L stainless steel	*Ag*-NPs	*E. coli*	[24]
Hydroxyapatite	Ag–TiO_2	*E. coli*	[25]

Table 1.
Major antibacterial materials.

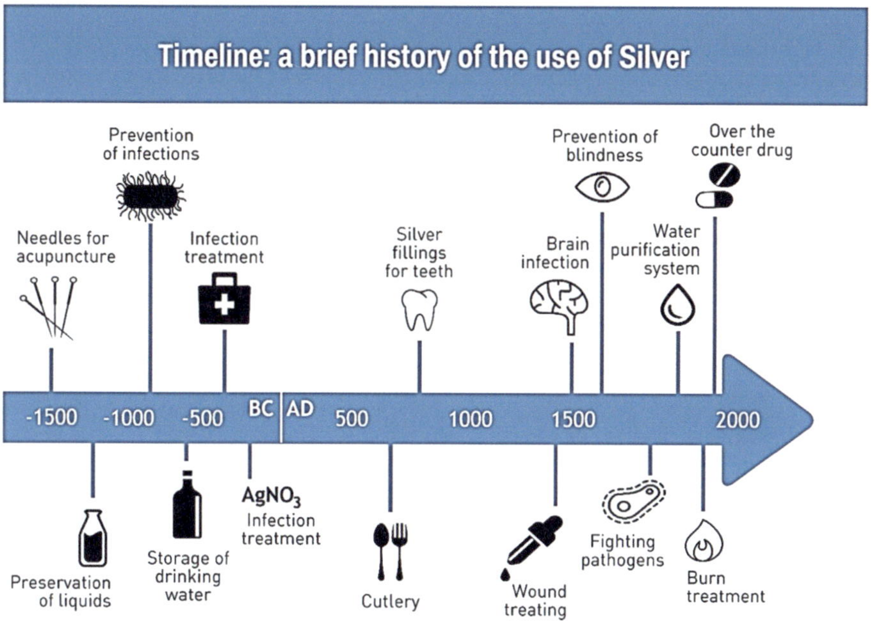

Figure 2.
Schematic representation of applications of silver-related products, along the course of human history (Adopted from Ref. [30] with permission).

2.1.2 Copper antimicrobial coating

Advancement in antimicrobial surfaces technology has resulted in the discovery of copper alloys (> 60% copper) and since 2008, being registered by the US Environmental Protection Agency (EPA) as the only solid antimicrobial material capable of destroying microbes [31–33]. Since then, copper and its alloy (with inherent antimicrobial property) used as frequently touched surfaces in hospitals and other hygiene critical environment have attracted the attention of the scientific community. For instance, compared to stainless steel surfaces, copper alloys exhibit excellent antimicrobial effectiveness against HCAI pathogens such as *Escherichia Coli* (*E. coli*) O157, *Methicillin Resistant Staphylococcus Aureus* (MRSA), *Clostridium difficile* (*C. difficile*), *influenza A virus, adenovirus* and *fungi* [34, 35]. Furthermore, in a recent study by Chatterjee et al. [36], copper-impregnated solid material surfaces in patient rooms exhibited antimicrobial effectiveness of 58% compared to standard hospital-laminated stainless steel surfaces.

2.2 Fabrication of superhydrophobic antibacterial surfaces

The unique water roll-off property of tropical plants, such as lotus leaves, has been observed by many scientists for centuries. However, it was only recently that advances in microscopy- and nanotechnology-enabled scientists to confirm the presence of micro-nanostructures to be responsible for the water roll-off property of these surfaces. Note that the micro-nanoscale structures enhance plants' ability to adapt and inhibit colonization and as evolutionary surviving strategy [37]. Again, it should be mentioned that the micro-nanostructures exist as long-chain palmitic (hexadecanoic) and stearic (octadecanoic) fatty acids [38], rendering them superhydrophobic.

In other words, the combined effects of micro-nanostructure and the low-surface energy fatty acid molecules result in superhydrophobicity. Inspired by nature, scientist now fabricates superhydrophobic surfaces in the laboratory by patterning micro-nano structures, followed by passivation with low-surface energy fatty acid molecules. Superhydrophobic surfaces have been fabricated using various strategies such as photolithography, sol-gel, plasma etching, anodization and chemical etching [39] to create micro-nano topographical patterns followed by passivation. Due to their water repellency, superhydrophobic surfaces find applications as self-cleaning-fabrics, anticorrosion, drag reduction and anti-biofouling surface [40]. In the context of antimicrobial surfaces, anti-biofouling surfaces are of particular interest; as in principle, they have the potential of presumably inhibiting initial bacterial attachment and subsequent biofilm formation. However, in practice, superhydrophobic surfaces have been observed to fail under long-term exposure in humid environments [41]. To this end, a dual-prong approach that incorporates bactericides in superhydrophobic coating holds promise not only in inducing both anti-biofouling and antibacterial properties, but also for improving their longevity. For instance, Chung et al. [42] fabricated silver-perfluorodecanethiolate coatings on silicon wafer with both superhydrophobic and antibacterial properties *via* precipitation method, using perfluorodecanethiol (PFDT) and silver as fluorinated and metal-thiolate complex precursors, respectively (**Figure 3(A)**). This silver-perfluorodecanethiolate superhydrophobic surface exhibited bacterial adhesion reduction factor of 77%. In a related work, Wang et al. [43] fabricated a superhydrophobic diamond films with both antibacterial and anti-biofouling properties using both hot filament chemical vapour deposition and sol-gel perfluorodecyltrichlorosilane (PFTS) process (**Figure 3(B)**). The perfluorodecyltrichlorosilane superhydrophobic diamond films exhibited bacterial adhesion reduction between 90 and 99%.

Among the antimicrobial surface technology discussed so far, mechano-bactericidal nanostructured surface is a very promising technology as it kills invading pathogen by physical contact, using its micro- and nano-topographical features. Mechano-bactericidal nanostructured surface also appears to provide a paradigm shift in the fabrication of next-generation novel antimicrobial surfaces. Contrary to conventional antimicrobial surfaces and coatings such as leachable antibacterial coatings (metallic silver and copper), and superhydrophobic antimicrobial surfaces,

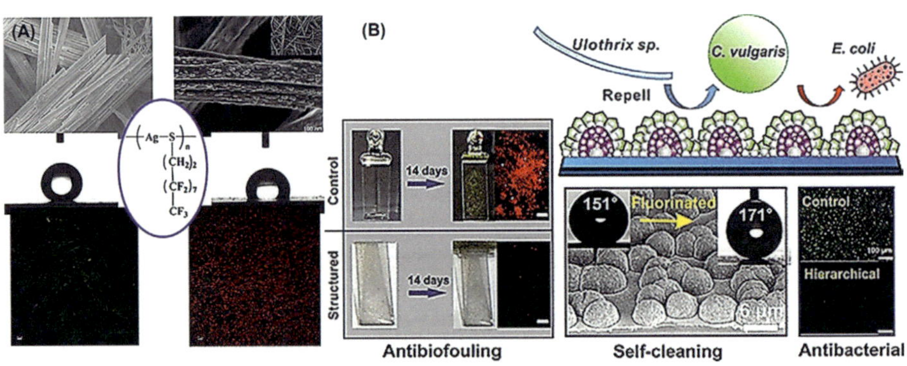

Figure 3.
Schematic representation of: (A) silver-perfluorodecanethiolate superhydrophobic coatings; (B) hierarchical structured diamond superhydrophobic films with anti-biofouling, self-cleaning and antibacterial properties (Adopted from Ref. [32, 33] with permission).

mechano-bactericidal surface does not suffer from coating instability, possible toxicity and antimicrobial resistance problems [13]. Thus, it is more environmentally sustainable and promises to revolutionize the entire antimicrobial surface technology space.

2.3 Fabrication of mechano-bactericidal nanostructured surface

In nature, many living things (including insects, plants, sharks geckos, etc.) develop defence mechanisms towards invading pathogens, despite pathogens' ability to employ complex adaptive mechanisms to colonize their surfaces [44]. However, it should be noted that most of these living things leverage on their micro- and nano-topographical patterns to induce antimicrobial killing, the so-called mechano-bactericidal nanostructured surfaces [44]. As noted in Section 1.0, the first mechano-bactericidal nanostructured surface (exhibited by cicada wings) was reported by Ivanova and co-workers [14]. In this pioneering work, Ivanova et al. showed that the micro-nano topographical patterns on the cicada wings induced mechanical stresses on the membrane envelop of the *P. aeruginosa* (gram-ve) bacterium, leading to the death of the latter under 30 minutes of contact. Following this work, many report on other wings (such as dragonfly and skin of geckos), and their synthetic analogues have been fabricated on metals, ceramics and polymers (**Figure 4**) [46]. Similar to other antimicrobial surfaces, mechano-bactericidal nanostructured surfaces are fabricated utilizing chemical and physical surface modification processes. These include alkaline hydrothermal process, reactive ion etching (RIE), thermal oxidation, anodization, chemical etching and glancing angle deposition (GLAD) [45]. In a typical hydrothermal process, the micro-nanoscale patterns of the metal (commonly pure titanium and Ti-6Al-4 V) is fabricated from an aqueous precursor sealed in stainless

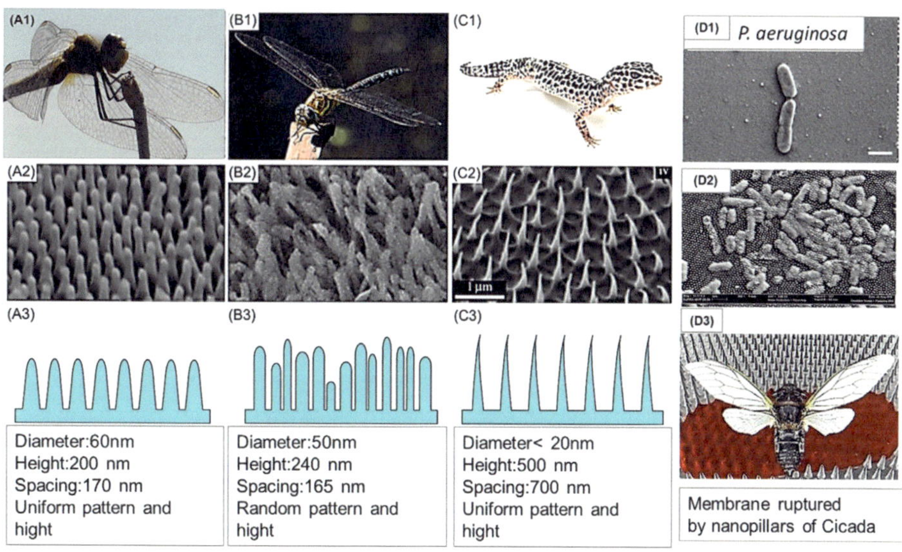

Figure 4.
Illustration of naturally occurring mechano-bactericidal surfaces: (A1–C1), cicada, dragonfly and gecko, respectively; (A2–C2), scanning electron microscopy (SEM) images of cicada wing, dragonfly wing and gecko skin; (A3–C3), schematic representation of nano topographical features on cicada wing, dragonfly wing and gecko skin; (D1–D3), Pseudomonas aeruginosa *ruptured by cicada wings (Adapted from Ref. [45] with permission).*

steel autoclave and operated above ambient temperature and atmospheric pressure (**Figure 5(A)**, Top). For instance, TiO_2 topographical nanostructures were achieved on titanium alloy, Ti-6Al-4 V using alkaline hydrothermal process [47]. Furthermore, by optimizing experimental parameters such as temperature, reaction time and acidic and alkaline concentration, topographical micro- and nanostructures similar to dragonfly insects were fabricated (**Figure 5(A)**, Bottom) [45].

Among the physical surface modification strategies for fabricating mechano-bactericidal nanostructured surfaces, thermal oxidation process appears paramount. Thermal oxidation involves optimization of temperature (within a range of 400 to 900°C) in a carbon-containing atmosphere to oxidize the metallic substrate and

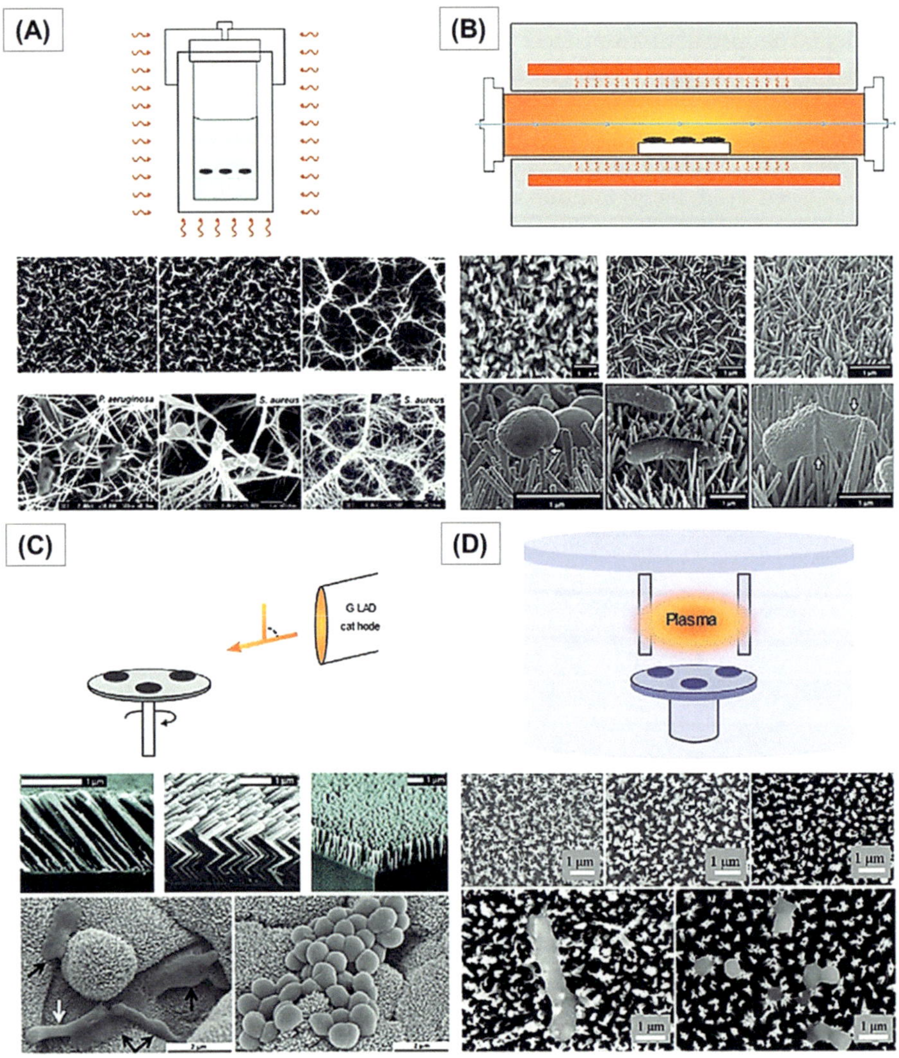

Figure 5.
Illustration of nano-topographical features obtained from nanofabrication techniques: (A) alkaline hydrothermal (AH); (B) thermal oxidation (TO); (C) glancing angle deposition (GLAD) and (D) reactive ion etching (RIE). Each panel illustrates a basic schematic of the technique, an example of features generated and evidence of bactericidal activity (Adapted from Ref. [45] with permission).

to grow the nanostructures topographical patterns (**Figure 5(B)**, Top) [45]. For example, thermal oxidation has been utilized to fabricate nanostructures analogue to the topographical features of the wings of dragonfly (**Figure 5(B)**, Bottom) [45].

Similar to thermal oxidation process is anodization; except that the oxidation process is achieved by electrochemical means. Anodization process has been used to generate such topographical features or mechano-bactericidal nanostructured surfaces. For instance, Agbe et al. [13] employed a one-step hard anodization process using phosphoric acid to fabricate mechano-bactericidal nanostructured surfaces that exhibited a pore diameter of 151 ± 37 nm, a cell diameter of 239 ± 53 nm, surface roughness (rms) of 2.9 ± 0.7 μm and bactericidal efficiency of 100% (**Figure 6**).

Other physical modification process for fabricating mechano-bactericidal nanostructured surfaces includes glancing angle deposition (GLAD) and reactive ion etching (RIE) [19]. Contrary to typical physical vapour deposition, where a stream of vapour-phase atom strikes and condenses upon a perpendicular substrate, forming dense and solid films, for GLAD, the substrate is flexible and can be tilted or rotated at glancing angle to create oblique deposition geometry [48]. In principle, the GLAD enables flexibility, as substrate is able to be manipulated, ensuring oblique angle deposition and growth of columnar architectures nanostructured on substrate's surfaces (**Figure 5(C)**, Top) [49]. Note that the main feature of GLAD is the "shadowing effect" or line-of-sight, which prevents incoming vapour from condensing into regions behind the nuclei, and thus, causing the nuclei to develop into columns that

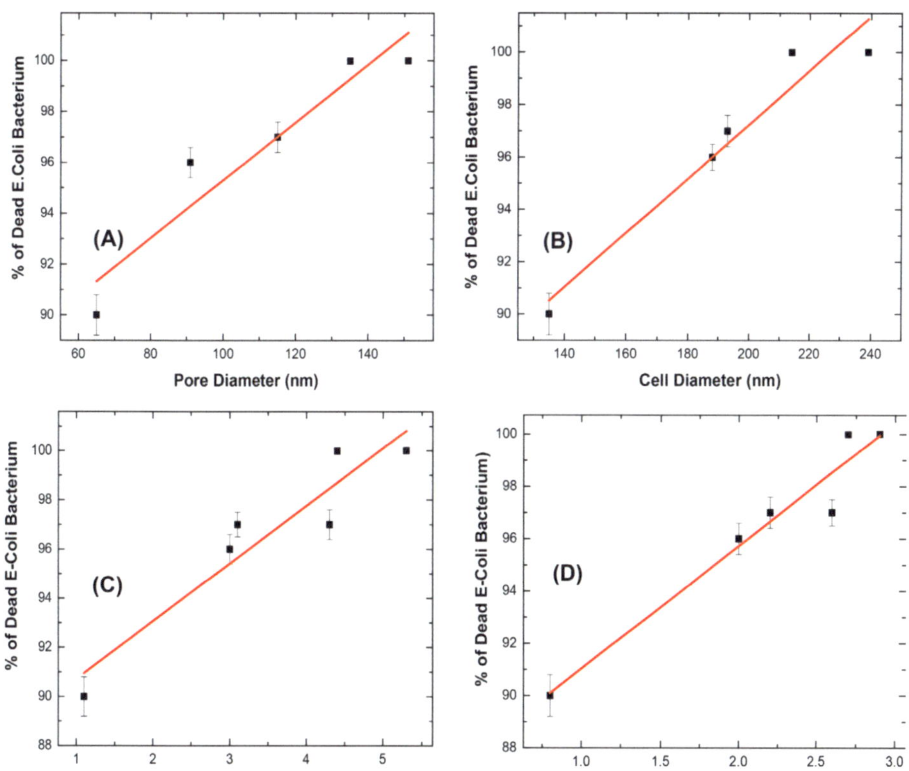

Figure 6.
Effects of nano-topographical features on E. coli *bacteria killing efficiency: (A) pore diameter, (B) cell diameter, (C) oxide thickness and (D) surface roughness (rms). (Adapted from Ref. [13] with permission).*

tilt towards the vapour source or line of sight [48]. Due to its versatility, the GLAD surface modification process has been used to fabricate precisely controlled micro- and nanostructures (**Figure 5(C)**, Bottom) [45]. Similar to the GLAD, the reactive ion etching (RIE) techniques have been effective in fabricating mechano-bactericidal nanostructured surfaces. These involve the use of dry etching reactive plasma and low vacuum systems, which result in partial removal of the top layer of the substrate, leading to fabrication of desired mechano-bactericidal nanostructured surfaces (**Figure 5(D)**, Top). For instance, Ishak et al. utilized the RIE process to fabricate micro-nano topographical features (**Figure 5(D)**, Bottom) [45].

Overall, mechano-bactericidal nanostructured surfaces are promising, innovative and biocompatible antimicrobial surface technology compared to the traditional antimicrobial coatings that are limited by coating instability, uncontrolled release profile, coating detachment problem and antimicrobial resistance problem [8]. Thus, mechano-bactericidal nanostructured surfaces are bioinspired and biocompatible, and have great potential for revolutionizing the antimicrobial surface technology space.

3. Killing mechanisms of mechano-bactericidal nanostructured surfaces

Although antibacterial mechanism of mechano-bactericidal nanostructured surface has been attributed to mechanical stress, the exact mechanism and the real-world application are still scientific challenge. For instance, while antimicrobial mechanism had initially been attributed to stretching, and subsequent rupture, of the bacterial membrane upon surface adhesion [50], others hypothesized the tearing of the bacterial membrane as bacteria attempt to move on the mechano-bactericidal surface (**Figures** 7 and **8**) [51].

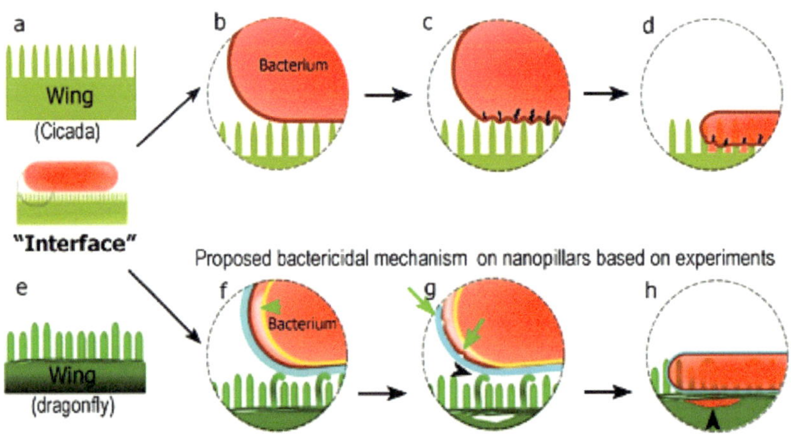

Figure 7.
Proposed mechanism of bactericidal activity of cicada wing (a) – (d). (a) Surface of the cicada wings, (b) A bacterium approaches surface, and experience compressive and adhesive forces under its weight. (c) Bacteria's membrane begins to rupture upon stretching. (d) Bacterium's cytoplasm leaks, leading to cell death. (e) Nanopillar density on dragonfly wings. (f) Taller nanopillars bends upon contact with bacterium. (g) The damaged bacterial membrane wrinkles and forms blebs (arrows). (h) Cytosol leaks and flows upon bacterium death. (Adapted from Ref. [50] with permission).

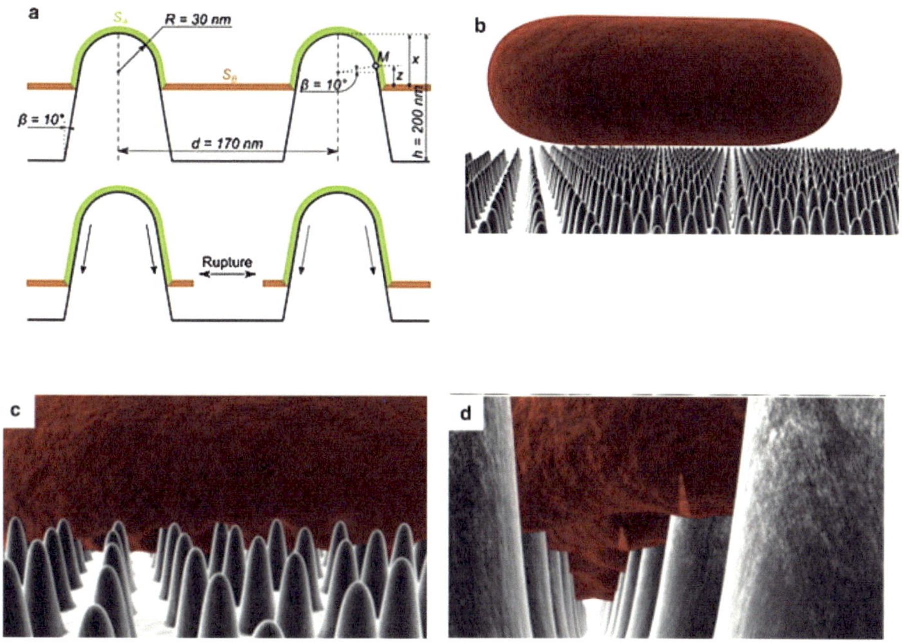

Figure 8.
(a). Illustration of bacterial outer layer adsorbed on cicada wing nanopillars. (b). Illustration of bacterial outer layer in contact with cicada wing nanopillars. (c). Illustration of a bacterial outer layer adsorbed on cicada: outer layer starts to rupture in regions between the pillars. (d). Illustration of a bacterial outer layer adsorbed on cicada: Bacterium collapses upon stretching on the nanopillar surface. Adapted from Ref. [51] with permission.

Meanwhile, a study by Jenkins et al. [52] disputed the membrane tearing hypothesis, claiming stress-mediated generation of reactive oxygen species (ROS) upon contact [52]. Furthermore, recent study suggests that the stretching or deformation of cell membrane was not sufficient to kill the bacteria because of sub-lethal mechanical injury, but rather due to subsequent programmed apoptosis-like death (ALD) of affected bacteria (**Figure 9**) [53].

From the foregoing, consensus on the antimicrobial mechanisms of mechano-bactericidal nanostructured surfaces is yet to be achieved. Thus, it is very pertinent, considering the complexity associated with bacterial-substrate interaction (bacteria type, surface chemistry, surface morphology, Van der Waals interactions, extra polymeric substance (EPS) and biofilm formation *via* quorum sensing), and multifaceted killing possibility [54, 55], that a multidisciplinary approach that combines mechanistic biophysical analysis, biochemical assays and computational modelling may be necessary for unravelling the mechanism. Interestingly, such approach has been well explored in multicellular eukaryotic cells. There is now growing evidence that biological systems are exceptionally sensitive not only to their chemical environment, but also to their mechanical environment. Therefore, forces and displacements experienced by a cellular system are crucial in modulating cellular response and behaviour. Ultimately, understanding exactly how bacteria respond to physical stimuli would not only aid in shedding light on antimicrobial mechanism but also provide significant implications for real world application of mechano-bactericidal nanostructured surface technology.

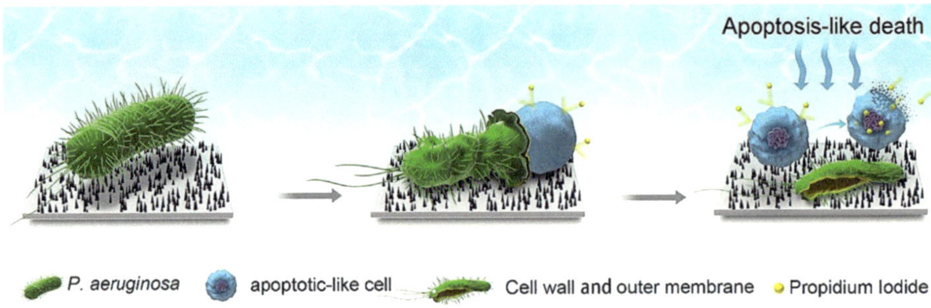

Figure 9.
Proposed mechanism for bactericidal action of Mechano-bactericidal nanostructured surface upon contact with gram-negative bacteria (Adapted from Ref. [53] permission with permission).

4. Clinical applications, challenges and future prospects

The next-generation antimicrobial surfaces, ideal for biomedical implants and frequently touched surface applications, must be of low cost and simplicity. Besides, they must be scalable, cost effective (in terms of value for end user), broad-spectrum effectiveness, durability and non-toxicity. Nonetheless, in spite of years of antimicrobial surface research, mechano-bactericidal nanostructured surfaces have not transition from the laboratory bench to the market [8]. Three challenges have resulted in the slow progression in the field of mechano-bactericidal nanostructured surface technology: *lack of durability* (the long-term stability and mechanical integrity of the nanostructured surfaces under various environmental conditions), *antibacterial resistance problem* and *the lack of standardized* testing protocols [56].

Despite the promising role of mechano-bactericidal nanostructured surfaces of being bioinspired and presumably killing microbes *via* mechanical stress, without releasing toxic-agents, the structural integrity of the nanostructured features (such as nanopillars, nanospikes and nano-ridges) seem to be weak unstable and temporary. It is unclear if these micro-nanotopographical features are progressively filled with debris and dead bacterial over time, leading to reduced killing ability over time. Thus, it is important that micro-mechanical tests, such as hardness, tribology, scratch and anti-adhesion studies, be performed on these mechano-bactericidal nanostructured surfaces to elucidate their structural integrity.

Another concern is biocompatibility issues and possibility of antibacterial resistance problem from the use of mechano-bactericidal nanostructured surfaces. Again, fate of mechano-bactericidal nanostructured surfaces on mammalian cells, particularly red blood cells (due to their small size ~6–8 μm), is unclear. In terms of resistance, it is common knowledge that HCAIs pathogens can develop resistance to antibiotics due to the tendency for microbes to develop antimicrobial resistance genes, induced by selection pressure. Sadly, this has resulted in the rise of multi-drug-resistant pathogens ("super bugs"), necessitating the need for novel and alternative antimicrobial arsenals to curb this rising global phenomenon. The rationale behind antimicrobial usage must be a balance between potential impacts on emergence of resistant microbial strains *vis-à-vis* preventing spread of pathogens [56]. However, because application of mechano-bactericidal surfaces in clinical settings is not wide spread (in contrast to antibiotics and antimicrobial agents), they are unlikely to develop antimicrobial resistance in the foreseeable future.

Finally, as a new line of research, the field of mechano-bactericidal nanostructured surface technology lacks *standardized* testing protocols. Hence, it is quite difficult validating the myriad of nanostructured mechano-bactericidal surfaces (micro-nanopatterns with different morphological features and processing technologies) available both in the market and in research and development phase [56]. For instance, protocols such as the ASTM 2180 [57], Japanese Standard JIS Z2801 [58], ISO 22196 and Live/dead staining [59], which were originally designed for leachable antimicrobial agents (characterized by diffusion of antimicrobial agents), are also used for validating antimicrobial effectiveness of mechano-bactericidal nanostructured surface. Furthermore, regarding mechano-bactericidal nanostructured surfaces, intended for applications on abiotic surfaces surrounding patients, novel dry seeding antimicrobial protocol that mimics the near dry ambient conditions, such as those reported by Agbe et al. [13], must be encouraged. In contrast, existing antimicrobial protocols require conditions of high humidity (> 90%) and temperature (~37°C), which do not necessarily reflect the near dry surface condition in real life [60]. Additionally, other protocols, such as Live and Dead staining assay, Agar zone of inhibition assay, Adhesion-based assay and Bioluminescent strains, are all not ideal for assessing antimicrobial effectiveness of mechano-bactericidal nanostructured surfaces. For instance, the Live/dead staining (e.g. LIVE/DEAD Bacterial Viability Kit (BacLightTM)), which is characterized by cell wall damage staining (following interaction with antimicrobial agents), does not provide insight into cellular structural integrity or elasticity and stiffness. Again, cell wall damage alone may not be sufficient grounds to justify cell death, because cells with damaged walls might be culturable [61]. It suffices to say that since antimicrobial mode of action for mechano-bactericidal nanostructured surfaces is presumably due to mechanical stresses, protocols that provide insight into the evolution of microbial cell death on mechano-bactericidal surface with time (capable of tracking cellular structural integrity with time, e.g. from pre-rupture, rupture, and post-rupture phases) such as atomic force microscopy-based force spectroscopy (AFM-FS) are urgently needed.

Furthermore, it is possible to leverage on the emerging field of mechano-biology and capabilities of advanced microscopy such as AFM-FS, fluorescence resonance energy transfer (FRET) technology and computational modelling (finite element analysis (FEA)). For instance, using AFM in dynamic mode, time-lapse imaging could be captured to observe changes in bacterial cell morphology over time. These presumably could help in elucidating the real-time mechanistic bacterial/mechano-bactericidal nanostructured surface interactions. Furthermore, while mechano-bactericidal nanostructured surfaces have been excellent in inactivating vegetative form of multidrug-resistant (MDR) bacteria, in particular, *P. aeruginosa*, *E. coli* and *Staphylococcus aureus* (*S. aureus.*) [62, 63], their effectiveness against biofilms, specifically, spore-forming bacteria such as *Clostridium difficile* (anaerobic) or *Bacillus subtilis* (aerobic), is still unknown. Nevertheless, such fundamental knowledge is critical if mechano-bactericidal surface technology is to be harnessed to solve HCAI and global antimicrobial resistance problems. In this regard, both experimental and quantitative models need to be explored on bacteria biofilm (*Bacillus subtilis*, *P. aeruginosa, E. coli* and *S. aureus*) to study how mechanical cues (stiffness, roughness, shear stress, deformation, etc.,) influence bacteria cell response, adhesion, proliferation and biofilm formation. Again, for real world application, it is important not only to understand antimicrobial mechanism of single cell bacteria but most importantly, multi-species biofilm colonies. Thus, it is hoped that an interdisciplinary approach rooted in mechano-biology, microbiology, biophysics, computational biology and biomaterial science and engineering will be utilized in investigating the mechanisms of novel and the next-generation mechano-bactericidal nanostructured surfaces for biomedical implants and touch surface applications.

5. Conclusion

Multidrug resistance pathogens such as *Staphylococcus aureus* (S.A.), *Pseudomonas aeruginosa* (P.A.) and *Escherichia-coli* (*E. coli*) can survive on biomedical implants and frequently touched surfaces. These surfaces can become potential reservoir for transmission of healthcare-associated pathogenic infections.

One potential strategy for breaking the chain of HCAI transmission is to engineer inherently antimicrobial surfaces that inhibit the attachment of pathogen, its proliferation and subsequent biofilm formation. Of particular focus in this review are mechano-bactericidal nanostructured surfaces. These surfaces hold promise for fabricating the next-generation antimicrobial surface technology due to their inherent antimicrobial nature, biocompatibility, non-toxicity and environmental sustainability nature.

However, challenges such as lack of long-term stability and mechanical durability of nanostructured surfaces, scalability, lack of standards and regulations and underlying killing mechanisms, need to be resolved. It is hoped that multi-disciplinary approach, including leveraging on the emerging field of mechano-biology, biophysics and computational models among others, would be adopted to help resolve issues bedevilling the full-scale application to enable the transition of mechano-bactericidal nanostructured antimicrobial surfaces from the bed side to the market.

Author details

Henry Agbe* and Samuel Kwofie
Department of Materials Engineering, College of Engineering,
Kwame Nkrumah University of Science and Technology, KNUST-Kumasi, Ghana

*Address all correspondence to: henry.agbe@yahoo.co.uk

References

[1] Vrana NE. Biomaterials and Immune Response: Complications, Mechanisms and Immunomodulation. Boca Raton, Florida, USA: Taylor & Francis Group, CRC Press; 2018

[2] Walker J. Decontamination in Hospitals and Healthcare. Cambridge, UK: Woodhead Publishing; 2019

[3] Schmidt MG. The role of antimicrobial surfaces in hospitals to reduce healthcare-associated infections (HAIs). In: Decontamination in Hospitals and Healthcare. Cambridge, UK: Woodhead Publishing; 2020. pp. 259-299

[4] MacLaurin A et al. Measuring and monitoring healthcare-associated infections: A Canadian collaboration to better understand the magnitude of the problem. Healthcare Quarterly (Toronto, Ont.). 2020;**22**(SP):116-128

[5] Organization WH. Report on the Burden of Endemic Health Care-Associated Infection Worldwide. Geneva, Switzerland: WHO; 2011

[6] Organization WH. Guidelines on Core Components of Infection Prevention and Control Programmes at the National and Acute Health Care Facility Level. Geneva, Switzerland: WHO; 2016

[7] Friedrich AW. Control of hospital acquired infections and antimicrobial resistance in Europe: the way to go. Wiener Medizinische Wochenschrift. 2019;**169**(1):25-30

[8] Ivanova EP, Crawford R. Antibacterial Surfaces. New York City, USA: Springer; 2015

[9] Tortora GJ, Funke BR, Case CL. Microbiology: An Introduction. Boston, USA: Addison-Wesley; 2010

[10] Ramos LS et al. Susceptibility of the *Candida haemulonii* complex to Echinocandins: Focus on both planktonic and biofilm life styles and a literature review. Journal of Fungi. 2020;**6**(4):201

[11] Hasan J et al. Antiviral and antibacterial nanostructured surfaces with excellent mechanical properties for hospital applications. ACS Biomaterials Science & Engineering. 2020;**6**(6):3608-3618

[12] Ng CT et al. Zinc oxide nanoparticles exhibit cytotoxicity and genotoxicity through oxidative stress responses in human lung fibroblasts and *Drosophila melanogaster*. International Journal of Nanomedicine. 2017;**12**:1621

[13] Agbe H, Sarkar DK, Chen X-G. Anodized aluminum surface with topography-mediated antibacterial properties. ACS Biomaterials Science & Engineering. 2022;**8**(3):1087-1095

[14] Hasan J et al. Selective bactericidal activity of nanopatterned superhydrophobic cicada *Psaltoda claripennis* wing surfaces. Applied Microbiology and Biotechnology. 2013;**97**(20):9257-9262

[15] Hasan J et al. Multi-scale surface topography to minimize adherence and viability of nosocomial drug-resistant bacteria. Materials & Design. 2018;**140**:332-344

[16] Hasan J, Crawford RJ, Ivanova EP. Antibacterial surfaces: The quest for a new generation of biomaterials. Trends in Biotechnology. 2013;**31**(5):295-304

[17] Morán G, Méallet-Renault R. Bacterial Pathogenesis and Antibacterial Control. London, UK: Intechopen; 2017

[18] Siedenbiedel F, Tiller JC. Antimicrobial polymers in solution and on surfaces: overview and functional principles. Polymers. 2012;**4**(1):46-71

[19] Glocker D, Ranade S. Medical Coatings and Deposition Technologies. Oxford, England, UK: John Wiley & Sons; 2016

[20] Chouirfa H et al. Review of titanium surface modification techniques and coatings for antibacterial applications. Acta Biomaterialia. 2018;**83**:37-54

[21] Ganewatta MS et al. Antibacterial and biofilm-disrupting coatings from resin acid-derived materials. Biomacromolecules. 2015;**16**(10):3336-3344

[22] Gargioni C et al. Self-assembled monolayers of copper sulfide nanoparticles on glass as antibacterial coatings. Nanomaterials. 2020;**10**(2):352

[23] Yusuf Y et al. Antibacterial properties of laser surface-textured TiO_2/ZnO ceramic coatings. Ceramics International. 2020;**46**(3):3949-3959

[24] Zhang S et al. A sol–gel based silver nanoparticle/polytetrafluorethylene (AgNP/PTFE) coating with enhanced antibacterial and anti-corrosive properties. Applied Surface Science. 2021;**535**:147675

[25] Reddy MP, Venugopal A, Subrahmanyam M. Hydroxyapatite-supported Ag–TiO_2 as *Escherichia coli* disinfection photocatalyst. Water Research. 2007;**41**(2):379-386

[26] Hill WR, Pillsbury DM. Argyria: The Pharmacology of Silver. New York, USA: Williams & Wilkins; 1939

[27] Oppermann R. In: Addicks L, editor. Silver in Industry. New York: Reinhold Publishing Corporation; 1940. 636 pages

[28] Barillo DJ, Marx DE. Silver in medicine: A brief history BC 335 to present. Burns. 2014;**40**:S3-S8

[29] Kowalczyk P et al. All that glitters is not silver—a new look at microbiological and medical applications of silver nanoparticles. International Journal of Molecular Sciences. 2021;**22**(2):854

[30] Möhler JS et al. Silver bullets: A new lustre on an old antimicrobial agent. Biotechnology Advances. 2018;**36**(5):1391-1411

[31] Mehtar S, Wiid I, Todorov S. The antimicrobial activity of copper and copper alloys against nosocomial pathogens and *Mycobacterium tuberculosis* isolated from healthcare facilities in the Western Cape: An *in-vitro* study. Journal of Hospital Infection. 2008;**68**(1):45-51

[32] Bryant C, Wilks SA, Keevil CW. Rapid inactivation of SARS-CoV-2 on copper touch surfaces determined using a cell culture infectivity assay. bioRxiv. 2021. 2021.01.02.424974

[33] Cortes AA, Zuñiga JM. The use of copper to help prevent transmission of SARS-coronavirus and influenza viruses. A general review. Diagnostic Microbiology and Infectious Disease. 2020;**98**:115176

[34] Grass G, Rensing C, Solioz M. Metallic copper as an antimicrobial surface. Applied and Environmental Microbiology. 2011;**77**(5):1541-1547

[35] Noyce J, Michels H, Keevil C. Potential use of copper surfaces to reduce survival of epidemic meticillin-resistant *Staphylococcus aureus* in the healthcare

environment. Journal of Hospital Infection. 2006;**63**(3):289-297

[36] Chatterjee P et al. Effectiveness of copper-impregnated solid surfaces on lowering microbial bio-burden levels in an acute care hospital. In: Open Forum Infectious Diseases. US: Oxford University Press; 2020

[37] Hasan J et al. Mimicking insect wings: The roadmap to bioinspiration. ACS Biomaterials Science & Engineering. 2019;**5**(7):3139-3160

[38] Ivanova EP et al. Molecular organization of the nanoscale surface structures of the dragonfly *Hemianax papuensis* wing epicuticle. PLoS One. 2013;**8**(7):e67893

[39] Safaee A, Sarkar DK, Farzaneh M. Superhydrophobic properties of silver-coated films on copper surface by galvanic exchange reaction. Applied Surface Science. 2008;**254**(8):2493-2498

[40] Brassard J-D, Sarkar DK, Perron J. Synthesis of monodisperse fluorinated silica nanoparticles and their superhydrophobic thin films. ACS Applied Materials & Interfaces. 2011;**3**(9):3583-3588

[41] Liu S et al. Dual-functional, superhydrophobic coatings with bacterial anticontact and antimicrobial characteristics. ACS Applied Materials & Interfaces. 2020;**12**(19):21311-21321

[42] Chung J-S et al. Silver-perfluorodecanethiolate complexes having superhydrophobic, antifouling, antibacterial properties. Journal of Colloid and Interface Science. 2012;**366**(1):64-69

[43] Wang T et al. Robust biomimetic hierarchical diamond architecture with a self-cleaning, antibacterial, and antibiofouling surface. ACS Applied Materials & Interfaces. 2020;**12**(21):24432-24441

[44] Lin N et al. Nanodarts, nanoblades, and nanospikes: Mechano-bactericidal nanostructures and where to find them. Advances in Colloid and Interface Science. 2018;**252**:55-68

[45] Ishak MI et al. Protruding nanostructured surfaces for antimicrobial and osteogenic titanium implants. Coatings. 2020;**10**(8):756

[46] Dickson MN et al. Nanopatterned polymer surfaces with bactericidal properties. Biointerphases. 2015;**10**(2):021010

[47] Damiati L et al. Impact of surface topography and coating on osteogenesis and bacterial attachment on titanium implants. Journal of Tissue Engineering. 2018;**9**:2041731418790694

[48] Hawkeye MM, Taschuk MT, Brett MJ. Glancing Angle Deposition of Thin Films: Engineering the Nanoscale. New Jersey, US: John Wiley & Sons; 2014

[49] Robbie K, Sit J, Brett M. Advanced techniques for glancing angle deposition. Journal of Vacuum Science & Technology B: Microelectronics and Nanometer Structures Processing, Measurement, and Phenomena. 1998;**16**(3):1115-1122

[50] Bandara CD et al. Bactericidal effects of natural nanotopography of dragonfly wing on *Escherichia coli*. ACS Applied Materials & Interfaces. 2017;**9**(8):6746-6760

[51] Linklater DP et al. Comment on "bactericidal effects of natural nanotopography of dragonfly wing on *Escherichia coli*". ACS Applied Materials & Interfaces. 2017;**9**(35):29387-29393

[52] Jenkins J et al. Antibacterial effects of nanopillar surfaces are mediated by cell impedance, penetration and induction of oxidative stress. Nature Communications. 2020;**11**(1):1626

[53] Zhao S et al. Programmed death of injured *Pseudomonas aeruginosa* on mechano-bactericidal surfaces. Nano Letters. 2022;**22**(3):1129-1137

[54] Catley TE, Corrigan RM, Parnell AJ. Designing effective antimicrobial nanostructured surfaces: Highlighting the lack of consensus in the literature. ACS Omega. 2023;**8**(17):14873-14883

[55] Linklater D, Ivanova EP. Challenges to the design and testing of antimicrobial nanostructured surfaces. Microbiology Australia. 2023;**44**(2):79-82

[56] Adlhart C et al. Surface modifications for antimicrobial effects in the healthcare setting: A critical overview. Journal of Hospital Infection. 2018;**99**(3):239-249

[57] ASTM E. Standard test method for determining the activity of incorporated antimicrobial agent (s). In: Polymeric or Hydrophobic Materials. West Conshohocken, PA, USA: ASTM International; 2012. p. 2180-07

[58] Japanese Industrial Standards Committee. Antimicrobial Products-Test for Antimicrobial Activity and Efficacy: JISZ 2801-2000. Japan; 2000

[59] Wiegand C et al. Critical physiological factors influencing the outcome of antimicrobial testing according to ISO 22196/JIS Z 2801. PLoS One. 2018;**13**(3):e0194339

[60] Knobloch JK-M et al. "Life-like" assessment of antimicrobial surfaces by a new touch transfer assay displays strong superiority of a copper alloy compared to silver containing surfaces. PLoS One. 2017;**12**(11):e0187442

[61] Shi L et al. Limits of propidium iodide as a cell viability indicator for environmental bacteria. Cytometry Part A: The Journal of the International Society for Analytical Cytology. 2007;**71**(8):592-598

[62] Nguyen SH et al. Natural antibacterial surfaces. In: Antibacterial Surfaces. New York City, USA: Springer; 2015. pp. 9-26

[63] Linklater DP et al. Nanopillar polymer films as antibacterial packaging materials. ACS Applied Nano Materials. 2022;**5**(2):2578-2591

Chapter 5

Cathodoluminescent Properties of Thin Silicon Films Deposited on Nanoscopic Cerium Dioxide Buffer Layer

Dmitry Evgenyevich Milovzorov

Abstract

Luminescent properties of silicon films deposited on glass substrate and cerium dioxide were investigated. Recombination of carriers through the defect or surface states was observed. Point defects were studied by using electron paramagnetic resonance spectroscopy. The crystal phase was tested by means of X-ray diffraction technique. There is a shift in energy for the position of spectral component to the high energy values when applying the external electric fields to the silicon film deposited on cerium dioxide layer. Electric field–assisted migration of atoms of oxygen from the buffer layer in silicon films and polarization of interface region result in the creation of additional defect states in the electronic structure. Raman spectroscopy investigations show the significant changes in spectral components for Si-Si and Ce-O dipoles. Color-switching effect can be used for application in the development of new controllable light-emitting device and its implementation in new color pixel technology.

Keywords: luminescence, silicon films, point defects, paramagnetic centers, color switching of spectra

1. Introduction

Light emission from silicon films was intensively investigated for several decades. Photoluminescence (PL) was observed at room temperatures in different types of silicon films as oxidized porous silicon [1–3], as in polycrystalline silicon films [4, 5]. Two dominated mechanisms were proposed to explain PL processes: quantum size effect in silicon nanocrystals, from one side, and molecular-like silicon assemblies inside the films, from another side. In previous works, there were several ways of PL description, for example, due to band-to-band indirect transitions [6], carrier recombination through the defect levels [7], or by means of additional surface states [8] in silicon. The correlation between the oxygen contents and PL intensity was studied in several works [9, 10]. However, rare works [11] are devoted to the possibility of changing PL spectral characteristics or color switching, but these silicon films may be very important for light emission devices' implementation to manage their spectra.

The evident advantage of electron probe over light laser beam is that the first has its spatial resolution in the range from 0.01 nm to 1 nm, but the same specification for the second method is in the range from 1 to 1000 μm. It is clear that the nanoscopic technological procedures and their testing can be realized by using only electron beams. The energy of electrons, usually, has to be used in the range of 15 keV by using the scanning electron microscopic equipment as electron source [12].

In the present work, we study cathodoluminescence (CL) of silicon films by using cathodoluminescent spectra, X-ray diffraction of silicon films, and Raman scattering spectra for detecting the crystalline fraction in films.

Color switching can be realized by applying external electric field. Also, the substrate's type plays a great role in generating local electric fields, which are appeared by external field. Moreover, it is necessary to investigate not only PL spectral characteristics but also the structural and bonding properties and nature of point defects in silicon film.

2. Cathodoluminescence measurements

We used electron beam gun of a scanning electron microscope (SEM JEOL) as the source of CL, and for the detection of CL signal, the double monochromator coupled with photomultiplier tube was used (Gatan MONOCL3 + 1sys). We set the acceleration voltage of SEM as the source of CL to 5 kV. The electron beam current was 1 nA. By using photomultiplier of CL, we measured CL of the silicon films at the range of 300 nm ~ 800 nm. The electron paramagnetic resonance spectrometer manufactured by Bruker Sci. company (USA) was used for spectroscopic measurements (**Figure 1**).

By lowering the temperature of deposition during silicon film preparation from gas mixture of silane diluted by hydrogen and silicon tetrafluoride, there is significant increase in structural properties of the film and its homogeneity. It is assumed that there is a structural transition at 250°C for gas hydrogen-to-silane ratio R = 2 and at 200°C for R = 3 and more.

Further decrease in deposition temperature from 300 to 220°C results in the appearance only nanocrystals with orientation (111). The other deposition condition for sample was flow rate ratio SiH_4:H_2:SiF_4 = 0.6:35:0.375, and the other conditions were the same as those for preparation of the previous silicon films. The preliminary

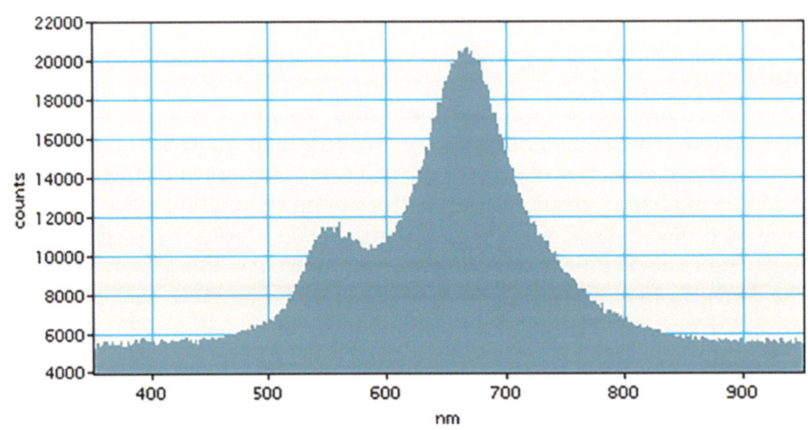

Figure 1.
Cathode luminescent spectrum of silicon film deposited at 220°C by PECVD method of film preparation.

cleaning operation for glass substrate processing was realized by molecular nitrogen plasma processing and, later, the hydrogen plasma processing. The duration of both procedures was 10 min.

Cathode luminescent spectrum of light emission from silicon film deposited at 220°C is shown in **Figure 1**. Several spectral peaks at 1.855 eV and 2.475 eV correspond to the carrier recombination through the band-to-band indirect optical transitions in nanocrystals and recombination through defect state (from point defect VO^-).

3. Silicon films deposited on cerium dioxide

In recent years, new techniques have helped to deposit various kinds of interface layers with different properties. The CeO layer was often used as a buffer layer for the deposition of crystalline silicon films [13]. It is known that the lattice constant for CeO (a = 0.541Å) crystal structure is approximately the same as that for silicon (the lattice mismatch 0.35%). **Figure 2** shows the Raman spectra of microcrystalline silicon films deposited on glass substrate and cerium dioxide buffer layer with relatively low and high contents of the crystalline phase. Cerium dioxide layer was grown by using the magnetron sputtering at the temperature of substrate 650°C. Also, the technical parameters of CVD facility, such as gap between electrodes and mesh of grids, can change the deposition conditions by changing the plasma characteristics in reactor.

In the preliminary test, silicon films, which were prepared by using PECVD method and were deposited on cerium dioxide nanocrystalline buffer layer, were tested by means of X-ray diffraction facility (see **Figures 3–5**). Using Scherrer formula [14], there was a possibility to estimate the crystal orientation of silicon nanocrystals and their average grain size. It was 177.3 Å (111).

The surface morphology and crystalline structures in the silicon film were detected by using the scanning electron microscope (Phillips Inc.). It is clear that the crystals' structures are widely spread in the film, and the thickness of the film is 75 nm. The

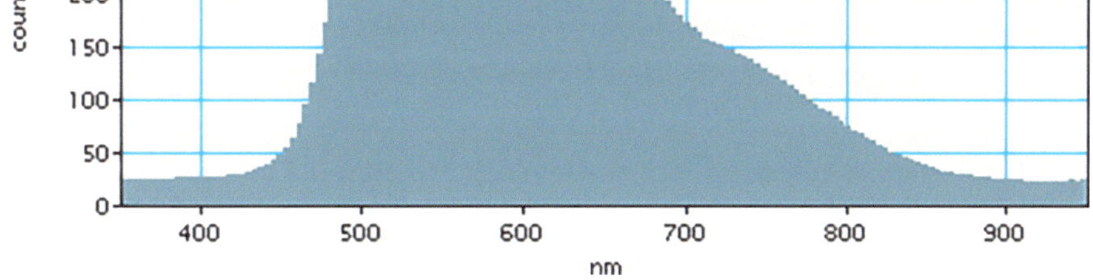

Figure 2.
Cathodoluminescent spectral characteristics of silicon film prepared at low temperature by its deposition on the buffer layer of cerium dioxide, which was deposited on glass substrate (a-Si/CeO_2/SiO_2). Gauss-shaped spectral peaks at 1.65 eV, 2.06 eV, and 2.5 eV were observed.

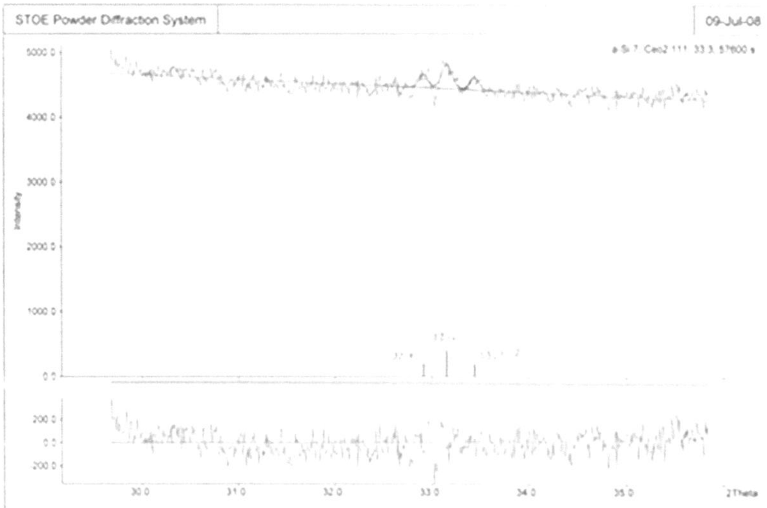

Figure 3.
XRD data for silicon film deposited on cerium oxide buffer layer. These experiments were carried out at the Moscow State University. The silicon crystal orientation is (111).

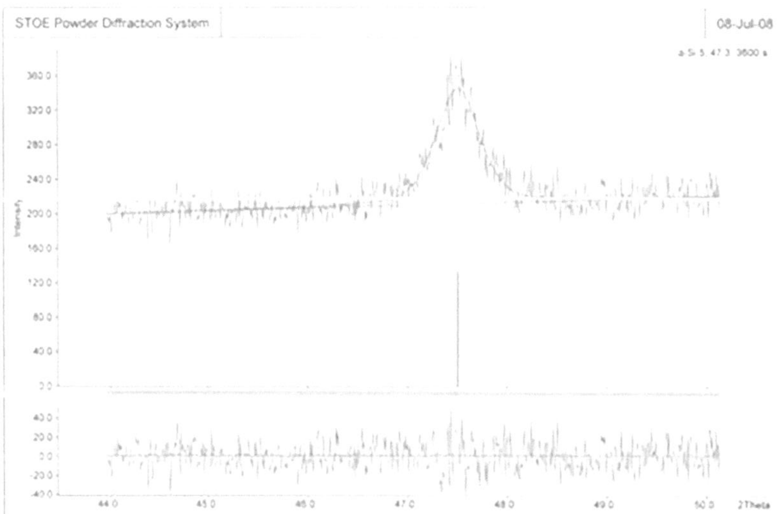

Figure 4.
XRD data for silicon film deposited on cerium oxide buffer layer. The crystal orientation is (220).

spatial distribution of silicon nanocrystals, which are gathered into conglomerates, is also seen (see **Figure 6**).

The role of oxygen incorporation into silicon net is most significant by using spectroscopic measurements of photoluminescent spectra. Electromigration process causes the oxygen to move by applying electric field in cerium dioxide due to the polarization of media. Atoms of oxygen penetrate into the interface region of the silicon film. According to work [11], the oxygen concentration is dramatically

Figure 5.
XRD data for cerium oxide buffer layer.

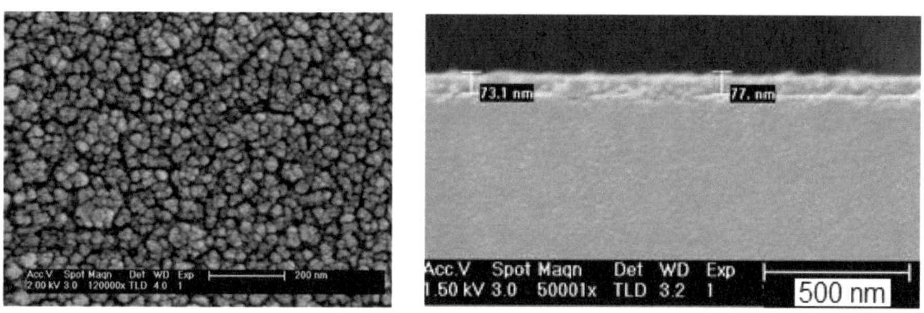

Figure 6.
Scanning electron microscopic photo of silicon film deposited on the buffer layer of cerium dioxide.

decreased over the period of time equals to several decades of minutes. It is illustrated by detecting the photoluminescent spectra with spectral peak at 650 nm. For the silicon films deposited on cerium dioxide buffered layer, the changes are seen in spectral pattern (see **Figure** 7), which are caused due to the shift of peak energy into more energetic range of values. The values of wave lengths of photoluminescent peaks are at 600 nm and 530 nm.

There are numerous nanocrystals with different polarized properties due to the variation of oxygen incorporation with silicon inside the film, nonspherical shape of nanocrystals, and broad deviation in their sizes. In any case, despite the traditional point of view that nanocrystal dipoles should have various dipole moments, which are distributed according to Gauss distribution function, the additional component of polarization of nanocrystal based on the quadruple moment should be distributed according to the χ^2 statistical distribution that causes the broadening of spectral pattern of peaks. χ^2 distribution has asymmetrical shape and is strictly distinguished from symmetrical Gauss function. Such asymmetry in spectral peaks is by measuring the cathodoluminescence (or photoluminescence [12]) photon signal from silicon films.

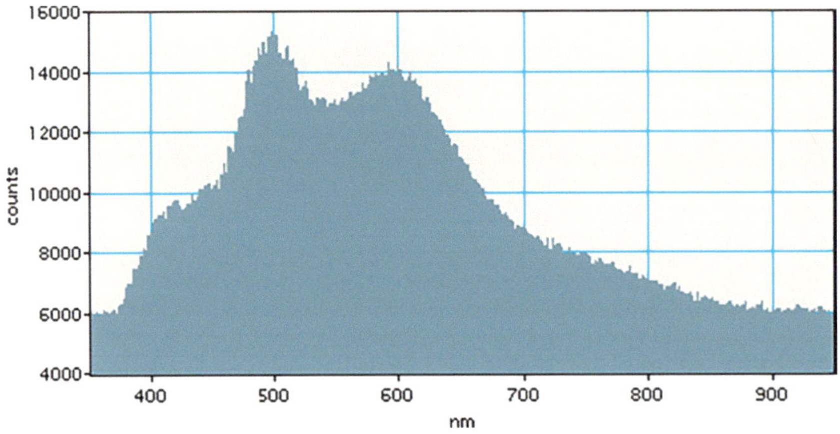

Figure 7.
Cathodoluminescent spectrum of silicon film deposited on the buffer layer of cerium dioxide by applied external electric field. It is seen, that the energetic positions of spectral Gauss shaped peaks are at 1.63 eV, 2.06 eV, 2.2 eV, 2.48 eV, and 2.95 eV. Integral intensity position is shifted to the high energy spectral lines.

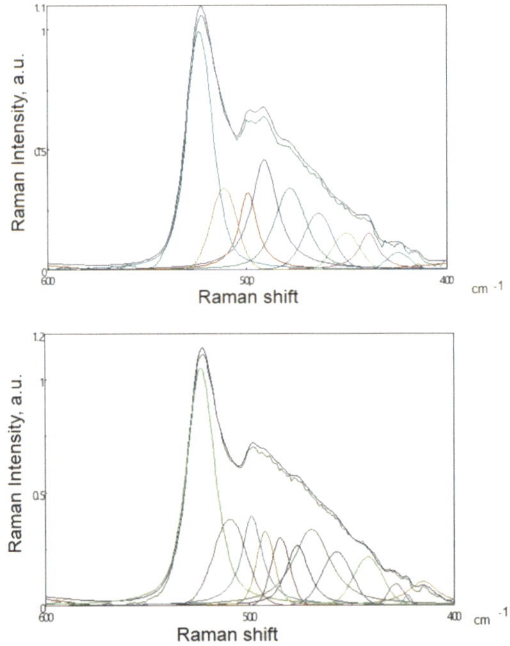

Figure 8.
Raman spectra of the silicon film deposited on cerium dioxide buffer layer without (upper graph) and by applied external electric field (low graph).

4. Raman measurements by applied external electric field

Raman measurements were carried out by means of laser spectrometer JRS1000 that consists of laser source, sample holder, and optical system for focusing the laser

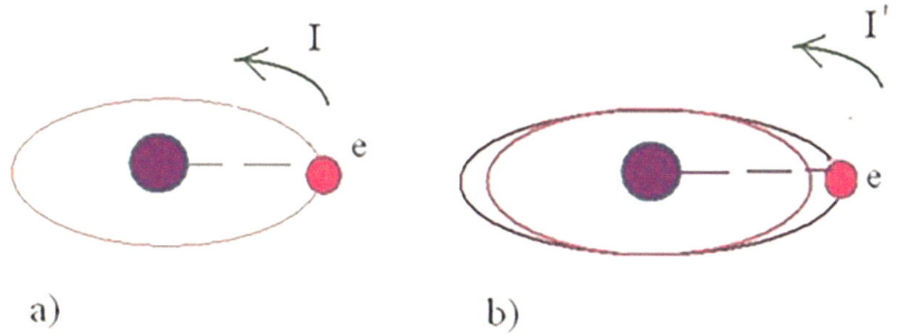

Figure 9.
The changes in polarization current by applied external electric field.

beam at the spot with a diameter of 1 μm. In addition, there is a microscope with microphotographer. The external electric field application is realized by using two metal pins adjusted on the surface of the sample by metal spring. The wave length of laser radiation was 514 nm.

The applied external electric field causes an increase in the magnitude of spectral peaks and the value of square under the spectral peaks raises (see **Figure 8**) for the silicon crystalline and nanocrystalline phases, and the crystalline phase of cerium dioxide. The external electric field perturbation moves the particles to the positions with minimal free energy. These changes in particle positions are related to the reorientations of vacancies [15], migration of dangling bonds [16], and even nanocrystals' grain boundary motion [17].

Raman spectra, which are shown in **Figure 8**, illustrate the structural changes that were done by applied external electric field inside the silicon film deposited on the buffer layer that consisted of cerium dioxide nanocrystals. The Swift-Hohenberg model [18] describes the pattern formation. Such theoretical methods can be used to analyze the change in grain boundaries and its moving between crystals and, also, to investigate the formation of defects in crystalline structures. According to Swift-Hohenberg model, it can be easy to reconstruct stationary solution for configuration with grain boundary between two crystals, which are located close to each other [17]. Order parameter is related to the formation hexagonal convective structures:

$$\frac{\partial\psi}{\partial t} = \nabla^2\left\{\left[\left(1+\nabla^2\right)^2 - \varepsilon\right]\psi + \psi^3\right\} + \zeta \tag{1}$$

where ε is the modified parameter of Releigh, ζ is the Gauss noise value, and the magnitude of ψ is $A_j(x,t) = A_{j0}\exp\left(\omega\left(\vec{Q}\right)t\right)\exp\left(i\vec{Q}\vec{x}\right)$; where A is an envelope of waves and x is the coordinate normal to the grain boundary. The complex amplitude is chosen since a phase change then corresponds to a spatial translation of the unstable mode. The solution of nonlinear Swift-Hohenberg equation also can be presented as linear combination of solutions if we derive the terms of ψ as partial solutions and further combining with them. The ψ local function can be transformed into a series according to perturbation theory as: $\psi = \psi_0 + \varepsilon\psi_1 + \varepsilon^2\psi_2 + \varepsilon^3\psi_3 + \varepsilon^4\psi_4 + \dots$, where the value ψ_0 is a steady-state solution, and the next values are the next order of perturbed decomposing procedure. The first order of solution can be expressed as $\psi_1 = P_1(x,t)\exp(ik) + Q_1(x,t)\exp(2ik) + R_1(x,t)\exp(3ik) + c.c.$ and amplitude

$A_j(t) = A_j(X,t)\left(1 + \sum_{j=1} \varepsilon^j Z_j\right)$, where Z_j is a renormalization constant. The amplitude $A_j(t)$ also can be written in spatial modulated form as $A_j(x,t) = A_{j0} \exp\left(\omega\left(\vec{Q}\right)t\right) \exp\left(i\vec{Q}\vec{x}\right)$; where $\vec{Q}$ is a perturbation vector. It is clear that the expression of the free energy F can be used by substituting ψ_1 local function:

$$F = \frac{\delta F_i}{\delta \psi}\bigg|_0 \left(w(x)\frac{\partial \psi}{\partial x} - \int w(x)\frac{\partial^2 \psi}{\partial x^2} dx\right) = \\ = 2\frac{\delta F_i}{\delta \psi}\bigg|_0 ((\sin(x) - \cos(x))(P_1'(x,t)\cos(k) + Q_1'(x,t)\cos(2k) + R_1'(x,t)\cos(3k))); \tag{2}$$

The linear functions are *P(x), Q(x), R(x)* of x variable. It is clear that free energy is a periodic function of spatial coordinate. Also, the crystal phase may be realized in the silicon film if there is a great value of free energy that is larger than the threshold value.

The energy of single dislocation was calculated by Peierls (1940), he explained the dislocation as a sum of chosen atom interactions with all neighboring atoms in upper and lower layers, and their displacements were composed into an elasticity value. The energy of moving displacements was calculated. It contributed into summing Peierls force, which is the exponential decay by increasing the thickness of considered interface layer. According to the work [17], the equation of grain boundary motion is as follows:

$$D\frac{\partial X_{gb}}{\partial t} = -p_{hex}\sin\left[2k_0 X_{gb}\sin(\theta/2)\right], \tag{3}$$

where D is a friction coefficient and p_{hex} is a Peierls force. The structural changes under the external electric field are shown in **Table 1**. There is an integration characteristic of spectral peaks that are generated by laser radiation's scattering on phonons in solid. These Raman peaks change their energetic position according to increasing electric field. The silicon crystalline and nanocrystalline–related spectral peaks to these fractions, respectively, increase by applied field, but the amorphous silicon phase–related spectral peaks decrease in their integral value or squire under each peak.

Wavenumber, cm^{-1}	0 V, Peak area	30 V, Peak area	Comments
524	19.9	22.8	Si-Si in c-Si
510–512	5.556	7.475	Si-Si with H
499	5.62	7.46	pure weak Lorentz nc-Si
490–492	11.3	4	a-Si:H
478–485	8.56	3.4	a-Si with dangling bonds, vacancies, O
464–470	4.58	10.6	Ce-O_2
450–458	2.56	4.24	Ce-O_2, crystalline

Table 1.
Peak area values by the different applied external electric fields.

By applying the electric field, there is a significant increase of the crystalline phase in the silicon film deposited on the buffer layer of cerium dioxide. During the reconstruction of the crystal phase, there are migration defects and reorientation of Si-O dipoles. Some of these dipoles are precipitates for the silicon grains that were made. Their reorientation causes the change in atomic fractions of various crystals. One of them will become larger than before, and the other will transform into small structures or even molecular assemblies.

5. Discussion

5.1 Theoretical assumption in explanation of cathodoluminescence

By the cathodoluminescence phenomena the electron beam irradiated the surface layer and causes the scattering of electrons by atoms of crystal structure and further excitation of atoms of media. Such kind of excitation stimulates the losses in kinetic energy of the electron that is retarding on the potential centers. After this, there is an emission of retarding radiation in broad spectral range. The cross section of scattering at the resonance maximum value of photon absorption is presented as [19]:

$$\sigma_{e-e} = \pi\lambda^2 g \frac{\Gamma_a \Gamma_b}{(E - E_0)^2 + \Gamma^2/4}; \tag{4}$$

where $\lambda = \frac{\sqrt{2}\pi\hbar}{\sqrt{mE}}$ is the de Broglie wavelength. As a result of electron-atom interaction, there is an energy transformation by generating polaritons in silicon media. The retarded electrons, which irradiate the surface of the thin film, emit continuous retarded radiation that is selected by resonant absorbing atom.

Electron repulsion on the electron cloud system of atom results in retarding of electron, and losses of its energy are in the resonant range for atomic system because retarding of electrons and repulsion forces cause the generation of retarding radiation. By resonant absorption process, the quantum of energy is absorbed by the atom. Such resonant absorption energy is related to the optical transitions inside the atom.

In low-energy E_1 of irradiated electrons, their retarding ability depends on velocity V_1, according to the linear dependence:

$$\frac{dE_1}{dz} = -\lambda V_1. \tag{5}$$

5.2 Luminescent properties of silicon films

Photoluminescent properties of silicon films were intensively studied for several decades. There are oxygen atoms inside the silicon films, which as impurity cause the significant changes to the optical properties of photoluminescent spectra. Variation in oxygen incorporation into silicon net causes the changes in dielectric function of media. Particularly, the variation in oxygen contents X of SiO_x from 0.5 to 2 causes the spread of dielectric function from few units to several dozens. The local fields, which appeared by oxygen incorporation, play a great role in generating spectral signal from the luminescent film. Silicon-oxide dipoles, which are created by oxygen incorporation, are widely investigated in recent decades [15, 16, 20, 21].

In addition, the color-switching effect of the spectral picture of photoluminescence emitted by the silicon film after oxygen processing that is studied in Ludwig et al. [11] experimentally confirms the changes in red-violet spectral ranges of photoluminescent emission. It is seen, according to their work, that the spectral picture changes from ultraviolet light of photoluminescence (from 400 nm to 500 nm) by low temperature of annealing the sample at 400–600°C to the red or even infrared spectral range (wavelength at 750–800 nm) by increasing the temperature up to 800–1000°C. Also, it is known that there are A-centers inside the silicon films [22] that disappeared by annealing to 1000°C and their existence inside the silicon film causes the photoluminescence in the range of wavelengths 600–800 nm. It is clear from the nonlinear laser spectroscopy experimental researches of other authors [23, 24] that the oxygen atoms leave the surface of silicon films by evaporation at a high temperature near 1000°C. In this way, there are defect levels inside the band gap of silicon due to oxygen incorporation, for example, Si-O and vacancy of Si plus Si-O-Si bridge. The first of these defects is related to the defect level at E_c-0.14 eV, but the second is at E_c-0.18 eV, where Ec is a bottom of the conductivity band in energy diagram of silicon.

The oxygen atomic assemblies can be converged one into another by annealing the sample, according to the work of Hofer and co-workers [25] such as molecular oxygen are dissociated and transformed into silicon-oxide bridges Si-O-Si by annealing to 450 K.

According to Sanderson's model, the charge of Si-O dipoles is varied from L. Pauling work [26] effective charge of silicon-oxygen dipole is 0.29 e. It slightly differs from silicon dioxide dipoles' effective charge that is 0.23 e according to Milovzorov [27]. This effective dipole charge can be estimated by using covalent radius r according to the formula: $r = r_0 - B^*\delta$, where r0 is the nonpolar radius and δ is the partial charge of the component's atom. Dipole moment can be written as: d = qd, where q is the charge and d is a distance. It is assumed that the distance is a sum of two covalent radii for two atoms (Si and O), which are composed of dipole $d = r_c^{Si} + r_c^{O}$. For evaluation the partial static charge on Si and O atoms as $e_s = 2(\delta_{Si}-\delta_O)$, and it is varied from 1.09 to 1.21 in units of electron charge, by index x increasing in SiO_x [28].

The most prominent way to solve this problem is using the quantum dipole moment, which was first proposed by Linus Poling [29]. According to my previous work [30], the quantum dipole moment μ can be calculated using the following formula:

$$\mu_{Si-O} = ex\int_V \psi_{3pSi}\psi_{2pO}d\nu = 4\pi ex\int_0^{2\pi}\int_0^{\xi}\int_0^{\pi}\Psi_{3p_xSi}\Psi_{2p_xO}r^2 d\phi dr d\theta \tag{6}$$

where $d\nu = 4\pi r^2 dr$, $0 < r \le \xi$, ζ is a covalent radius of atom of oxygen, and x is a bond length.

According to the estimation, the dipole moment for nonpertubed Si-O bond value equals to 1.12 D.

For atomic orbital Slater of silicon and oxygen atoms, the formula for dipole moment can be written as [30]:

$$\mu(r) = 230ea_0\exp\left(\frac{-3.655r}{a_0}\right)\left[0.27\left(\frac{r}{a_0}\right)^5 + 0.37\left(\frac{r}{a_0}\right)^4 + 0.41\left(\frac{r}{a_0}\right)^3 + 0.37\left(\frac{r}{a_0}\right)^2 + 0.18\left(\frac{r}{a_0}\right) + 0.05\right]$$

where covalent radius is r = 1.6 Å; $r/a_0 = 3$, dipole moment equals to the value of $\mu_{Si-O} = 1.12$ D.

By deformation of film due to acoustic or electric field influence, the additional term will appear in formula:

$$\mu_{Si-O} = \mu_{Si-O}{}^{0} + \frac{\partial \mu_{Si-O}}{\partial x}\partial x. \tag{7}$$

The second term depends on charge spatial distribution in dipole. **Figure 9** shows the changes in polarization current due to the influence of external electric field on one electron system (see **Figure 9**).

Polarization of nanocrystal, also, is significant for the linear and nonlinear optical properties.

Nonlinear polarization associated with the phonons can be expressed as:

$$P = \frac{1}{V}\sum_{i=1}^{N} <\mu>_i = \frac{N}{V}\left[\alpha_0 + \left(\frac{\partial \alpha}{\partial Q}\right)_0 <Q> + \left(\frac{\partial^2 \alpha}{\partial^2 Q}\right)_0 <Q>^2\right]E \tag{8}$$

The vibrational wave can be expressed as: $<Q> = \frac{Q}{2}\exp(i(\omega_v t - k_v z) + c.c.$ The probability of interband transition exponentially increases by lowering the size of nanocrystals [31].

By using the operator approach, it is easy to write the operator divergence as:

$$\left(div\vec{D}\right)\psi = -\frac{1}{c}\left(\frac{\partial}{\partial t}\sum_{s=1}^{n}\frac{\varepsilon_S}{4\pi}\Delta(X - X_S)\right)\psi \tag{9}$$

By multiplying the equation for the wave function of electron to conjugate wave function and repeating this procedure for complex conjugate equation multiplied to ordinary wave function, and, after this, finding the subtract between these values, we can find the expression:

$$\psi^*\left(div\vec{D}\right)\psi - \psi\left(div\vec{D}\right)\psi^* = -\frac{1}{c}\left(\frac{\partial}{\partial t}\sum_{s=1}^{n}\frac{\varepsilon_S}{4\pi}\Delta(X - X_S)\right)(\psi^*\psi - \psi\psi^*); \tag{10}$$

Using Maxwell equation for polarized media, we can find the expression: $div\vec{D} = div\left(\vec{E} + 4\pi\vec{P}\right) = div\vec{E} + 4\pi div\vec{P} = 4\pi\left(\rho + div\vec{P}\right)$

The time evolution of polarization for a chain consisting of two and three quantum nanoscopic systems is studied by Wang et al. [32].

The current density can be written according to the quantum mechanics approach as:

$$J_P = \left(\psi^*\left(div\vec{P}\right)\psi - \psi\left(div\vec{P}\right)\psi^*\right) = -\frac{1}{4\pi c}\left(\frac{\partial}{\partial t}\sum_{s=1}^{n}\frac{\varepsilon_S}{4\pi}\Delta(X - X_S)\right)(\psi^*\psi - \psi\psi^*); \tag{11}$$

Therefore, the repolarization current can be expressed in the following form:

$$I = \int_S J ds = \int_S \left(\psi^*\left(div\vec{P}\right)\psi - \psi\left(div\vec{P}\right)\psi^*\right)ds = -\frac{1}{16\pi^2 c}\left(\int_{S'}\sum_{s=1}^{n}\varepsilon_S\frac{\partial}{\partial t}\Delta(X - X_S)(\psi^*\psi - \psi\psi^*)ds'\right) \tag{12}$$

The integrative characteristic of current is transformed into a simple form and depends on polarization of media and frequencies according to which the current is distributed:

$$I(t) = \frac{1}{4\pi}\int_{-\infty}^{\infty}\sum_{s=1}^{n}\varepsilon_s\frac{\partial}{\partial t}\Delta(X - X_S)\exp(-i\omega\tau)d\tau =$$

$$\frac{1}{4\pi}\sum_{s=1}^{n}\varepsilon_s\int_{-\infty}^{\infty}\frac{\partial}{\partial t}\Delta(X - X_S)\exp(-i\omega\tau)d\tau = \frac{1}{4\pi}\sum_{s=1}^{n}\varepsilon_s\frac{\partial}{\partial t}\int_{-\infty}^{\infty}\Delta(X - X_S)\exp(-i\omega\tau)d\tau =$$

$$= \frac{1}{4\pi}\sum_{s=1}^{n}\varepsilon_s\left(\omega_S{}^2 + 1^{`}\right)^{1/2}\exp(-i(\omega - \omega_S)t) \tag{13}$$

After applying external electric field, the resulting field can be cast as:

$$\vec{E}_{\sum} = \vec{E}_0 - \frac{\vec{P}_1 - \vec{P}_2}{\varepsilon_0} \tag{14}$$

where polarizations' values are $P_1, P_2.$

For a single nanoscopic dipole (nanocluster or nanocrystal), there is an expression:

$$\frac{\partial\mu}{\partial t} = \frac{\partial x}{\partial t}q + \frac{\partial q}{\partial t}x = \bullet xq + Ix \tag{15}$$

The first term describes changing in own frequencies of the system, but the second is a current of changing quantity of charges (for nanosized clusters or nanocrystals).

Each polarization value for the film is a sum of all dipoles' moments: $\vec{P}_n = \sum_{k=1}^{N}\vec{\mu}_k.$

The differences between values of various polarizations are:

$$\vec{P}_n - \vec{P}_l = \sum_{k=1}^{N}\overrightarrow{(\mu_{kn}} - \vec{\mu}_{kl}) = \sum_{k=1}^{N}\int\frac{\overrightarrow{\partial(\mu_{kn} - \vec{\mu}_{kl})}}{\partial t}dt = \sum_{k=1}^{N}\sum_{\nu=1,2,3}\int(\bullet x_\nu q + Ix_\nu)dt = \Delta P +$$

$$+\sum_{k=1}^{N}\sum_{\nu=1,2,3}\int(Ix_\nu)dt = \Delta P + \frac{1}{4\pi}\sum_{s=1}^{n}\varepsilon_s\left(\omega_S{}^2 + 1^{`}\right)^{1/2}\exp(-i(\omega - \omega_S)t);. \tag{16}$$

The spectral decomposition of resulted field of polarized film can be written as:

$$\vec{E}_{\sum}(\omega) = \vec{E}_0 + \frac{\vec{P}_2 - \vec{P}_1}{\varepsilon_0} = \vec{E}_0 + \frac{1}{4\pi\varepsilon_0}\sum_{s=1}^{n}\varepsilon_s\left(\omega_S{}^2 + 1^{`}\right)^{1/2}\exp(-i(\omega - \omega_S)t); \tag{17}$$

By electro-dipole approach, the energy interaction as a function of time is the following:

$$V(t) = -\vec{\mu}\vec{E}(t); \tag{18}$$

Polarization has the view:

$$P_{\mu}{}^{(1)}(t) = \frac{i}{\hbar} N \sum_{n} \int_{-\infty}^{t} d\tau \langle [\mu_m(t), \tilde{\mu_n}(\tau)] \rangle E_n(t+\tau) = \sum_{n} \int_{-\infty}^{t} d\tau K_{mn}{}^{(1)}(t) E_n(t+\tau); \quad (19)$$

where the response function

$$K_{mn}{}^{(1)}(t) = \frac{i}{\hbar} N \langle [\mu_m(t), \tilde{\mu_n}(\tau)] \rangle; \quad (20)$$

$$\begin{aligned} P_{\mu}{}^{(1)}(t) &= \sum_{n} \int_{-\infty}^{\infty} E_n(\omega) \exp(-i\omega t) d\omega \int_{-\infty}^{t} d\tau K_{\mu n}{}^{(1)}(\tau) \exp(-i\omega\tau) \\ &= \varepsilon_0 \sum_{n} \int_{-\infty}^{\infty} \chi_{\mu n}{}^{(1)}(\omega) E_n(\omega) \exp(-i\omega t) d\omega; \end{aligned} \quad (21)$$

$$\begin{aligned} \chi'_{nn}(\omega) &= \frac{N}{\varepsilon_0 \hbar} \left| \mu_{ge}{}^{n} \right|^2 \frac{\omega_0 - \omega}{(\omega_0 - \omega)^2 + \Gamma^2}; \\ \chi''_{nn}(\omega) &= \frac{N}{\varepsilon_0 \hbar} \left| \mu_{ge}{}^{n} \right|^2 \frac{\Gamma}{(\omega_0 - \omega)^2 + \Gamma^2}; \end{aligned} \quad (22)$$

More common results were received by Mexican researches [33] and can be expressed in the form of Yang-Mills-Maxwell equation:

$$\frac{\partial(s_{matter} + s_{radiation})}{\partial t} + div\left(\vec{J} + \vec{J}_i^{\alpha}\right) = -\frac{\partial \mu^{\alpha}}{\partial t} + div\,\vec{\lambda}^{\alpha}; \quad (23)$$

where in the left side of the equation are the changing of the spin and currents. But in the right side are located derivation of magnetic moment values over time and divergence of polarization.

The photoluminescent radiation and Raman scattering spectroscopy optical response depend on the characteristics of polarization of nanocrystals breaking the symmetry, variation of shape, and quantity of charged particles on the surface of the nanocrystals. For polycrystalline silicon media, the nanocrystals are included in amorphous silicon that is in the grain boundary region between the nanocrystals. According to calculation of Theiss [34] for nanocrystal with diameter 10 nm the quantity of atoms is 26,183, the surface/volume ratio is 600 m^2/cm^{-3}, quantity of atoms in surface layer is 2199 atoms, fraction of silicon atoms is 8.4%, but for nanocrystal with diameter 5 nm these parameters are the following: 3273, 1200 m^2/cm^{-3}, 550 atoms and 16.8%.

Absorption spectral characteristics usually are given by spectral superposition of a series of Gaussian convolutions of Lorentzian or Voigtian spectral peaks [35]. Such contributions to the line shape are the inhomogeneous effects arising from the different surroundings near point defects or impurities. This broadening of spectra is related to the compositional materials or media with great density of defects.

For the nanocrystalline silicon film by applying the external electric field, there is deformation wave in grain boundary region, migration charged impurities (atoms of oxygen and hydrogen), and even restructured behavior of crystal network inside the nanocrystal. By increasing the external field value up to the critical value of the crystalline structure will destroyed and Raman spectral peak for silicon around 520 cm^{-2} will disappeared. However, there is an opportunity to reconstruct again the crystal structure for the best performance.

On the other side, the preparation of silicon film on the buffer layer of cerium dioxide results in a decrease in the thickness of amorphous transitional layer between the silicon dioxide substrate and silicon nanocrystal. The lattice constants of crystalline cerium dioxide and bulk silicon are differed very few. Due to the electromigration process, the light atoms (compared to heavy atoms of Ce) of oxygen can move through the buffer layer to the silicon film and insist the sharp jump of polarization of film. The current-voltage characteristics for such kind of silicon films become nonlinear.

6. Conclusion

Cathodoluminescent properties of silicon films were studied for various types of films, particularly, nanocrystalline silicon films prepared on glass substrate or amorphous silicon films deposited on cerium buffer layer. Structural changes inside the silicon film are investigated by external applied electric field. The electromigration of point defects, dangling bonds, or hydrogen atoms results in an increase in the crystalline fraction in the film. The energy of interaction of external electric field with the atomic system in solids can be distributed in the numerous interatomic bonds according to their own electric dipole field interactions, which are related to their own spectral characteristics.

Acknowledgements

Author Dmitry Milovzorov thanks co-workers from Moscow State University for help in X-ray diffraction experiments and fruitful discussions. Also, the author is grateful to Prof. Pavel Todua, Director of Scientific Research Center of Surface and Vacuum of Russia, and his co-worker, Prof. V. Mitukhlyaev for help in realization of the cathodoluminescence experiments.

Author details

Dmitry Evgenyevich Milovzorov[1,2]

1 ROSNOU Russian New University, Moscow, Russian Federation

2 Fluens Technology Group Ltd., Moscow, Russian Federation

*Address all correspondence to: milovzorovde@stud.rosnou.ru

References

[1] Kanemitsu Y. Slow decay dynamics of visible luminescence in porous silicon: Hopping of carriers confined on a shell region in nanometer-size Si crystallites. Physical Review B. 1993;**48**:12357-12360

[2] Parkinson M, Bayliss S, Clarke D, Law D, Beamson G. An analysis of photoluminescence voltage quenching and band structure of nanoporous silicon. Thin Solid Films. 1998;**326**: 194-200

[3] Schoisswohl M, von Bardeleben J, Bratus V, Munder H. Defects in luminescent and non-luminescent porous Si. Thin Solid Films. 1995;**255**: 163-165

[4] Tamura H, Ruckschioss M, Wirschem T, Veprek S. Origin of the green/blue luminescence from nanoscrystalline silicon. Applied Physics Letters. 1994;**65**:1537-1539

[5] Tong S, Liu X, Bao X. Study of photoluminescence in nanocrystalline silicon/amorphous silicon multilayers. Applied Physics Letters. 1995;**66**: 469-471

[6] Milovzorov D, Inokuma T, Kurata Y, Hasegawa S. Relationship between structural and optical properties in polycrystalline silicon films prepared at low temperature by PECVD. Journal of Electrochemical Society. 1998;**145**: 3615-3620

[7] Delerue C, Allan G, Lannoo M. Theoretical aspects of the luminescence of porous silicon. Physical Review B. 1993;**48**:11024-11034

[8] Nayefeh M, Rigakis N, Yamani Z. Photoexcitation of Si-Si surface states in nanocrystallites. Physical Review B. 1997;**56**:2079-2082

[9] Kanemitsu Y, Okamoto S. Resonantly excited photoluminescence from porous silicon: Effects of surface oxidation on resonant luminescence spectra. Physical Review B. 1997;**56**:R1696-R1699

[10] Inokuma T, Wakayama Y, Muramoto T, Aoki R, Kurata Y, Hasegawa S. Optical properties of Si clusters and Si nanocrystallites in high-temperature annealed SiO_x films. Journal of Applied Physics. 1998;**83**:2228-2234

[11] Ludwig M, Augustian A, Hummel R. Color-switching effect of photoluminescent silicon after spark-processing in oxygen. Semiconductor Science and Technology. 1997;**12**: 981-986

[12] Park SY, Lee SW, Kang JS, Lee HR, Joo MY, Kim JK, et al. Cathodoluminescence Properties of Silicon Thin Films Crystallized by Electron Beam Exposure, Conference Paper July 2013. DOI: 10.1109/ IVNC.2013.6624766

[13] Jones T, Croke E, Garland C, Marsh O, McGill T. Journal of Vacuum Science and Technology. 1998;**B16**:2686

[14] B.D. Cullity Elements X-Ray Diffraction. Reading: Addison-Wesley; 1978. p. 102

[15] Dobaczewski I, Andersen O, Rubaldo L, Gorscinski K, Markevich VP, Peaker AP, et al. Saddle point for oxygen reorientation in the vicinity of a silicon vacancy. Physical Review B. 2003;**67**: 195204-195208

[16] Biswas R, Pan BC, Ye YY. Metastability of amorphous silicon from silicon network rebonding. Physical Review Letters. 2002;**88**(20):205502

[17] Boyer D, Vinals J. Weakly nonlinear theory of grain boundary motion in patterns with crystalline symmetry. Physical Review Letters. 2002;**89**(N5): 055501-055504

[18] Cross M, Hohenberg P. Pattern formation outside of equilibrium. Reviews of Modern Physics. 1993;**65**: 851-1124

[19] Breit G, Wigner E. Physical Review. 1936;**49**:519-536

[20] Lanoo M, Allan G. A cluster plus effective medium tight-binding study of SiO_x system. Solid State Communications. 1978;**28**:733-739

[21] Strathis J, Cartier E. Conductance measurements on P_b centers at the (111) Si: SiO_2 interface. Journal of Applied Physics. 1996;**80**:3915-3921

[22] Shimasaki M, Show Y, Iwase M, Izumi T, Ichinobe T, Nosaki S, et al. Applied Surface Science. 1996;**92**: 617-620

[23] Daum W, Krause H, Reichel U, Ibach H. Nonlinear optical spectroscopy at silicon interfaces. Physica Scripta. 1993;**49**:513-518

[24] Hofer U, Morgen P, Wurth W, Umbach E. Metastable molecular precursor for the dissociative adsorption of oxygen on Si (111). Physical Review Letters. 1985;**55**(27):2979-2982

[25] Hofer U, Puschmann A, Coulman D, Umbach E. Absorption of molecular oxygen on Si (111). Surface Science. 1989;**211**(212):948-958

[26] Pauling L. The nature of silicon-oxide bonds. American Mineralogist. 1980;**65**:321-323

[27] Milovzorov D. Optical properties of thin silicon films. In: Narayana L, Kalipali D, editors. Fundamental Science and Application of Silicon Photonics. London, UK: INTECH; 2018. pp. 1-24. (in publication)

[28] He L, Wang D, Hasegawa S. A study of plasma-deposited amorphous SiO_x: H ($0 \leq x \leq 2$) films using infrared spectroscopy. Journal of Non-Crystalline Solids. 2000;**261**:67-71

[29] Poling L, Wilson EB. Quantum Mechanics with Applications to Chemistry. Dover, NY; 1985. p. 272

[30] Milovzorov D. Memory cell with photoacoustic switching. In: Lai WY, Pau S, Lopez OD, editors. Nanofabrication: Technologies, Devices, and Applications. 2005. pp. 427-437

[31] Orlova E, Harrison P, Zheng WM, Haisall MP. The influence of localization in quantum well on the time of life states of impurities' levels. Physics and Technology of Semiconductors. 2005; **39**:70

[32] Wang C, Yakimenko I, Zozoulenko I, Berggren K. Journal of Applied Physics. 1998;**84**(n5):2684-2689

[33] B. Berche, E. Medina, Classical Yang-Mills Theory in Condensed Matter Physics, arXiv:1210.0105v1 [cond-matter], 29 September 2012

[34] Theiss W. Surface Science Reports. 1997;**29**:91-192

[35] Cannas M. Point defects in amorphous SiO_2: Optical activity in the visible, UV and vacuum-UV spectral regions [PhD thesis]. Physics, University of Palermo; 1998. 145 p

Section 3

State-of-the-Art Applications

Chapter 6

Nanostructured Diamond Sensors Detect Emerging Contaminants

Xiaoxi Yuan, Feng Yang and Hongdong Li

Abstract

Emerging contaminants have induced severe risks to the aquatic environment, and the continuous accumulation of various residues has forced aquatic ecosystems to deteriorate. The accumulation of emerging contaminants through food chains eventually causes serious harm to human production activities and life safety. Therefore, effective trace detection of emerging contaminants is needed to ensure water quality and safety. Electrochemical sensors are outstanding for portability, simple operation, high sensitivity, low detection limit, and fast detection speed. They could be used for trace detection of emerging contaminants in liquid environments. This study presented the performance of various nanostructured boron-doped diamond electrodes for the trace detection of emerging contaminants. The diamond electrodes were modified in different ways, including nanostructure, tip curvature, surface terminal, and modification regulation. The effects of various modifications on the electrochemical properties of diamond electrodes were discussed. Moreover, key modification strategies were studied to improve the sensitivity, selectivity, robustness, and real water performance of diamond electrodes in trace detection of emerging contaminants. Specific examples included the detection of endocrine-disrupting chemicals (e.g., nonylphenol), antibiotics (e.g., minocycline and doxorubicin), and persistent organic pollutants (e.g., polychlorinated biphenyls). Diamond-based water quality sensors could detect, analyze, track, and manage emerging contaminants quickly at trace levels.

Keywords: nanostructured diamond, sensor, detect, emerging contaminants, electrochemical

1. Introduction

Emerging contaminants, relative to traditional pollutants (e.g., chemical oxygen demand and ammonia nitrogen), refer to those toxic or hazardous chemicals that have been discovered recently or ignored before and pose risks to the ecological environment or human health, but have not yet been included in management or have not been effectively prevented and controlled due to insufficient management measures. Emerging contaminants mainly include persistent organic pollutants (POPs), endocrine-disrupting chemicals (EDCs), pharmaceuticals and personal care products (PPCPs), perfluorinated compounds (PFCs), antibiotics, and microplastics [1]. At present, governments and research institutions worldwide have paid much attention to the control of emerging contaminants. However, the management of emerging contaminants has faced many challenges owing to incomplete environmental risk assessment, imperfect regulatory system, and other factors. Water pollution has always been a major environmental problem in the world. With accelerated industrialization and urbanization, emerging contaminants, such as PPCPs, industrial additives, and EDCs are continuously discharged into environmental water bodies [2]. The development of efficient and sensitive detection technologies is significant for monitoring and assessing emerging contaminants in environmental waters. Studies show emerging contaminants, even at low environmental concentrations, could severely impact organisms owing to their accumulation in food chains. Therefore, trace detection and detection limit improvement are serious issues faced by researchers.

The commonly used methods for detecting emerging contaminants include chromatography, mass spectrometry, and electrochemical methods. Gas-phase mass spectrometry and high-performance liquid chromatography are widely used detection methods with high sensitivity and accuracy. However, they have high requirements for sample pretreatment, expensive equipment, complex testing processes, and strong professional operability. These limitations restrict the further development of chromatography and mass spectrometry. Electrochemical sensing is superior over traditional methods in terms of sensitivity, price, and operation. More importantly, electrochemical sensing could be used for real-time *in-situ* detection. It has gradually become a research hotspot worldwide and has incomparable advantages to other detection methods [3].

Compared with traditional metal electrodes and other sp^2 carbon electrodes, conductive diamond has unique electrochemical properties, including low background current, wide potential window, weak molecular adsorption, stable microstructure, and fast electron transfer kinetics. Iwaki et al. [4] introduced diamond into electrochemistry by using an ion-implanted diamond as an electrode. Pelskov et al. [5] explored the photoelectrochemical properties of polycrystalline diamond films. Later, Fujishima et al. [6] studied the application of boron-doped diamond (BDD) films in photoelectrodes. Over the past four decades, conductive diamonds (e.g., BDD, nitrogen-doped diamond, phosphorus-doped diamond, and hydrogen-terminated undoped diamond) have been widely used in various electrochemical applications [7].

Conductive diamond perform well in detecting emerging contamination. Gan et al. [8] prepared a carbon-black nanoparticle-modified BDD electrode to detect the emerging contaminant estradiol, which achieved a nanomolar detection limit. Diamond electrodes could improve electrochemical performances through diamond nanostructures (e.g., nanotextures, nanowires, nanoneedles, nanotrees, nanoparticles, and porous diamond). Nanostructured diamond could be achieved through masking and plasma etching. Surface nanostructures enlarge surface areas and improve detection capability [9]. The electrochemical activity of the interface is also related to the surface state. This article elaborates on the construction of diamond electrodes with different surface and interface structures for trace detection of emerging contaminants.

2. Nanostructured diamond sensor detects emerging contaminants

Diamond is a kind of wide band gap semiconductor with unique electrical, physical, chemical, and mechanical properties. As a material with excellent performance, it has a wide range of applications in practical fields, such as detecting emerging contaminants. At present, the growth mechanism of diamonds is very mature, and researchers pay more attention to the nanostructured processing and surface modification of diamonds. Diamond is the hardest substance in nature; thus, it is difficult to achieve the preparation of nanostructured diamond by simple mechanical processing. It is urgent to find a simple and easy method to solve the current preparation problem. At the same time, combining traditional diamond material with nanomaterials and biomaterials could fully expand the field of detecting emerging contaminants by diamond sensors. Our team is committed to the research of BDD to detect emerging contaminants. We comprehensively improve the detection performance of nanostructured diamond materials. The following describes specific application examples.

2.1 Nanopyramid diamond electrochemical detection of EDC—nonylphenol

Nonylphenol (4-NP) is an important fine chemical raw material and intermediate, and is mainly used to produce surfactants, antioxidants, textile printing and dyeing auxiliaries, pesticide emulsifiers, and resin and rubber stabilizers [10]. It is known as the "sperm killer". Yuan Xiaoxi et al. used hydrogen plasma to etch doped diamonds from top to bottom to form nanopyramid diamond electrodes [11]. Hydrogen plasma treatment has many advantages (e.g., simple operation, high efficiency, and lack of pretreatment) and is a new idea in the field of nanostructured processing of superhard carbon materials. The nanopyramid diamond-doped sensor has high stability and strong specificity in detecting 4-NP. The detection limit of the sensor reaches the sub-nanomolar level (0.26 nM). The sensitivity of the detection limit is attributed to the three-dimensional nanopyramid morphology that provides more active sites. Moreover, the optimized hydrophobic interface of the electrode hydrogen terminal plays an important role. The hydrogen terminal is hydrophobic and 4-NP could easily reach the electrode surface and undergo an oxidation reaction, thus facilitating trace detection.

2.2 Diamond aptamer sensor for detecting EDC—bisphenol A

Bisphenol A (BPA) is widely used in the manufacture of plastic (milk) bottles, sippy cups for infants, and the inner coatings of food and beverages (milk powder) [12]. BPA is distributed widely in mineral water bottles, medical devices, and the lining of food packaging. BPA could cause endocrine disorders and threaten the health of fetuses and children. It could induce precocious puberty, and obesity due to metabolic disorders [13].

Ma et al. used ion sputtering to uniformly modify gold films on the diamond surface and then annealed at high temperature to form gold nanoparticles in uniform particle size (about 8 nm). The aptamer and mercaptohexanol self-assembled on the gold nanoparticles. Owing to the specific binding between BPA and the aptamer, the sensor could be used in trace detection of BPA. The actual usable area of the electrode was enlarged five times after modification with gold nanoparticles, and the large specific surface area provided sufficient binding sites for aptamer modification. The aptamer composite sensor as-prepared has high stability, strong specificity and a detection limit as low as 7.2 fM, and is reusable. It is one of the most sensitive BPA sensors available [14]. The excellent detection performance is attributed to the synergistic coupling effect among diamond, gold particles, the aptamer, and mercaptohexanol.

2.3 Diamond aptasensor for detection of persistent organic pollutants—polychlorinated biphenyls

Polychlorinated biphenyls (PCBs) are persistent organic pollutants that could be used as additives in insulating oils, heat carriers, and many industrial products (e.g., various rubbers, coatings, resins, ceramic glazes, fire retardants, and pesticides) [15]. PCBs are extremely insoluble in water, but are highly soluble in fats and organic solvents. PCBs are extremely chemically stable and highly resistant to acid, alkali, and oxidation, and they do not decompose and often accumulate in biological fats. It has been realized globally that PCBs have high toxicity, low lethal doses, high bioaccumulative, and persistent hazards [16]. Many countries have begun to reduce or stop the production of PCBs since 1973. However, PCBs are quite stable and released slowly, and the final storage point of PCBs entering the environment is mainly riverine sediments. PCBs have strong aggregation properties in the body and are gradually enriched through food chains. PCBs, after entering the environment, will remain in the environment for a long time and hardly degrade. Water and soil contaminated by PCBs could not be recovered for a long time. Yuan et al. [17] prepared a boron-doped diamond (BDD) substrate, and increased the density of gold nanoparticles and modified aptamers through secondary sputtering annealing, thus achieving the trace detection of PCB-77 molecules.

A linear relationship within 1.0×10^{-15}–1.0×10^{-11} M and with a detection limit as low as 0.32 fM was achieved with electrochemical impedance spectroscopy. The aptamer sensor has high sensitivity, specificity, stability, and reusability. Testing in physical samples, such as milk proved the aptamer sensor worked stably in more complex environments and had good practical value.

2.4 BDD with high graphite content detection of antibiotic minocycline

Minocycline is a long-term, effective semi-synthetic tetracycline with an antibacterial spectrum similar to that of tetracyclines. It has the strongest antibacterial effect among tetracyclines and is widely used to treat bacterial infections [18]. However, an overdose of minocycline could lead to many side effects. The most common side effects include nausea, vomiting, dizziness, vestibular dysfunction, photosensitivity, headaches, pigmentation, and skin rashes. More serious adverse reactions were also frequently reported, such as drug allergy, autoimmune syndrome, hepatotoxicity, pneumonitis, and carcinogenesis [19]. Therefore, it is necessary to detect minocycline contents using highly sensitive and selective methods.

Wan et al. [20] grew a BDD film with high graphite content. The electrochemical behavior of minocycline hydrochloride on the electrode was studied through cyclic voltammetry, electrochemical impedance spectroscopy, and differential pulse voltammetry. The linearity was high in the concentration range of 0.02–100 μM, and the detection line reached 9.8 nM. In this work, the synergistic effect of BDD and D/CNWs was used to achieve the high sensitivity, high selectivity, anti-interference, and repeatability of the electrode in the detection of minocycline hydrochloride.

2.5 Nanocrystalline graphite diamond composite electrodes for detecting antibiotic acetaminophen

Antibiotic acetaminophen (APAP) is one of the most effective pain and fever relievers. However, the amount of APAP-containing production wastewater is increasing following the intensified usage. The APAP content in drinking water increases through various pathways (e.g., water circulation and discharge), leading to serious water safety issues [21]. Excessive APAP intake by the human body could cause heart and gastrointestinal diseases. Statistics show that severe cases could end up in death. Therefore, real-time monitoring of APAP in drinking water is very important [22].

Wang et al. prepared a nanocrystalline graphite/doped diamond composite electrode using chemical vapor deposition. The sp^3 structure of diamond on the (111) crystal plane under high-temperature conditions in a boron-rich environment was transformed into nanocrystalline graphite in the sp^2 graphite phase. Density functional theory calculations showed the APAP adsorbing energy of graphite was lower than that of diamond, indicating that graphite has a stronger adsorption effect. Combined with the excellent performance of the diamond electrode, the nanocrystalline graphite/diamond composite electrode structure was used to sensitively detect APAP. The linear detection range was 0.02–50 μM, and the detection limit was as low as 5 nM [23]. The nanocrystalline graphite and BDD in the composite film had a strong interfacial bonding force, but did not fall off easily, and showed good application potential even in harsh environments.

2.6 Nickel nanoparticle on boron-doped diamond electrode detecting antibiotic azithromycin

Antibiotics have attracted wide attention as emerging contaminants in recent years. As one of the highest concentrations of antibiotics in aqueous substrates, azithromycin (AZI) has side effects causing numerous serious and significant damage to humans and environment [24]. Therefore, there is an urgent need to develop efficient and sensitive analytical sensing devices for environmental monitoring. In our recent study, we developed a Ni nanoparticle-modified boron-doped diamond (NiNPs-BDD) electrode to achieve high-performance electrochemical detection of AZI. The NiNPs-BDD electrode consists of Ni nanoparticles embedded on the bottom of square pits with submicron-sized appearing on the surface of boron-doped diamond (BDD), which showed higher sensitivity compared with unmodified BDD. The electrode demonstrates linear ranges at concentration scales of 0.1–100 μM in phosphate buffer solutions with a pH of 7.5, and the detection limits of Ni-BDD was 0.148 μM. The NiNPs-BDD electrode shows good repeatability, stability, and washability with respect to the most traditional electrodes.

2.7 Porous diamond electrodes for detecting clenbuterol

Clenbuterol (CLB) in medical doses applies to the treatment of respiratory diseases. It could also effectively increase muscle mass and reduce body fats. Therefore, CLB is abused by athletes as a stimulant to increase muscle mass, and in the livestock industry. CLB with a long half-life could remain in edible meat products and accumulate in body tissues after being eaten [25]. Excessive intake of CLB could produce toxic side effects, such as palpitation, muscle tremors, and vomiting [26]. For the sake of food safety and human health, the use of CLB in animal husbandry is prohibited, and the abuse of drugs that may cause serious irritation to the human body is strongly opposed.

Ma et al. [27] prepared a gold nanoparticle porous diamond electrode in one step through wet annealing. A composite structure of gold nanoparticles and columnar nanopores (average diameter of ~60 nm) was formed on the BDD surface (Au-pBDD). At low concentrations of CLB, Au-pBDD has a distinct oxidation peak with a detection limit as low as 66 nM, which could not be detected by the two-dimensional BDD membrane electrode. The excellent detection performance of the composite electrode is attributed to the coupling effect between porous diamond and gold nanoparticles.

3. Conclusions

Our team is devoted to research on the detection of emerging contaminants using doped diamonds. Starting from the structure-activity relationship of doped diamond, we propose the multiscale principle of "multilevel morphology construction—electronic structure regulation—electrode interface optimization—synergistic coupling" to comprehensively improve the detection performance of diamond materials. This principle is of great significance to both basic research and practical application. The high detection sensitivity, wide linear range, repeatability, and high stability, make diamond materials suitable for trace detection of various emerging contaminants. Moreover, vapor-deposited diamond is cheap and suitable for industrial promotion. Developing real-time detection equipment for emerging contaminants in drinking water, and ensuring the safety of drinking water are major issues of livelihood. The successful conversion of research findings will directly benefit thousands of households.

Acknowledgements

We acknowledge the support of Program of Science and Technology Development Plan of Jilin Province of China (Nos. 20220508136RC and 02201ZYTS318).

Conflict of interest

The authors declare no conflict of interest.

Author details

Xiaoxi Yuan[1*], Feng Yang[1] and Hongdong Li[2]

1 Institute for Interdisciplinary Quantum Information Technology, Jilin Engineering Normal University, Changchun, P.R. China

2 State Key Laboratory of Superhard Materials, College of Physics, Jilin University, Changchun, P.R. China

*Address all correspondence to: xxyuan@jlenu.edu.cn

References

[1] Hir ZAM, Abdullah AH. Hybrid polymer-based photocatalytic materials for the removal of selected endocrine disrupting chemicals (EDCs) from aqueous media: A review. Journal of Molecular Liquids. 2022;**361**:119632. DOI: 10.1016/j.molliq.2022.119632

[2] Wee SY, Aris AZ. Endocrine disrupting compounds in drinking water supply system and human health risk implication. Environment International. 2017;**106**:207-233. DOI: 10.1016/j.envint.2017.05.004

[3] dos Santos RP, Cruz WD, Magalhães KF, de Araújo DM, Medeiros MC, Martínez-Huitle CA, et al. Electrochemical degradation of a commercial formulation of the insecticide pyriproxyfen using boron-doped diamond anode. Journal of the Electrochemical Society. 2020;**167**:146510. DOI: 10.1149/1945-7111/abc4bb

[4] Iwaki M, Sato S, Takahashi K, Sakairi H. Electrical conductivity of nitrogen and argon implanted diamond. Nuclear Instruments and Methods in Physics Research. 1983;**209**:1129-1133. DOI: 10.1016/0167-5087(83)90930-4

[5] Pelskov YV, Sakharova AY, Krotova M, Bouilov LL, Spitsyn BV. Photoelectrochemical properties of semiconductor diamond. Journal of Electroanalytical Chemistry and Interfacial Electrochemistry. 1987;**228**:19-27. DOI: 10.1016/0022-0728(87)80093-1

[6] Patel K, Hashimoto K, Fujishima A. Photoelectrochemical investigations on boron-doped chemically vapour-deposited diamond electrodes. Journal of Photochemistry and Photobiology A: Chemistry. 1992;**65**:419-429. DOI: 10.1016/1010-6030(92)80023-o

[7] Yang N, Yu S, Macpherson JV, Einaga Y, Zhao H, Zhao G, et al. Conductive diamond: Synthesis, properties, and electrochemical applications. Chemical Society Reviews. 2019;**48**:157-204. DOI: 10.1039/c7cs00757d

[8] Gan P, Compton RG, Foord JS. The voltammetry and electroanalysis of some estrogenic compounds at modified diamond electrodes. Electroanalysis. 2013;**25**:2423-2434. DOI: 10.1002/elan.201300362

[9] Wang Q, Subramanian P, Li M, Yeap WS, Haenen K, Coffinier Y, et al. Non-enzymatic glucose sensing on long and short diamond nanowire electrodes. Electrochemistry Communications. 2013;**34**:286-290. DOI: 10.1016/j.elecom.2013.07.014

[10] Zhao Y, Ji J, Wu Y, Chen S, Xu M, Cao X, et al. Nonylphenol and its derivatives: Environmental distribution, treatment strategy, management and future perspectives. Chemosphere. 2024;**352**:141377. DOI: 10.1016/j.chemosphere.2024.141377

[11] Yuan X, Gao N, Gao X, Qiu D, Xu R, Sun Z, et al. Nanopyramid boron-doped diamond electrode realizing nanomolar detection limit of 4-nonylphenol. Sensors and Actuators B: Chemical. 2019;**281**:830-836. DOI: 10.1016/j.snb.2018.11.011

[12] Czarny-Krzymińska K, Krawczyk B, Szczukocki D. Bisphenol A and its substitutes in the aquatic environment: Occurrence and toxicity assessment. Chemosphere.

2023;**315**:137763. DOI: 10.1016/j.chemosphere.2023.137763

[13] E. EFSA Panel on Food Contact Materials, P. Aids, Lambré C, Baviera JMB, Bolognesi C, Chesson A, Cocconcelli PS, et al. Re-evaluation of the risks to public health related to the presence of bisphenol A (BPA) in foodstuffs. EFSA Journal. 2023;**21**:e06857. DOI: 10.2903/j.efsa.2023.6857

[14] Ma Y, Liu J, Li H. Diamond-based electrochemical aptasensor realizing a femtomolar detection limit of bisphenol A. Biosensors and Bioelectronics. 2017;**92**:21-25. DOI: 10.1016/j.bios.2017.01.041

[15] Lv Y, Ma J, Yu Z, Liu S, Yang G, Liu Y, et al. Fabrication of covalent organic frameworks modified nanofibrous membrane for efficiently enriching and detecting the trace polychlorinated biphenyls in water. Water Research. 2023;**235**:119892. DOI: 10.1016/j.watres.2023.119892

[16] Idowu IG, Megson D, Tiktak G, Dereviankin M, Sandau CD. Polychlorinated biphenyl (PCB) half-lives in humans: A systematic review. Chemosphere. 2023;**345**:140359. DOI: 10.1016/j.chemosphere.2023.140359

[17] Yuan X, Jiang Z, Wang Q, Gao N, Li H, Ma Y. Polychlorinated biphenyl electrochemical aptasensor based on a diamond–gold nanocomposite to realize a sub-femtomolar detection limit. ACS Omega. 2020;**5**:22402-22410. DOI: 10.1021/acsomega.0c02846

[18] Jain N, Jain GK, Ahmad FJ, Khar RK. Validated stability-indicating densitometric thin-layer chromatography: Application to stress degradation studies of minocycline. Analytica Chimica Acta. 2007;**599**:302-309. DOI: 10.1016/j.aca.2007.08.020

[19] Wrightson WR, Myers SR, Galandiuk S. Analysis of minocycline by high-performance liquid chromatography in tissue and serum. Journal of Chromatography B: Biomedical Sciences and Applications. 1998;**706**:358-361. DOI: 10.1016/s0378-4347(97)00647-6

[20] Wan L, Liang Y, Yang Y, Zhang Q, Cui Z, Liu J, et al. Electrochemical detection of phenacetin with graphite/boron-doped diamond electrode grown at high methane concentration. Diamond and Related Materials. 2024;**144**:110965. DOI: 10.1016/j.diamond.2024.110965

[21] Juang R-S, Hsieh C-T, Lin T-A, Shao Y-C, Gandomi YA. Reduced graphene oxide decorated with poly (diallyldimethylammonium chloride) for sensitive detection of acetaminophen in water and human urine samples. Colloids and Surfaces A: Physicochemical and Engineering Aspects. 2023;**677**:132426. DOI: 10.1016/j.colsurfa.2023.132426

[22] Wang S, Chen F, Li Z, Tao H, Qu L, Li J, et al. A graphene oxide/Zn-metal organic framework electrochemical sensor for acetaminophen detection. Surfaces and Interfaces. 2023;**39**:102910. DOI: 10.1016/j.surfin.2023.102910

[23] Wang P, Yuan X, Cui Z, Xu C, Sun Z, Li J, et al. A nanometer-sized graphite/boron-doped diamond electrochemical sensor for sensitive detection of acetaminophen. ACS Omega. 2021;**6**:6326-6334. DOI: 10.1021/acsomega.0c06141

[24] Niyitanga T, Khan MQ, Ahmad K, Khan RA. Fabrication of an azithromycin sensor. Biosensors. 2023;**13**:986. DOI: 10.3390/bios13110986

[25] Li Y, Zhang H, Cui Z, Liu S, Xu J, Jia C, et al. Chemical staining enhanced Enzyme-linked immunosorbent assay for

sensitive determination of clenbuterol in food. Food Chemistry. 2023;**400**:134012. DOI: 10.1016/j.foodchem.2022.134012

[26] Rajput JK. Nanosensors: A smart remedy for early detection of clenbuterol contamination in food. Food Chemistry. 2023;**426**:136569. DOI: 10.1016/j.foodchem.2023.136569

[27] Ma Z, Wang Q, Gao N, Li H. Electrochemical detection of clenbuterol with gold-nanoparticles-modified porous boron-doped diamond electrode. Microchemical Journal. 2020;**157**:104911. DOI: 10.1016/j.microc.2020.104911

Chapter 7

Development of Nanostructured $BiVO_4$ Photoelectrodes for the Degradation of Pharmaceutical Pollutants in Wastewaters

Prabhakarn Arunachalam, Mabrook S. Amer, Maged N. Shaddad and Abdullah M. Al-Mayouf

Abstract

Global environmental problems, such as energy shortages and water pollution, threaten sustainable development. In recent years, pharmaceuticals have been recognized as emerging water pollutants. Water splitting using photoelectrochemical (PEC) technology has attracted considerable attention for its eco-friendly and promising characteristics. Photoelectrocatalytic oxidation has been widely studied in order to remove them from contaminated water. This chapter describes recent advances in competitive photoelectrocatalytic integrated photoelectrodes utilizing $BiVO_4$ and other cocatalytic materials. Further, we discuss the development of photoelectrocatalytic devices for the efficient degradation of pharmaceutical pollutants (ciprofloxacin, sulfamethoxazole, and tetracycline hydrochloride) in wastewater. This photoelectrode nanostructure is designed to capture more photons, reduce electron-hole recombination, and utilize more solar energy. Therefore, we examine the important properties of $BiVO_4$-based materials for efficient degradation of PEC pollutants in this book chapter. Our book chapter discusses key challenges and future prospects in wastewater treatment and offers new insights.

Keywords: bismuth vanadate, pharmaceutical pollutants, antibiotics, photoelectrochemical, metal oxide

1. Introduction

Globally, urbanization and population growth are the major causes of water contamination. The leather, paper, pharmaceutical, and textile industries consume enormous amounts of water for their production and discharge large amounts of waste [1, 2]. In particular, industrial waste generated during the production of prevalent antibiotic drugs like tetracycline (TC), sulfamethoxazole, oxytetracycline, and ibuprofens threatens human and aquatic life. Water bodies discharge these waste products directly into them, which are persistent and nonbiodegradable [3, 4]. The majority of pharmaceutical pollutants can be found in wastewater and groundwater. These pollutants can be harmful to marine life, and drinking water contaminated with them can cause chronic health problems, such as increased antibiotic-resistant bacteria strains [5]. Due to the worldwide rise of multidrug-resistant bacteria, antibiotic resistance has become the most pressing concern, making it imperative to find appropriate methods for removing antibiotics from pharmaceutical wastewater [6]. TC antibiotics have recently been eliminated from water sources using new technologies [6]. In addition to electrochemical oxidation, heterogeneous photocatalysis has also evolved for water treatment, but both are energy-intensive and inefficient [7]. In order to remove these contaminants more efficiently, photoelectrocatalytic degradation is a promising eco-friendly technique. During this process, hydroxyl radicals, superoxide radicals, and holes are generated to oxidize organic compounds into CO_2 and water [8]. The PEC technology, which combines electrochemistry and photocatalysis, is becoming increasingly popular because of its high energy efficiency [9].

In combination with PEC water splitting method, pharmaceutical pollutant degradation can be used to solve energy and environmental problems [10–13]. Interestingly, combining PEC oxidation with antibiotics degradation is an effective approach of utilizing solar energy for improving energy and environmental quality [13]. This process produces radicals using both light and electrical energy, as well as using photocatalysis and electrolysis [14]. Therefore, the solar energy can be effectively transformed and kept by utilizing the photoinduced holes of the photoanode to oxidize pharmaceutical pollutants and the photoelectron migration to the cathode for H_2 production. Furthermore, there is no doubt that fossil fuels can make this system cost-efficient at the industrial scale. In addition, it is crucial that solar-assisted photoelectrolysis reactors can be manufactured very cheaply in order to compete on a commercial scale. A number of attempts have been made in recent years to remove waste drugs using PEC technology. In a recent study, the Liu group demonstrated 81% removal efficiency of TC antibiotics using gold-sensitized TiO_2 nanotube arrays [15]. Similarly, Li et al. constructed a nanostructured n-ZnO/p-Cu_2O/n-TNA electrode for breakdown of tetracycline by PEC, achieving 85% removal of tetracycline after four repeated experiments [16]. Due to the only absorption of UV light by TiO_2-based electrodes, they are not very efficient in converting solar energy into electricity. It would be helpful to expand these studies to different kinds of semiconductor photocatalysts. For effective PEC systems, photoelectrodes should have a small bandgap, high conversion efficiency, and good aqueous durability. In order to meet these requirements, solar light can be used as a source of photon energy; however, visible light active photocatalysts are necessary for the photoanodic material. On this regard, semiconductor photoelectrodes such as WO_3 [17], Ag_3VO_4 [18], $BiVO_4$ [19, 20], Fe_2O_3 [21], CuS [22], and Ag_3PO_4 [23] have been investigated for PEC processes.

Its nontoxic nature, stability, and favorable band alignments make BiVO4 semiconductor material especially suitable for solar water splitting [24]. As a result, there are several inherent difficulties; among them are the high carrier recombination rates, the slow oxygen evolution rates, as well as the photocorrosion of metals [25]. Since cocatalysts are low overpotentials and durable and can arrest surface holes quickly, they can significantly improve PEC performance [26]. Adding metal to electrodes can improve electrode performance as well, but its origin and application depend on the metal [24]. In a recent study by Qiu et al., tandem solar cells with a single perovskite cell were found to be able to split water unaided with an efficiency of 6.2% for more than 10 hours [27]. As a result, several modifications have been made to the composition of bismuth in the Earth's crust in order to make a low-cost metal vanadate based on this method [28, 29]. Due to their slow electronic properties and low solar conversion efficiencies, $BiVO_4$ photoanodes cannot be used commercially in PEC systems at present. In particular, the photoactive semiconductor material must be infused with a powerful, stable, and inexpensive water oxidation catalyst. Prussian blue (PB)-type electrocatalysts were recently used as supporting cocatalysts and hole-storage layers in PEC water splitting to protect $BiVO_4$ anodes that are subject to photocorrosion [30, 31]. Numerous methods have been demonstrated to address these concerns, such as ion doping [32–34], nanostructuring [35], surface variation with passivation layers or electrocatalysts [36, 37], and combinatorial synthesis [38]. It should be noted that $BiVO_4$ surfaces are capable of encapsulating oxygen evolution catalysts (OECs) such as CoPi [39], IrOx [40], FeOOH [41], and NiOOH [42]. Following this, PEC technology was also used to reduce pharmaceutical pollutants. Further, according to Liu et al. (2021), an F-doped $BiVO_4$@NiFeLDH

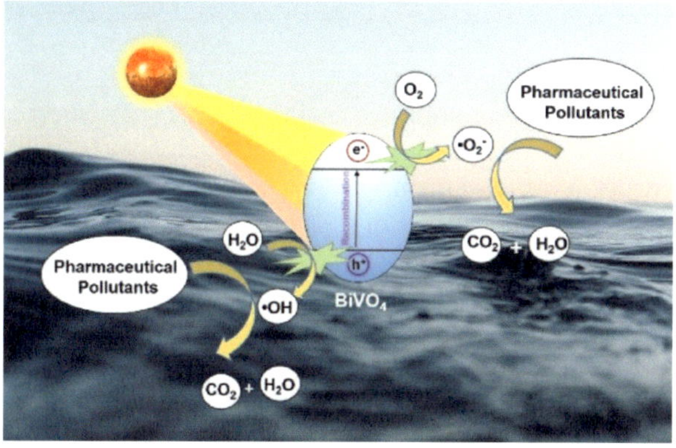

Figure 1.
Schematic representation of bismuth vanadate nanomaterials for PEC pharmaceutical pollutants degradation.

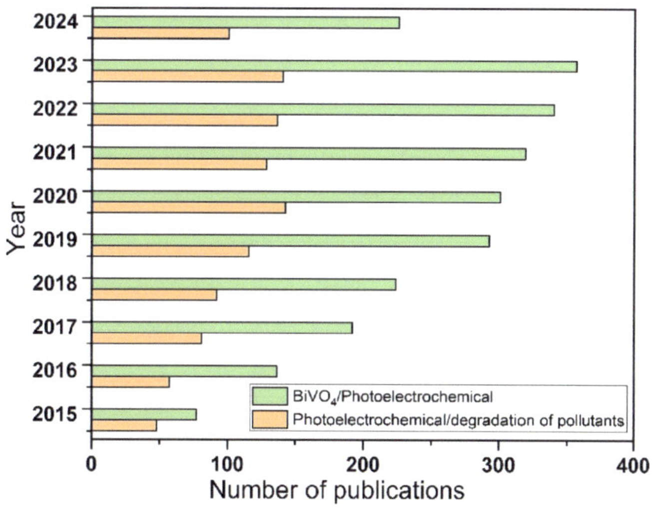

Figure 2.
Statistical histogram of the Web of Science searching results on published literature about $BiVO_4$ photoanodes from 2015 to 2024.

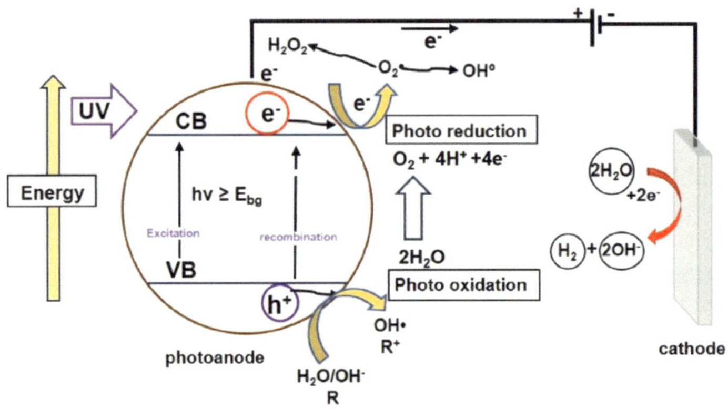

Figure 3.
Schematic representation of surface reactions that occur during photoelectrocatalytic reactions using photoanodes.

photoanode is being developed to remove antibiotic TCs [43]. Surface and interface engineering can be used to design composite photoelectrodes based on $BiVO_4$ that improve light absorption but also promote oxidation [44–46]. Finally, this book chapter investigates how metal-doped $BiVO_4$ and OECs can be employed to eliminate antibiotics efficiently (**Figures 1** and **2**).

In the last ten years, for instance, $BiVO_4$ has become more popular for PEC applications according to the Web of Science statistics (**Figure 3**) (obtained with the keywords listed in the legend). Moreover, it is possible to develop commercially viable $BiVO_4$ photoanodes using different modification strategies. As part of the book chapter, we discuss recent progress regarding earth-abundant electrocatalysts' potential role in PEC degradation of pollutants. We also present various examples of functional and complete PEC degradation of pollutants, along with a critical evaluation of the ongoing disputes in this field. We also discuss future challenges related to $BiVO_4$-based photoanodes.

2. Fundamentals of photoelectrocatalysis

PEC combines electrolytic and photocatalytic steps, which has gotten a lot of attention because it retards recombination and prolongs electron-hole lifetimes. This process involves ejecting an electron from the fully filled valence band (VB) of a semiconductor into the empty conductive band (CB), which generates a positively charged vacancy or hole (h^+_{VB}) [47]. Irradiation of light with higher energies than bandgaps on a semiconductor photocatalyst causes electrons to move to the CB (e^-_{CB}), leaving a positive vacancy in the VB (h^+_{VB}), which can yield the OH radical by oxidizing H_2O/OH^- (Eqs. 1–3) [48]. The photocatalyst's radical production capacity is decreased by the rapid recombination of electron-hole (e^-_{CB}/h^+_{VB}) pairs whose lifetimes are measured in nanoseconds [49]. Using an external potential, the e^-_{CB}/h^+_{VB} pair's recombination rate can be altered [50]. In PEC, a catalyst is decorated over a conductive support matrix called a photoanode. Light irradiation produces photo-excited electrons that are transferred to the cathode by the external positive potential applied to the photoanode. Through the reaction with H_2O/OH^-, the h^+_{VB} produces more radicals.

$$\text{Photoanodes} + h\nu \rightarrow \text{e}^{-}{}_{CB} + \text{h}^{+}{}_{VB} \quad (1)$$

$$\text{h}^{+}{}_{VB} + \text{H}_2\text{O} \rightarrow HO^{\bullet} + \text{H}^{+} \quad (2)$$

$$\text{h}^{+}{}_{VB} + OH^{-} \rightarrow HO \quad (3)$$

$$\text{h}^{+}{}_{VB} + \text{R} \rightarrow \text{R}^{+}{}_{ox} \quad (4)$$

$$\text{e}^{-}{}_{CB} + \text{O}_2 \rightarrow \text{O}_2{}^{-\bullet} \quad (5)$$

Furthermore, organic contaminants (R) could be directly oxidized by the light-induced h^{+}_{VB}, as shown in Eq. (4), further increasing the degradation efficiency. Additionally, superoxide anion is also formed by electron excited electrons interacting with oxygen molecules (Eq. 5). In the literature [51, 52], applied potential is considered a major factor contributing to PEC efficiency.

As a consequence, electrons can be transferred into the external circuit when the photoanode is applied instead of oxygen molecules. Thus, the photoinduced hole or OH• will remain at the electrode surface. It should be noted that the rate of photogenerated carrier recombination is limited, though there is the possibility of improving the efficiency of the oxidation at the semiconductor-electrolyte interface [53, 54]. There are four major physiochemical steps essential for the PEC reaction to take place. Among the processes described here are (i) capturing photons; (ii) creating/separating electron-hole pairs; and (iii) transporting separated charge carriers via charge transfer; all of which must be catalytically active. The last two processes involve electron/hole pairs that can either recombine in bulk, so it is crucial to have an effective charge separation and higher carrier transport; (iv) surface chemical reactions.

3. Integrated photoelectrochemical processes

PEC water splitting combined with antibiotics degradation should result in better solar energy utilization and protection of the environment [10, 11]. Using this method, pollutants (such as pharmaceutical pollutants) are oxidized in the photoanode, while the photoelectrons migrate to the cathode, which then generates hydrogen, allowing solar energy to be converted and stored efficiently. PEC water splitting offers two advantages: It splits the water and it degrades organic pollutants. As a result of this process, both hydrogen fuel and harmful antibiotics are removed and detoxified, thus simultaneously addressing both energy and environmental problems. Using an anode that oxidizes pollutants (like pharmaceuticals), a cathode that generates hydrogen can convert and store solar energy effectively. Using this PEC method, carbon emissions and other forms of pollution can be reduced [55]. In PEC, organic pollutants are decomposed using photoelectrodes. Further, in PEC cells, semiconductor photocatalysts absorb most of the solar spectrum. In particular, $BiVO_4$ semiconductor materials have a favorable band alignment, a nontoxic nature, and a structural stability that makes them capable of splitting solar water [24]. It is necessary to resolve several intrinsic difficulties, such as high carrier recombination rates and slow oxygen evolution rates [25]. Due to their low overpotentials and high durability, cocatalysts can drastically improve PEC performance [26].

4. BiVO4-based photoanode materials

As a semiconductor with a relatively small optical bandgap, $BiVO_4$ is of considerable interest in PEC systems, as its CB edge is relatively close to 0 V in comparison with RHE. It is theoretically possible to generate a maximum AM 1.5 photocurrent response of 7.5 mA/cm^2 on $BiVO_4$ electrodes and absorb photons with energies above the bandgap to perform STH energy conversion at a rate of 9%. Moreover, bare $BiVO_4$ photoanodes have very low visible-light absorption intensity, resulting in very low photocatalytic performance and a photocurrent density of only 1 mA/cm^2. Since photoinduced holes and electrons are recombined at a high surface recombination temperature, a high voltage is needed to obtain high conversion efficiency. Although $BiVO_4$ photoanodes have an acceptable CB energy level, short carrier lifetimes, and sluggish oxygen evolution kinetics, they are not suitable for oxidizing PEC water. It is therefore very desirable to develop $BiVO_4$-based composite photoelectrodes by surface/interface engineering, which promotes coupled oxidation reactions and enhances light absorption.

Furthermore, by doping $BiVO_4$ electrodes with metal, the performance can be improved, but its origins and applications vary depending on the metal. It has been demonstrated that Zr doping over $BiVO_4$ results in a significant reduction in overall resistance as well as an increase in PEC activity [46]. In addition, it was shown that Zr^{4+} can improve performance if Bi^{3+} sites are substituted with Zr^{4+}. Photocatalytic properties are enhanced by nanostructured electrodes when the depletion layer width is increased and the minority carrier travel distance is reduced. It has been possible to enhance the functional properties of $BiVO_4$ photoanodes by using heterostructure strategies with Fe_2O_3. The synergistic interface between $BiVO_4$ (which offers good light-gathering properties) and WO_3, which enhances charge carrier transport, improved $BiVO_4$'s PEC behavior [56, 57]. Additionally, different semiconductor combinations have been effective in water splitting reactions, such as $Si/TiO_2/BiVO_4$ [58], $SnO_2/BiVO_4$ [59], WO_3/Fe_2O_3 [60], and Si/Fe_2O_3 [61]. According to Cai et al., Fe_2O_3 nanoparticles are an efficient supporting catalyst for the photocatalytic degradation of organic pollutants [62]. Notably, $BiVO_4$ photoanodes have also been decorated successfully with Fe-based cocatalysts, such as FeOOH [41] and Ni-FeO_x [63].

Recent studies have demonstrated the electrochemical features of silver phosphate and nickel hydroxyl phosphate over $BiVO_4$ photoanodes [64–66]. As reported recently, our group demonstrated the cooperative effect of Mo incorporation and vacuum annealing on Bi_2O_3 PEC behavior in light-driven water oxidation [67]. Recently, the Co-based WOC has gained extensive attention due to its higher efficiency under near-neutral pH conditions, and its lower cost to produce [68]. Additionally, photoinduced holes are transported from $BiVO_4$ to $Co^{3+}/^{2+}$ species through the water oxidation reaction, lowering the overpotential significantly [26, 69, 70]. In recent years, several studies have shown that Co-based catalysts are effective in water splitting when applied to WO_3, Fe_2O_3, and ZnO photoanodes [71–73]. Furthermore, cocatalyst incorporation resulted in significant enhancements for $BiVO_4$ photoanodes. In a study by Long et al., powdered composites of $BiVO_4/Co_3O_4$ on conducting substrates enhanced photon-to-current conversion efficiency fourfold compared to bare $BiVO_4$ [74]. Afterward, with Co-Pi catalyzed W-loaded $BiVO_4$, Zhong et al. reported a photocurrent of 1.4 mA/cm^2 at 1.23 V_{RHE} [75]. In addition, the CoPi-catalyzed W:$BiVO_4$ photoelectrodes showed that bulk recombination limits the total PEC behavior of Co-Pi catalyzed water electrolysis catalysts. It is still necessary to correct the bulk recombination limitation process in order to achieve effective $BiVO_4$-based photoelectrodes. Abdi et al. demonstrated 90% external quantum efficiency for CoPi-catalyzed $BiVO_4$ photoanodes. Further, they demonstrated that slow electron transport is a crucial issue to be studied in $BiVO_4$ electrodes to rule out bulk recombination. Furthermore, Abdi et al. were able to attain 1.7 mA/cm^2 at 1.23 V_{RHE} for cocatalyzed bare $BiVO_4$ [76]. Two competing theories have recently emerged regarding the mechanism behind photocurrent enhancement in CoPi-loaded photoelectrodes. Barroso et al. reported that better band bending leads to photogenerated carriers separating and extending the carrier life in semiconductors [77]. Alternatively, Klahr et al. found that there is no change in band bending and instead suggested that Co-Pi collects photogenerated holes from semiconductors at the surface, preventing carrier recombination [78]. According to Zachäus et al., the intensity-modulated photocurrent spectroscopy was applied to analyze the dynamics of the surface carriers of Co-Pi–loaded $BiVO_4$ photoelectrodes. In a detailed investigation, Co-Pi OEC loading reduced $BiVO_4$ recombination behavior by a factor of 10–20 without significantly influencing charge transfer kinetics. Moreover, $BiVO_4$ photoanodes are restricted by surface recombination instead of charge transfer because CoPi is used to passivate their surfaces [79]. Under AM 1.5G, Lichterman's group used $CoO_x/BiVO_4$ photoelectrodes to achieve 1.49 mA/cm^2 at 1.23 V_{RHE} [80]. Upon deposition of Co_3O_4 for $BiVO_4$, Wang et al. demonstrated 17-fold improvements in PEC water electrolysis characteristics [70]. Recently, Hegnaer et al. combined photoactive $BiVO_4$ materials with electrocatalytic materials such as CoFe-PB electrocatalysts. Furthermore, by cathodically shifting $BiVO_4$'s onset potential by 0.8 V, the mixture increases its photovoltage by 0.45 V. Moreover, $BiVO_4$/CoFe-PB photoelectrodes exhibit a sixfold enhancement in photocurrent compared to RHE at 1.23 V [81]. In $BiVO_4$ semiconductors, surface engineering via element doping is highly expected to alter the band structure due to its strong oxidation capacity. For example, F^- ions are more electronegative and have smaller atomic radius than O_2^- ions, which favors lattice doping for positively shifting VB positions and enhancing charge separation. Furthermore, improved surface reaction kinetics can be achieved by decorating excellent OECs with interface engineering [82]. A recent investigation by Liu et al. examined the simultaneous removal of antibiotic TCH from wastewater integrated with the generation of pure H_2 using an F-doped $BiVO_4$@NiFe-LDH core-shell electrode. At 1.23 V, the maximal photocurrent response of the F-$BiVO_4$@NiFe-LDH photoanode was approximately sixfold enrichment than that of the $BiVO_4$ photoelectrode pristine. Further, the composite photoanode was effective in degrading TCH in PEC, more efficiently than previously reported (**Figure 4**). Further, the composite electrode showed efficacy when applied to the degradation of TCH in PEC, achieving 86% degradation in 2 h and no attenuation after four cycles (**Figure 4f**). Therefore, the composite materials fabricated from $BiVO_4$ provide an efficient, sustainable, and sunlight-driven method for reducing organic pollutants while generating hydrogen.

Adding metals to $BiVO_4$ improves its performance in PEC but also promotes charge separation, reducing the likelihood of charge recombination. During the last few years, our group developed ZrO_2- and α-Fe_2O_3-loaded $BiVO_4$ electrode materials. To produce monoclinic ZrO_2 and α-Fe_2O_3 nanoparticles, sequentially depositing Zr and Fe precursors onto $BiVO_4$ was followed by thermal treatment. Thus, the electrode produced five times higher water oxidation photocurrent than pristine $BiVO_4$ [46]. Afterward, Ikeda et al. demonstrated the effects of Zr doping on the structural and catalytic features of $BiVO_4$ powder [83]. By integrating Zr, monoclinic scheelite structures (ms-$BiVO_4$) were formed. By doping Zr into ms-$BiVO_4$ powder, PEC properties were enhanced. After a certain point, doping caused structural changes,

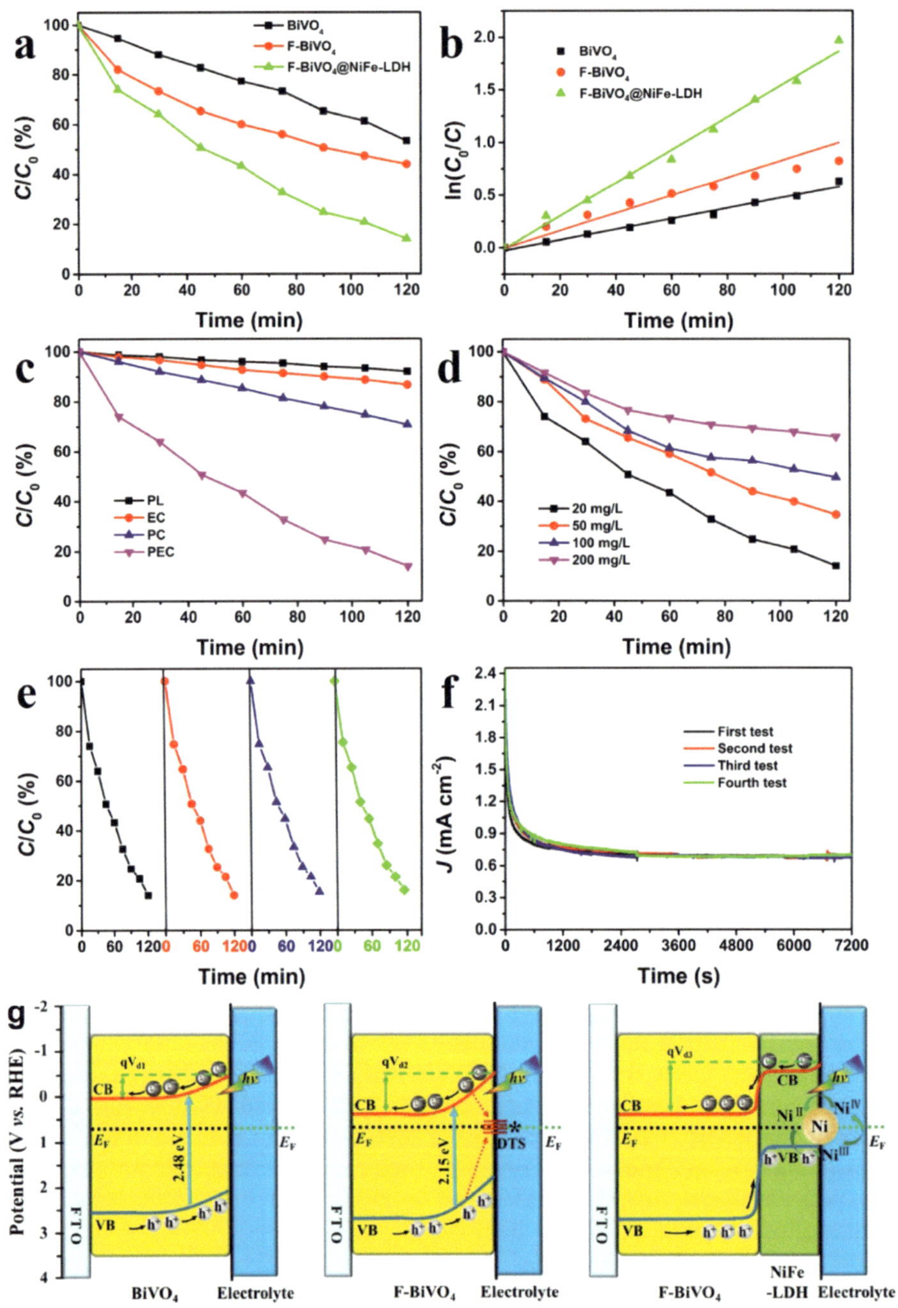

Figure 4.
(a) PEC degradation of TCH by constructed electrodes ($BiVO_4$, F-$BiVO_4$, and F-$BiVO_4$@NiFe-LDH). (b) TCH removal kinetics. (c) Concentration-time profiles for different TCH degradation approaches (PC = photocatalysis, PL = photolysis, EC = electrocatalysis) by means of F-$BiVO_4$@NiFe-LDH photoanodes. (d) PEC degradation of TCH at varying concentrations. (e) Durability of the fabricated composite electrode for TCH degradation. (f) Current-time TCH oxidation profiles. (g) Schematic band diagrams of the fabricated photoanodes [43].

resulting in a reduction in PEC activity [57]. We prepared AgNi-OH-Pi/Zr-$BiVO_4$ by electrochemically depositing NiOH-Pi, Ag-Pi, and AgNi-OH-Pi layers onto Zr-$BiVO_4$ electrodes and then using PEC to transform them. It was demonstrated that the modified photoanodes demonstrated excellent PEC functionality of 3.14 mA/cm^2 and long-term durability of more than 60 hours. A higher efficiency can be credited to more efficient charge separation and excellent surface characteristics [65]. Further, OECs incorporated over $BiVO_4$ electrodes can enhance sluggish water oxidation and photocorrosion problems [64]. Besides improving electrode photostability and charge transfer, coating OECs with $BiVO_4$ can also prevent photocorrosion and slow water oxidation. Additionally, OECs can reduce the bandgap of $BiVO_4$. This allows for more efficient sunlight absorption and creates more efficient photoelectrode materials. Moreover, NiFeOOH is considered a highly efficient PEC water splitter since it has nanomorphologies, high light absorption capacity, and a low overpotential. In a study by Rong et al, NiFeOOH enhanced OERs in carbon nanotubes [84]. In a similar study, Fang et al. demonstrated that NiFeOOH and Co-Pi catalysts can be used to modify $BiVO_4$ surfaces to enhance water splitting in PEC [85]. In a recent study, we synthesized different groupings of Zr:$BiVO_4$/NiFeOOH by electrodeposition and PEC alteration to determine their optimal catalyst combination for antibiotics degradation beyond PEC water oxidation [86]. As a result, Zr:$BiVO_4$/NiFeOOH heterojunction photoanodes are more durable and water oxidation occurs more efficiently. Compared to Zr:$BiVO_4$, the developed Zr:$BiVO_4$/NiFeOOH photoelectrode exhibited a photocurrent enhancement of 1.27 mA/cm^2 at 1.23 V_{RHE}, which was fivefold higher than Zr:$BiVO_4$ (**Figure 5**). Interestingly, the developed heterojunctioned electrodes have been found to be capable of degrading TCH, riboflavin (Rf), and ciprofloxacin (Cip) efficiently in PEC. Compared to pristine Zr:$BiVO_4$ photoelectrodes, the TCH degradation rate achieved 96% in 1 hour. In this way, Zr:$BiVO_4$ photoelectrodes altered with OEC decoration improved PEC degradation beyond water splitting.

Bismuth sulfide (Bi_2S_3) has a low bandgap of 1.4 eV, allowing it to create heterostructured photoelectrodes by integrating with $BiVO_4$ [87]. It has been shown that heterostructured electrodes of $BiVO_4$ and Bi_2S_3 can be used for PEC water splitting. Through hydrothermal synthesis, Liu et al. established a heterojunctioned electrode comprised of $BiVO_4$/Bi_2S_3 nanowires and achieved a photocurrent density of 0.51 mA/cm^2 [88]. Also, Zhuo and colleagues found that walnut-like heterostructured photoanodes made of $BiVO_4$/Bi_2S_3 enhanced photocatalytic performance [89]. In one study, Chen et al. developed heterostructured electrodes composed of $BiVO_4$/Bi_2S_3 on conducting films, generating 1.43 mA/cm^2 of photocurrent [90]. Recent years have seen PEC approaches applied to the degrading of waste drugs. It would be necessary to develop a new photoelectrode made from $BiVO_4$ through surface/interface engineering, improved adsorption of light, better charge transfer, and enhanced coupled oxidative reactions to defeat this problem.

Based on the above background, we present an electrochemical fabrication method that combines electrodeposition, in-situ PEC ion exchange. In this study, we evaluate the durability of hybrid Zr:$BiVO_4$@Co to Zr:$BiVO_4$@Bi_2S_3-CoS heterostructure film electrodes under PEC conditions for water oxidation reaction and TCH degradation [91]. CoS is integrated into Zr:$BiVO_4$ to boost charge separation and OER rates. Compared to parent Zr:$BiVO_4$ films, the heterostructured Zr:$BiVO_4$@Bi_2S_3-CoS films showed nearly 3.5-time enhancement at 1.23 V vs. RHE (**Figure 6**). In addition

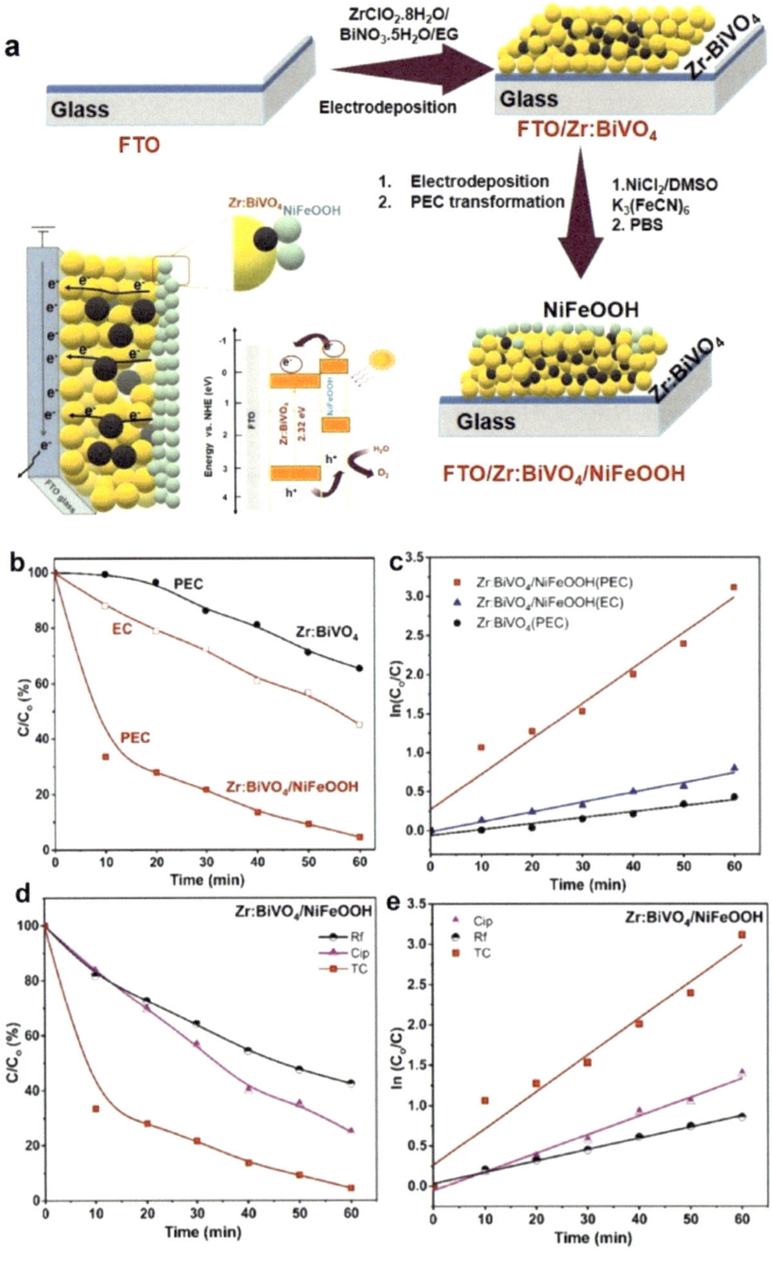

Figure 5.

(a) Schematic illustration of Zr:BiVO$_4$/NiFeOOH electrodes on an FTO obtained via a PEC ion-exchange approach, (b) concentration-time profiles for different TCH degradation approaches using Zr:BiVO$_4$/NiFeOOH photoelectrodes, (c) TCH removal kinetics, (d) PEC degradation of different pharmaceutical pollutants, (e) and its corresponding kinetic plots [87].

to enhanced optical absorption, Zr:BiVO$_4$ and Bi_2S_3 heterojunction photoanodes have excellent surface and transport properties. The composite electrodes also degraded TCH through PEC effectively. In 30 minutes, the degradation efficiency of the TCH exceeded that of the pristine BiVO$_4$ electrode by 96%.

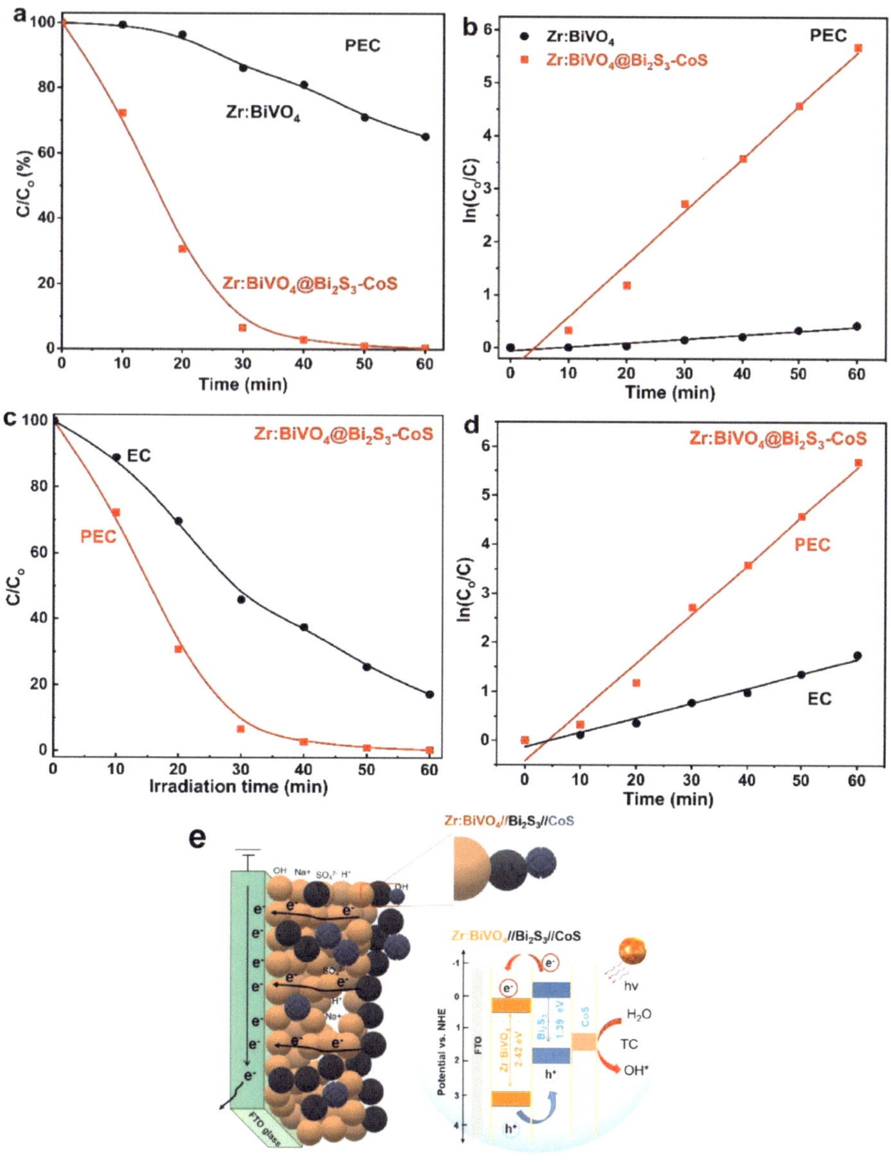

Figure 6.
PEC degradation of TCH for fabricated electrodes (a), kinetic plots (b), concentration-time profiles for a variety of TCH degradation approaches with Zr:$BiVO_4$@Bi_2S_3-CoS photoelectrodes (c), (d) and its corresponding kinetic plots, (e) schematic representation of charge transfer pathways in Zr:$BiVO_4$@Bi_2S_3-CoS electrodes.

5. Conclusions

Our book chapter explains the fundamental principles of PEC water splitting and pollutant degradation on $BiVO_4$ materials, along with the scientific challenges related to this process. It is ideal for water splitting to occur via bare semiconductors with bandgap energies overlapping the redox potentials of water reduction and oxidation, so that photogenerated charge carriers have an appropriate overpotential for HER and OER. With the development of $BiVO_4$ as a photoanode, practical PEC devices can be designed using this material. There have been a few publications on innovative approaches to $BiVO_4$ modification, but much more attention is needed to advance these approaches. $BiVO_4$ photoanodes from the last few decades are reviewed, along with current initiatives looking at integrated approaches to PEC. In addition to PEC solar-to-chemical energy conversion, there are a variety of ways to improve semiconductor photocatalytic materials' photocatalytic properties. Many semiconductor photocatalytic materials have inherent limitations, such as easy recombination between electrons and holes, a slow OER at the surface, and low light absorption. In addition, we reviewed a number of popular modifications involving surface engineering and interface engineering. By combining these engineering strategies, $BiVO_4$-based photocatalyst properties have been greatly enhanced. This chapter reviews recent advancements in sunlight-driven, efficient, sustainable methods of degrading pharmaceutical pollutants in combination with hydrogen production. It is certain that progress in developing highly active photoanode materials will continue, as well as the certainty of bandgap design in the near future.

Acknowledgements

The authors extend their appreciation to the Deputyship for Research & Innovation, "Ministry of Education" in Saudi Arabia for funding this research work through the project number (IFKSUDR_E151).

Author details

Prabhakarn Arunachalam[1*], Mabrook S. Amer[1], Maged N. Shaddad[2] and Abdullah M. Al-Mayouf[1]

1 Electrochemical Sciences Research Chair, Department of Chemistry, Science College, King Saud University, Riyadh, Saudi Arabia

2 Department of Chemistry, College of Science and Humanities in Al-Kharj, Prince Sattam Bin Abdulaziz University, Al-Kharj, Saudi Arabia

*Address all correspondence to: parunachalam@ksu.edu.sa

References

[1] Levin-Reisman I, Ronin O, Gefen I, Braniss I, Shoresh N, Balaban NQ. Antibiotic tolerance facilitates the evolution of resistance. Science. 2017;**355**:826-830

[2] Huang S, Zhang Q, Liu P, Ma S, Xie B, Yang K, et al. Novel up-conversion carbon quantum dots/α-FeOOH nanohybrids eliminate tetracycline and its related drug resistance in visible-light responsive Fenton system. Applied Catalysis B: Environmental. 2020;**263**:118336

[3] Wei X, Feng H, Li L, Gong J, Jiang K, Xue S, et al. Synthesis of tetragonal prismatic γ-In_2Se_3 nanostructures with predominantly {110} facets and photocatalytic degradation of tetracycline. Applied Catalysis B: Environmental. 2020;**260**:118218

[4] Zhu B, Song D, Jia T, Sun W, Wang D, Wang L, et al. Self-assembly of $Bi_2Sn_2O_7$/β-Bi_2O_3 S-Scheme heterostructures for efficient visible-light-driven photocatalytic degradation of tetracycline. ACS Omega. 2021;**6**:1647-1656

[5] Bhujbalrao R, Anand R. Deciphering determinants in ribosomal methyltransferases that confer antimicrobial resistance. Journal of the American Chemical Society. 2019;**141**:1425-1429

[6] Wang H, Chen T, Chen D, Zou X, Li M, Huang F, et al. Sulfurized oolitic hematite as a heterogeneous Fenton-like catalyst for tetracycline antibiotic degradation. Applied Catalysis B: Environmental. 2020;**260**:118203

[7] Das R, Vecitis CD, Schulze A, Cao B, Ismail AF, Lu X, et al. Recent advances in nanomaterials for water protection and monitoring. Chemical Society Reviews. 2017;**46**:6946-7020

[8] Garcia-Segura S, Brillas E. Applied photoelectrocatalysis on the degradation of organic pollutants in wastewaters. Journal of Photochemistry and Photobiology C: Photochemistry. 2017;**31**:1-35

[9] Suhaimi NAA, Shahri NNM, Samat JH, Kusrini E, Lim JW, Hobley J, et al. Domination of methylene blue over rhodamine B during simultaneous photocatalytic degradation by TiO_2 nanoparticles in an aqueous binary solution under UV irradiation. Reaction Kinetics, Mechanisms and Catalysis. 2022;**135**:511-527

[10] Liu J, Li J, Li Y, Guo J, Xu SM, Zhang R, et al. Photoelectrochemical water splitting coupled with degradation of organic pollutants enhanced by surface and interface engineering of $BiVO_4$ photoanode. Applied Catalysis B: Environmental. 2021;**278**:119268

[11] Liu J, Li J, Shao M, Wei M. Directed synthesis of SnO_2@ $BiVO_4$/Co-Pi photoanode for highly efficient photoelectrochemical water splitting and urea oxidation. Journal of Materials Chemistry A. 2019;**7**:6327-6336

[12] Davies KR, Allan MG, Nagarajan S, Townsend R, Dunlop T, McGettrick JD, et al. Solar light-driven simultaneous pharmaceutical pollutant degradation and green hydrogen production using a mesoporous nanoscale WO_3/$BiVO_4$ heterostructure photoanode. Journal of Environmental Chemical Engineering. 2023;**11**:110256

[13] Xia L, Li J, Bai J, Li L, Zeng Q, Xu Q, et al. Preparation of a $BiVO_4$ nanoporous

photoanode based on peroxovanadate reduction and conversion for efficient photoelectrochemical performance. Nanoscale. 2018;**10**:2848-2855

[14] Peleyeju MG et al. Photoelectrocatalytic water treatment systems: Degradation, kinetics and intermediate products studies of sulfamethoxazole on a TiO_2–Exfoliated graphite electrode. RSC Advances. 2017;**7**:40571-40580

[15] Liu Y, Yao Q, Wu X, Chen T, Ma Y, Onga CN, et al. Gold nanocluster sensitized TiO_2 nanotube arrays for visible-light driven photoelectrocatalytic removal of antibiotic tetracycline. Nanoscale. 2016;**8**:10145-10151

[16] Li J, Lv S, Liu Y, Bai J, Zhou B, Hu X. Photoeletrocatalytic activity of an n-ZnO/p-Cu_2O/n-TNA ternary heterojunction electrode for tetracycline degradation. Journal of Hazardous Materials. 2013;**262**:482-488

[17] Umukoro EH, Peleyeju MG, Ngila JC, Arotiba OA. Towards wastewater treatment: Photo-assisted electrochemical degradation of 2-nitrophenol and orange II dye at a tungsten trioxide-exfoliated graphite composite electrode. Chemical Engineering Journal. 2017;**317**:290-301

[18] Chemelewski WD, Mabayoje O, Mullins CB. SILAR growth of Ag_3VO_4 and characterization for photoelectrochemical water oxidation. Journal of Physical Chemistry C. 2015;**119**:26803-26808

[19] Sinclair TS, Gray HB, Müller AM. Photoelectrochemical performance of $BiVO_4$ photoanodes integrated with [NiFe]-Layered double hydroxide nanocatalysts. European Journal of Inorganic Chemistry. 2018;**2018**:1060-1067

[20] Dong L et al. Sunlight responsive $BiVO_4$ photocatalyst: Effects of pH on L-cysteine-assisted hydrothermal treatment and enhanced degradation of ofloxacin. Catalysis Communications. 2011;**16**:250-254

[21] Li J, Zhou J, Hao H, Li W. Controlled synthesis of Fe_2O_3 modified Ag-010$BiVO_4$ heterostructures with enhanced photoelectrochemical activity toward the dye degradation. Applied Surface Science. 2017;**399**:1-9

[22] Ma Q et al. Construction of CuS/TiO_2 nano-tube arrays photoelectrode and its enhanced visible light photoelectrocatalytic decomposition and mechanism of penicillin G. Electrochimica Acta. 2018;**283**:1154-1162

[23] Bi Y, Ouyang S, Umezawa N, Cao J, Ye J. Facet effect of single-crystalline Ag_3PO_4 sub-microcrystals on photocatalytic properties. Journal of the American Chemical Society. 2011;**133**:6490-6492

[24] Wang Z, Huang X, Wang X. Recent progresses in the design of $BiVO_4$-based photocatalysts for efficient solar water splitting. Catalysis Today. 2019;**335**:31-38

[25] Sivula K, Le Formal F, Grätzel M. Solar water splitting: Progress using hematite (α-Fe_2O_3) photoelectrodes. ChemSusChem. 2011;**4**:432-449

[26] Park Y, McDonald KJ, Choi K-S. Progress in bismuth vanadate photoanodes for use in solar water oxidation. Chemical Society Reviews. 2013;**42**:2321-2337

[27] Qiu Y, Liu W, Chen W, Zhou G, Hsu PC, Zhang R, et al. Efficient solar-driven water splitting by nanocone $BiVO_4$-perovskite tandem cells. Science Advances. 2016;**2**:e1501764

[28] Guo W, Chemelewski WD, Mabayoje O, Xiao P, Zhang Y, Mullins CB. Synthesis and characterization of CuV_2O_6 and $Cu_2V_2O_7$: Two photoanode candidates for photoelectrochemical water oxidation. Journal of Physical Chemistry C. 2015;**119**:27220-27227

[29] Cardenas-Morcoso D, Peiro-Franch A, Herraiz-Cardona I, Gimenez S. Chromium doped copper vanadate photoanodes for water splitting. Catalysis Today. 2017;**290**:65-72

[30] Jiang H, Xu Y-T, Wang T, Zhu P-L, Yu S, Yu Y, et al. Nickel hexacyanoferrate flower-like nanosheets coated three dimensional porous nickel films as binder-free electrodes for neutral electrolyte supercapacitors. Electrochimica Acta. 2015;**166**:157-162

[31] Hegner FS, Cardenas-Morcoso DC-M, Gimenez S, Lopez N, Galan-Mascaros JR. Level alignment as descriptor for semiconductor/catalyst systems in water splitting: The case of hematite/cobalt hexacyanoferrate photoanodes. ChemSusChem. 2017;**10**:4552-4560

[32] Abdi FF, Han L, Smets AHM, Zeman M, Dam B, van de Krol R. Efficient solar water splitting by enhanced charge separation in a bismuth vanadate-silicon tandem photoelectrode. Nature Communications. 2013;**4**:2195

[33] Kay A, Cesar I, Gratzel M. Benchmark for water photooxidation by nanostructured α-Fe_2O_3 films. Journal of the American Chemical Society. 2006;**128**:15714-15721

[34] Zandi O, Klahr BM, Hamann TW. Highly photoactive Ti-doped α-Fe_2O_3 thin film electrodes: Resurrection of the dead layer. Energy & Environmental Science. 2013;**6**:634-642

[35] Osterloh FE. Inorganic nanostructures for photoelectrochemical and photocatalytic water splitting. Chemical Society Reviews. 2013;**42**:2294-2320

[36] Steier L, Herraiz-Cardona I, Gimenez S, Fabregat-Santiago F, Bisquert J, Tilley SD, et al. Understanding the role of underlayers and overlayers in thin film hematite photoanodes. Advanced Functional Materials. 2014;**24**:7681-7688

[37] Ma Y, Kafizas A, Pendlebury SR, Le Formal F, Durrant JR. Photoinduced absorption spectroscopy of CoPi on $BiVO_4$: The function of CoPi during water oxidation. Advanced Functional Materials. 2016;**26**:4951-4960

[38] Jiang C, Wang R, Parkinson BA. Combinatorial approach to improve photoelectrodes based on $BiVO_4$. ACS Combinatorial Science. 2013;**15**:639-645

[39] Pilli SK, Furtak TE, Brown LD, Deutsch TG, Turner JA, Herring AM. Cobalt-phosphate (Co-Pi) catalyst modified Mo-doped $BiVO_4$ photoelectrodes for solar water oxidation. Energy & Environmental Science. 2011;**4**:5028-5034

[40] Ye H, Park HS, Bard AJ. Screening of electrocatalysts for photoelectrochemical water oxidation on W-doped $BiVO_4$ photocatalysts by scanning electrochemical microscopy. Journal of Physical Chemistry C. 2011;**115**:12464-12470

[41] Seabold JA, Choi KS. Efficient and stable photo-oxidation of water by a bismuth vanadate photoanode coupled with an iron oxyhydroxide oxygen evolution catalyst. Journal of the American Chemical Society. 2012;**134**:2186-2192

[42] Peng Y, Du M, Zou X, Jia G, Santoso SP, Peng X, et al. Suppressing photoinduced charge recombination at the $BiVO_4$|| NiOOH junction by sandwiching an oxygen vacancy layer for efficient photoelectrochemical water oxidation. Journal of Colloid and Interface Science. 2022;**608**:1116-1125

[43] Liu J, Li J, Li Y, Guo J, Xu SM, Zhang R, et al. Photoelectrochemical water splitting coupled with degradation of organic pollutants enhanced by surface and interface engineering of $BiVO_4$ photoanod. Applied Catalysis B: Environmental. 2020;**278**:119268

[44] Zango ZU, Lawal MA, Usman F, Sulieman A, Akhdar H, Eisa MH, et al. Promoting the suitability of graphitic bio-carbon nitride and metal oxide nanoparticles: A review of sulfonamides photocatalytic degradation. Chemosphere. 2024;**351**:141218

[45] SureshR,GnanasekaranL,RajendranS, SotoMoscoso M, Chen WH, Show PL, et al. Application of nanocomposites in integrated photocatalytic techniques for water pollution remediation. Environmental Technology and Innovation. 2023;**31**:103149

[46] Shaddad MN, Ghanem MA, Al-Mayouf AM, Gimenez S, Bisquert J, Herraiz-Cardona I. Cooperative catalytic effect of ZrO_2 and α-Fe_2O_3 nanoparticles on $BiVO_4$ photoanodes for enhanced photoelectrochemical water splitting. ChemSusChem. 2016;**9**:2779-2783

[47] Peleyeju MG, Arotiba OA. Recent trend in visible-light photoelectrocatalytic systems for degradation of organic contaminants in water/ wastewater. Environmental Science: Water Research & Technology. 2018;**4**:1389e1411

[48] Carraway ER, Hoffman AJ, Hoffmann MR. Photocatalytic oxidation of organic acids on quantum-sized semiconductor colloids. Environmental Science & Technology. 1994;**28**:786-793. DOI: 10.1021/es00054a007

[49] Hoffmann MR, Martin S, Choi W, Bahnemann DW. Environmental applications of semiconductor photocatalysis. Chemical Reviews. 1995;**95**:69-96

[50] Esquivel K, Arriaga LG, Rodríguez FJ, Martínez L, Godínez LA. Development of a TiO_2 modified optical fiber electrode and its incorporation into a photoelectrochemical reactor for wastewater treatment. Water Research. 2009;**43**:3593-3603

[51] Selcuk H, Sene JJ, Anderson MA. Photoelectrocatalytic humic acid degradation kinetics and effect of pH, applied potential and inorganic ions. Journal of Chemical Technology and Biotechnology. 2003;**78**:979-984

[52] Wang N, Li X, Wang Y, Quan X, Chen G. Evaluation of bias potential enhanced photocatalytic degradation of 4-chlorophenol with TiO_2 nanotube fabricated by anodic oxidation method. Chemical Engineering Journal. 2009;**146**:30-35

[53] Gogate PR, Pandit AB. A review of imperative technologies for wastewater treatment I: Oxidation technologies at ambient conditions. Advances in Environmental Research. 2004;**8**(3-4):501-551

[54] Ikehata K, Naghashkar NJ, El-Din MG. Degradation of aqueous pharmaceuticals by ozonation and advanced oxidation processes: A review. Ozone Science and Engineering. 2006;**28**:353-414

[55] Pagliaro M, Konstandopoulos AG, Ciriminna R, Palmisano G. Solar

hydrogen: Fuel of the near future. Energy & Environmental Science. 2010;**3**:279-287

[56] Siripala W, Ivanovskaya A, Jaramillo TF, Baeck S, McFarland EW. A Cu_2O/TiO_2 heterojunction thin film cathode for photoelectrocatalysis. Solar Energy Materials and Solar Cells. 2003;**77**:229-237

[57] Shi X, Choi Y, Zhang K, Kwon J, Kim DY, Lee JK, et al. Homochiral D 4-symmetric metal–organic cages from stereogenic Ru (II) metalloligands for effective enantioseparation of atropisomeric molecules. Nature Communications. 2014;**5**(5):4775

[58] Grigioni I, Stamplecoskie KG, Selli E, Kamat PV. Dynamics of photogenerated charge carriers in $WO_3/BiVO_4$ heterojunction photoanodes. Journal of Physical Chemistry C. 2015;**119**:20792-20800

[59] Jung H, Chae SY, Shin C, Min BK, Joo O-S, Hwang YJ. Effect of the $Si/TiO_2/BiVO_4$ heterojunction on the onset potential of photocurrents for solar water oxidation. ACS Applied Materials and Interfaces. 2015;7:5788-5796

[60] Chatchai P, Murakami Y, Kishioka S-y, Nosaka AY, Nosaka Y. $FTO/SnO_2/BiVO_4$ composite photoelectrode for water oxidation under visible light irradiation. Electrochemical and Solid-State Letters. 2008;**11**:H160-H163

[61] Sivula K, Formal FL, Grtzel M. WO_3–Fe_2O_3 photoanodes for water splitting: A host Scaffold, guest absorber approach. Chemistry of Materials. 2009;**21**:2862-2867

[62] Cai P, Zhou S-M, Ma D-K, Liu S-N, Chen W, Huang S-M. Fe_2O_3-modified porous $BiVO_4$ nanoplates with enhanced photocatalytic activity. Nano-Micro Letters. 2015;7:183-193

[63] Thorne JE, Li S, Du C, Qin G, Wang D. Energetics at the surface of photoelectrodes and its influence on the photoelectrochemical properties. Journal of Physical Chemistry Letters. 2015;**6**:4083-4088

[64] Shaddad MN, Arunachalam P, Labis J, Hezam M, Al-Mayouf AM. Fabrication of robust nanostructured (Zr) BiVO4/nickel hexacyanoferrate core/shell photoanodes for solar water splitting. Applied Catalysis B: Environmental. 2019;**244**:863-870

[65] Shaddad MN, Arunachalam P, Alothman AA, Beagan AM, Alshalwi MN, Al-Mayouf AM. Synergetic catalytic behavior of AgNi-OH-pi nanostructures on Zr: $BiVO_4$ photoanode for improved stability and photoelectrochemical water splitting performance. Journal of Catalysis. 2019;**371**:10-19

[66] Shaddad MN, Cardenas-Morcoso D, Arunachalam P, García-Tecedor M, Ghanem MA, Bisquert J, et al. Enhancing the optical absorption and interfacial properties of BiVO4 with Ag_3PO_4 nanoparticles for efficient water splitting. The Journal of Physical Chemistry C. 2018;**122**:11608-11615

[67] Shaddad MN, Arunachalam P, Hezam M, AL-Saeedan NM, Gimenez S, Bisquert J, et al. Unprecedented solar water splitting of dendritic nanostructured Bi_2O_3 films by combined oxygen vacancy formation and Na_2MoO_4 doping. International Journal of Hydrogen Energy. 2021;**46**:23702-23714

[68] Kanan MW, Nocera DG. In situ formation of an oxygen-evolving catalyst in neutral water containing phosphate and Co^{2+}. Science. 2008;**321**:1072

[69] McAlpin JG, Surendranath Y, Dinca M, Stich TA, Stoian SA, Casey WH, et al. EPR evidence for Co (IV) species

produced during water oxidation at neutral pH. Journal of the American Chemical Society. 2010;**132**:6882

[70] Wang J, Osterloh FE. Limiting factors for photochemical charge separation in $BiVO_4/Co_3O_4$, a highly active photocatalyst for water oxidation in sunlight. Journal of Materials Chemistry A. 2014;**2**:9405-9411

[71] Seabold JA, Choi KS. Effect of a cobalt-based oxygen evolution catalyst on the stability and the selectivity of photo-oxidation reactions of a WO_3 photoanode. Chemistry of Materials. 2011;**23**:1105

[72] Zhong K, Sun J, Inumaru H, Gamelin DR. Solar water oxidation by composite catalyst/α-Fe_2O_3 photoanodes. Journal of the American Chemical Society. 2009;**131**:6086

[73] Zhong DK, Cornuz M, Sivula K, Gratzel M, Gamelin DR. Photo-assisted electrodeposition of cobalt–phosphate (Co–Pi) catalyst on hematite photoanodes for solar water oxidation. Energy & Environmental Science. 2011;**4**:1759

[74] Long MC, Cai WM, Kisch H. Visible light induced photoelectrochemical properties of n-$BiVO_4$ and n-$BiVO_4$/p-Co_3O_4. Journal of Physical Chemistry C. 2008;**112**:548

[75] Zhong DK, Choi S, Gamelin DR. Near-complete suppression of surface recombination in solar photoelectrolysis by "Co-Pi" catalyst-modified W:$BiVO_4$. Journal of the American Chemical Society. 2011;**133**:18370-18377

[76] Abdi F, van de Krol R. Nature and light dependence of bulk recombination in Co-Pi-catalyzed $BiVO_4$ photoanode. Journal of Physical Chemistry C. 2012;**116**:9398

[77] Barroso M, Mesa CA, Pendlebury SR, Cowan AJ, Hisatomi T, Sivula K, et al. Dynamics of photogenerated holes in surface modified α-Fe_2O_3 photoanodes for solar water splitting. Proceedings of the National Academy of Sciences. 2012;**109**:15640-15645

[78] KlahrB,GimenezS,Fabregat-SantiagoF, Bisquert J, Hamann TW. Electrochemical and photoelectrochemical investigation of water oxidation with hematite electrodes. Energy & Environmental Science. 2012;**5**:7626

[79] Zachäus C, Abdi F, Peter LMP, van de Krol R. Photocurrent of $BiVO_4$ is limited by surface recombination, not surface catalysis. Chemical Science. 2017;**8**:3712-3719

[80] Lichterman MF, Shaner MR, Handler SG, Brunschwig BS, Gray HB, Lewis NS, et al. Enhanced stability and activity for water oxidation in alkaline media with bismuth vanadate photoelectrodes modified with a cobalt oxide catalytic layer produced by atomic layer deposition. Journal of Physical Chemistry Letters. 2013;**4**:4188

[81] Hegner FS, Herraiz-Cardona I, Cardenas-Morcoso D, Lopez N, Galan-Mascaros JR, Gimenez S. Cobalt hexacyanoferrate on $BiVO_4$ photoanodes for robust water splitting. ACS Applied Materials & Interfaces. 2017;**9**:37671-37681

[82] Qiu Y, Pan Z, Chen H, Ye D, Guo L, Fan Z, et al. Current progress in developing metal oxide nanoarrays-based photoanodes for photoelectrochemical water splitting. Scientific Bulletin. 2019;**64**:1348-1380

[83] Ikeda S, Kawaguchi T, Higuchi Y, Kawasaki N, Harada T, Remeika M, et al. Effects of zirconium doping into a monoclinic scheelite $BiVO_4$ crystal

on its structural, photocatalytic, and photoelectrochemical properties. Frontiers in Chemistry. 2018;**6**:266

[84] Rong F, Zhao J, Yang Q, Li C. Nanostructured hybrid NiFeOOH/CNT electrocatalysts for oxygen evolution reaction with low overpotential. RSC Advances. 2016;**6**:74536-74544

[85] Fang G, Liu Z, Han C. Enhancing the PEC water splitting performance of $BiVO_4$ co-modifying with NiFeOOH and Co-Pi double layer cocatalysts. Applied Surface Science. 2020;**515**:146095

[86] Arunachalam P, Shaddad MN, Amer MS, Abdulhadi AQ. Enhanced photoelectrochemical water splitting coupled with pharmaceutical pollutants degradation on Zr: BiVO4 photoanodes by synergetic catalytic activity of NiFeOOH nanostructures. Alexandria Engineering Journal. 2024;**99**:64-75

[87] Wan Y, Han M, Yu L, Yi G, Jia J. 3D Bi 2 S 3 salix leaf-like nanosheet/TiO_2 nanorod branched heterostructure arrays for improving photoelectrochemical properties. CrystEngComm. 2016;**18**:1577-1584

[88] Liu C, Li J, Li Y, Li W, Yang Y, Chen Q. Epitaxial growth of Bi 2 S 3 nanowires on $BiVO_4$ nanostructures for enhancing photoelectrochemical performance. RSC Advances. 2015;**5**:71692-71698

[89] Zhou Z, Li Y, Lv K, Wu X, Li Q, Luo J. Fabrication of walnut-like $BiVO_4$@ Bi_2S_3 heterojunction for efficient visible photocatalytic reduction of Cr (VI). Materials Science in Semiconductor Processing. 2018;**75**:334-341

[90] Chen HQ, Lin LY, Chen SL. Direct growth of $BiVO_4/Bi_2S_3$ nanorod array on conductive glass as photocatalyst for enhancing the photoelectrochemical performance. ACS Applied Energy Materials. 2018;**1**:6089-6100

[91] Arunachalam P, Shaddad MN, Amer MS, Alsalman AM, Madhavan J. Cooperative catalytic behavior of CoS and Bi 2 S 3 nanoparticles on Zr: $BiVO_4$ photoanodes for enhanced photoelectrochemical sulfite oxidation coupled with pharmaceutical pollution degradation. Environmental Science: Nano. 2024;**11**:2668-2682

Chapter 8

Nanomaterials Based Sensors: From Preparation to Versatile Applications

Wessam S. Omara, Maram A. AlGhamdi and Noha Taymour

Abstract

Nowadays, nanomaterials are considered a magic tool in many industrial applications. They offer smart solutions for problems related to health, the environment, and energy. Because of their tiny size and precise characters, nanomaterials represent a unique and vital component in microelectromechanical systems (MEMS) and nanoelectromechanical systems (NEMS)-based technologies. One of the most important applications of these technologies is sensors. Sensors are divided into two main constituents, the first is the device and the second is the sensing materials. This chapter explains the main sensor components. It highlights the different classes of sensing materials. The main categories of sensing techniques are summarized. It also explains the relation between sensing materials and proposed sensing techniques, with a focus on materials properties at the nanoscale and their selective response to certain stimuli. Finally, it concludes with the challenges and prospective of the sensor industry and market.

Keywords: nanomaterials, sensing techniques, nanosensors, selectivity, stimuli, response time

1. Introduction

In contemporary research, nanotechnology has garnered significant attention, giving a promising future for both interdisciplinary and applied sciences. The unique properties of nanoscale materials, characterized by their diminutive size and precise attributes, enable their application across a broad spectrum of fields. Nanoparticles can be synthesized from a variety of chemical compositions, including both synthetic and naturally derived materials. The most prevalent types include metals, metal oxides, silicates, non-oxide ceramics, polymers, organic compounds, carbon-based materials, and biomolecules. These nanoparticles can adopt various morphologies, such as spheres, fibers, wires, platelets, and tubes. The optical characteristics of nanoparticles are particularly noteworthy, as their absorption and emission wavelengths can be finely tuned through adjustments in particle size and surface modifications. When the dimensions of nanoparticles fall below the critical wavelength of light, they can achieve transparency. Furthermore, nanocomposites that combine

IntechOpen

organic polymers with inorganic nanoparticles in the nanoscale range represent a novel class of materials, exhibiting unique properties that differentiate them from their conventional counterparts [1]. For metals, with decreasing size of nanoparticles, the sintering and melting temperatures also decrease. Nanoparticles may be incorporated into a solid matrix to provide better thermal conduction.

Among noble-metal nanomaterials, gold nanoparticles (AuNPs) have received considerable attention, whereas; it is biologically safe and exhibit no apparent toxicity [2]. Gold nanoparticles have been found to be useful in different applications, and some potential applications are still being reviewed [3]. Some of these applications are directly related to their optical properties, where the sensitivity of the plasmon resonance frequency toward the refractive index of the surroundings make them suitable candidates for sensing applications [4]. So, the addition of Au NP was supposed to enhance both the optical and electrical properties of PANI; hence the sensitivity of the sensor should be enhanced accordingly [5, 6].

One of the main challenges that may be faced in the proper treatment of infectious diseases is the perfect diagnosis, it is a critical point in treatment when physicians have many different epidemics with similar symptoms [7]. Since the development of biosensors based on high selectivity and sensitivity become necessary to detect the antigen, suggest and recommend a suitable treatment and apply the therapy. This selectivity depends on the presence of an enzyme or antibody corresponding to this substrate (antigen). Enhancing the responsiveness and overall performance of enzyme electrodes remains a central objective in the field of biosensor research [8].

A biosensor is an analytical instrument that integrates a biological sensing component with a transducer to convert the concentration of a target analyte into a measurable signal, such as an optical, micromechanical, or electrical output. These devices offer portability, user-friendliness, and high specificity, making them compatible with advanced data processing technologies. As a result, biosensors hold immense potential for applications across diverse domains, including pharmaceuticals, healthcare, environmental pollution monitoring, food processing, and agricultural product analysis. The development of biosensors has the capacity to revolutionize these industries by providing rapid, accurate, and cost-effective analytical tools for a wide range of applications [9, 10].

Based on the prepared sensing materials, there are many techniques that vary from optical, electrochemical, mechanical, etc. [11]. Optical biosensors measure changes in light absorbed or emitted as a result of a biochemical reaction [12]. Optical-based detection provides a large number of subclasses based on absorption, reflection, refraction, dispersion, infrared, Raman, chemiluminescence, fluorescence, and phosphorescence [13]. Depending on the prepared sensing materials, there are many techniques that vary from optical, electrochemical, mechanical, ... etc. [14]. Optical biosensors measure changes in light absorbed or emitted as a result of a biochemical reaction. Surface Enhanced Raman Spectroscopy (SERS) is considered one of the most powerful spectroscopic tools as an optical biosensor, which is a surface-sensitive technique that enhances Raman scattering where a rough metal surface or by nanostructures such as plasmonic-magnetic or silica nanotubes acts as adsorbed surface to entrap target molecules. The enhancement factor can be magnified as much as 1010 to 1011, in which this technique may detect single molecules [10]. The electrochemical sensing technique receives great attention does you to its ability to record tiny changes in conductivity related current, voltage, or resistance of a conducting material [15]. Conducting polymers are recognized for their biocompatibility with biological molecules in neutral aqueous environments. These polymers can undergo reversible electrochemical doping and undoping, which results in notable alterations in their conductivity and spectroscopic characteristics. Such changes can serve as indicators for the detection of biochemical reactions, making these materials valuable in various analytical applications [16]. The field of cantilever-like sensors is still in its formative stages, having originated in the mid-1990s. The cantilever surface bending resulting from molecular adsorption on the surface of the microcantilever represents the adsorption-induced surface stress variation [17]. Moreover, the cantilever frequency is also changed according to mass loading. Thanks to the invention of the atomic force microscope AFM in 1986, the pending and deflection of the microcantilever can be detected easily. Microcantilever surface functionalization plays a vital role in sensor selectivity which is achieved by the immobilization of bioreceptors on the modified surface [18]. Recently, conducting polymers were used as coating material to enhance the adsorption and immobilization of target molecules [19]. The AFM functions are not only as a miniaturized phonograph where images are obtained by raster scanning the surface with an AFM probe in microcantilever signal read out, it is also the most useful technique to monitor the value added by using nucleic acid aptamers or bioreceptors to immobilize target biomolecules on different polymer surface. In addition, it helps biologists to study cell surface topography and offers detailed information about changes that occurred in infected cells [14, 20].

2. Classification of sensing nanomaterials

There are four main classes of materials based on their different structure and properties; Metallic, ceramic, polymeric, and composite materials. The classification of materials is usually based mainly on their structure properties. Each class has one or more subclasses depending on the variation of its properties under certain conditions or processing [21].

Composites consist of at least two distinct types of materials. Another category of materials is advanced materials, which are utilized in high-tech applications, including:

- *Semiconductors*: Semiconductors are materials—either inorganic or organic—that can regulate their electrical conductivity based on factors such as chemical structure, temperature, light exposure, and the presence of dopants. The term "semiconductor" reflects their conductivity level, which falls between that of metals like copper and gold, and insulators such as glass.

- *Biomaterials*: A biomaterial is defined as any substance engineered to interact with biological systems for medical purposes. These materials must be nontoxic and compatible with bodily tissues, ensuring they do not elicit adverse biological reactions.

- *Smart materials*: Smart materials are designed to detect and respond to environmental changes in predetermined ways.

- *Nanomaterials*: Nanomaterials possess structural characteristics on the nanometer scale, with some engineered at the atomic or molecular level [22].

Sensing nanomaterials can be categorized into more than eight classes depending on their structure, properties, and response in the presence of analyte. **Figure 1**

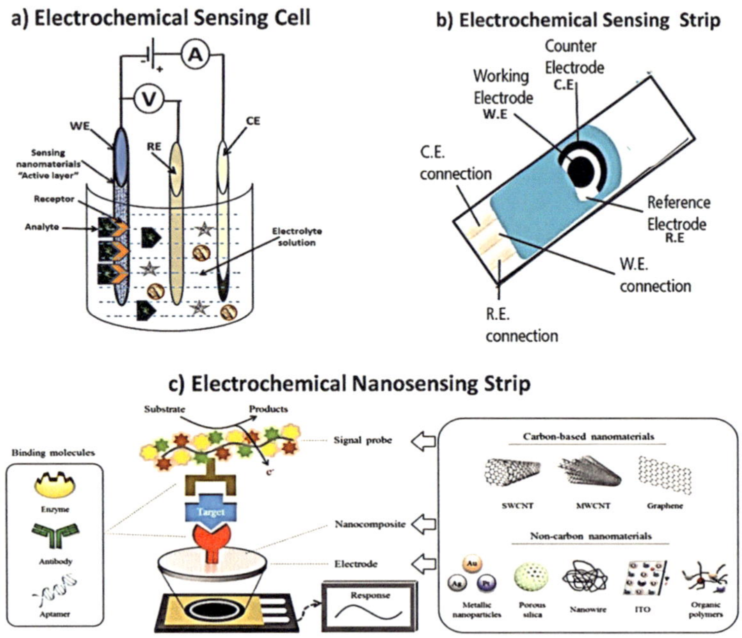

Figure 1.
The electrochemical sensing component (a) electrochemical sensing cell, (b) electrochemical sensing strip, and (c) electrochemical nanosensor strip.

presents the electrochemical sensing component and setup. These classes include; metallic, magnetic, ceramic, polymeric, carbon, quantum dots, hydrogel, and in/organic nanocomposite [23], **Figure 2** presents different types of sensing nanomaterials. The chemical, optical, and electrical properties of sensing materials control their

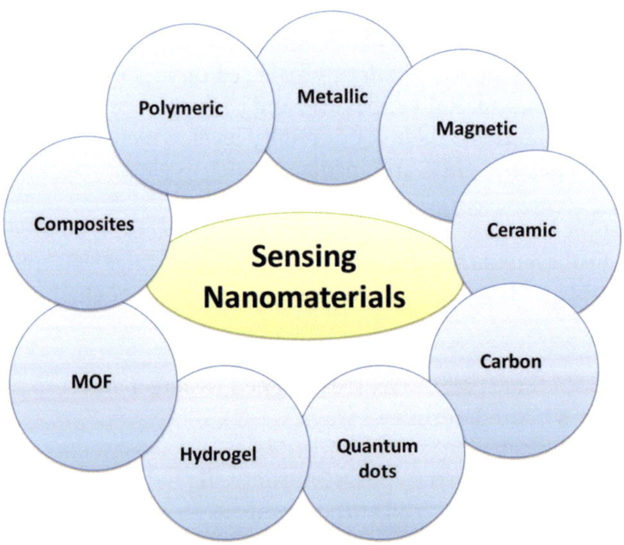

Figure 2.
The different classes of sensing nanomaterials.

suitability for specific sensing applications. Also, sensing media or the surrounding environments determine the processing of sensing materials that are integrated into the sensing device [24].

3. Preparation of nanomaterials

There are two main approaches for preparing nanomaterials.

The first one is the Top-Down Approach, and the second is the bottom-up approach [25]. **Figure 3** summarizes the main synthesis methods included in each approach.

3.1 Top-down approach

In the top-down approach, suitable raw materials are reduced to nanoscale particles through lithographic techniques like milling, sputtering, and grinding [25, 26].

3.2 Bottom-up approach

The bottom-up approach employs a range of chemical and biological methods to synthesize nanoparticles through the self-assembly of atoms. This process involves the formation of new nuclei that subsequently grow into particles at the nanoscale [25].

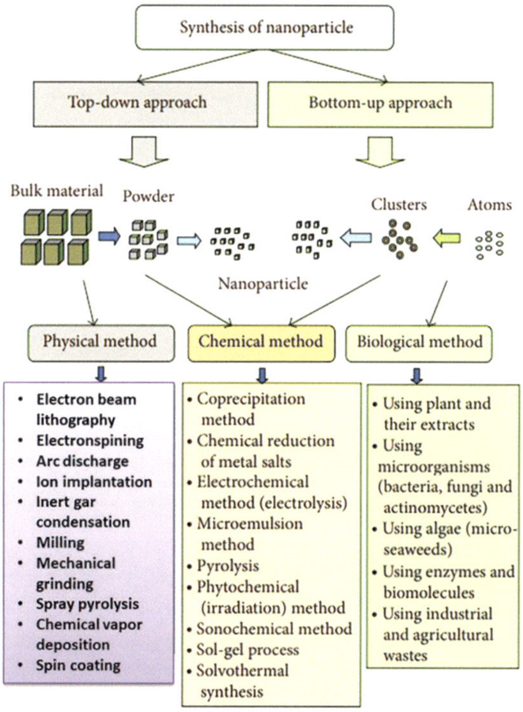

Figure 3.
Different approaches to preparing nanomaterials.

4. Sensing nanomaterials

As mentioned earlier, sensing nanomaterials are divided into many classes based on their structure and properties on the nanoscale.

4.1 Metallic nanoparticles

Most metals have completely new properties in nanoscale [27]. Metal nanoparticles, such as gold (Au) and silver (Ag), have been widely utilized in the design of nanosensors for sensing applications due to their optical and electrochemical properties that are changeable and measurable as a response in the presence of stimuli [28]. Their optical properties could be enhanced or inhibited in the presence of different concentrations of target molecules [29]. At the same time, the conductivity and other electrical properties of these metallic nanoparticles could be increased or decreased as a sign of the presence of a certain concentration of a specific analyte [29]. Most sensing devices possess a gold layer as an inert surface for further treatment and functionalization with chemically functionalized compounds [30]. Gold has various shapes and different properties on the nanoscale; such as nanosphere, nanorod, nanostar, etc. [27], **Figure 4** presents the different shapes and sizes of Au nanoparticles. For example, gold nanoparticles AuNps could be used in optical or electrochemical detection of food poisoning, water and contamination, environmental pollution, and early detection of diseases in the medical field [31], As mentioned before, the sensing materials and techniques usually depend on the type of target molecule and the sensing media one of the interested nanosensing applications in the biomedical field is periodontitis detection.

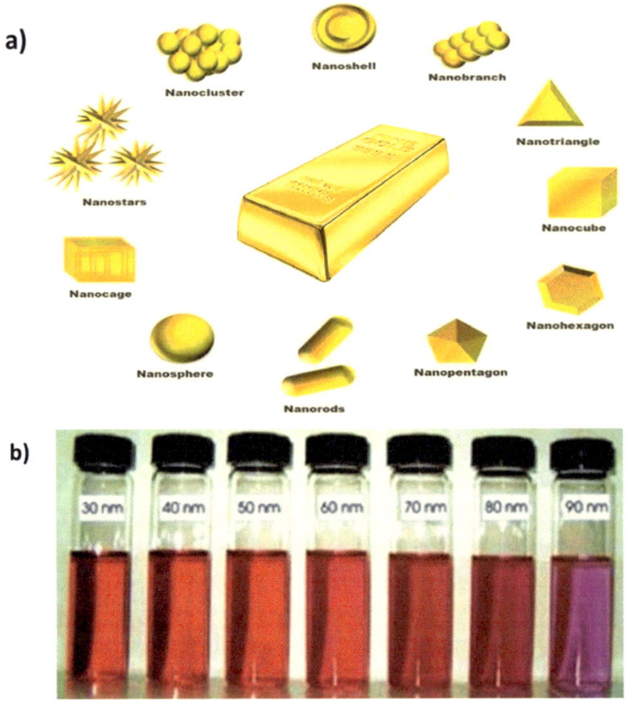

Figure 4.
(a) Different shapes of Au nanoparticles, and (b) colors of different sizes of Au nanoparticles.

Researchers have used Au nanoparticles with specific antibodies or aptamers to detect periodontal disease biomarkers. Silver nanoplate-based multicolorimetric sensors are also employed to detect alkaline phosphatase (ALP) and interleukin-1beta (IL-1β) in saliva [32].

4.2 Magnetic nanoparticles

Magnetic nanoparticles (MNPs) have drawn a lot of interest in the development and fabrication of sensors and biosensors for a variety of uses, such as biology, food, medicine, and the environment. MNPs might be integrated into the transducers or distributed throughout the specimen, where they would be drawn onto the active sensing surface of the sensors or biosensors by external magnetic fields [33]. Flexible magnetic field nanosensors, which provide a number of benefits like comfort, real-time health monitoring, durability, motion sensing, and seamless interaction with other sensors, hold great promise for wearable electronics. They are anticipated to transform wearable technology, spur innovation across a range of industries, and improve the user experience as a whole [34]. Iron oxide (Fe_3O_4) nanoparticles are the most famous nanosensor used in different sensing applications, where researchers have developed magnetic nanoparticle-based biosensors that can detect periodontal pathogens, such as Porphyromonas gingivalis and Aggregatibacter actinomycetemcomitans, in clinical samples [35, 36].

4.3 Carbon-based nanomaterials

Carbon-based nanomaterials, including carbon nanotubes (CNTs) and graphene nanosheets, have also been explored for their sensing applications because of their optical electrochemical properties. Due to their meritorious qualities, which include large specific surface area, simple functionalization, biocompatibility, long-term chemical stability, high thermal and electrical conductivity, electrocatalytic activity, and optical qualities [37]. **Figure 5** presents different forms of carbon-based

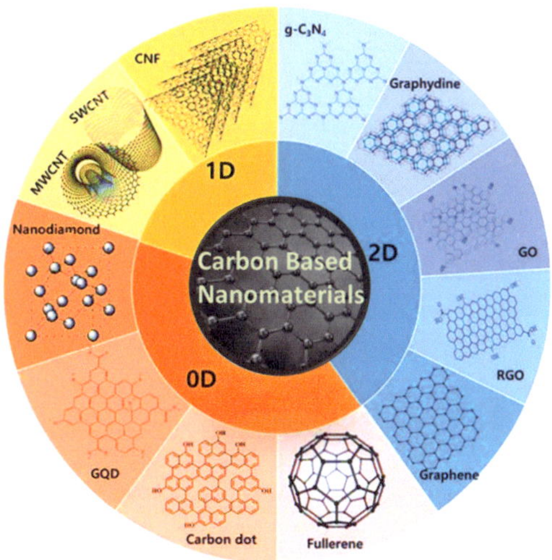

Figure 5.
Different carbon-based sensing nanomaterials.

materials. These materials play a vital role in the entrapment of target molecules in the case of environmental sensors for the detection of pesticides, heavy metals, organic dyes phenols, and other chemical contaminants [38]. Carbon nanomaterials in quantum dots scale have good optical properties that support their applications in optical sensors and bioimaging [39]. Researchers have reported the use of CNT-based sensors for the detection of salivary biomarkers associated with periodontitis, such as alkaline phosphatase and lactate dehydrogenase [40].

4.4 Quantum dots

Quantum dots (QDs) are semiconductor nanocrystals whose size is less than 10 nm and they possess unique optical properties, such as size-tunable emission spectra and high photostability. These characteristics have led to the development of QD-based fluorescent sensors that have been used widely in food pathogens, pesticides, infection, cancer, and pollution detection [41]. For this application and others in energy, electronics, and medicine, in 2023 the Nobel Prize in chemistry was awarded to Moungi Bawendi, Louis Brus, and Alexei Ekimov for their work on tiny quantum dots. In dentistry, QDs play a vital role in the detection of salivary proteins, enzymes, and nucleic acids associated with periodontitis, offering improved sensitivity and multiplexing capabilities. Fu et al. [39] reviewed various nanomaterials such as metal/metal oxide nanoparticles, carbon nanotubes, and quantum dots for detecting oral diseases, including periodontitis, **Figure 6** presents different types of quantum dots nanomaterials and their electronic structure. Quantum dots also provide a reliable tool for molecular imaging techniques [39].

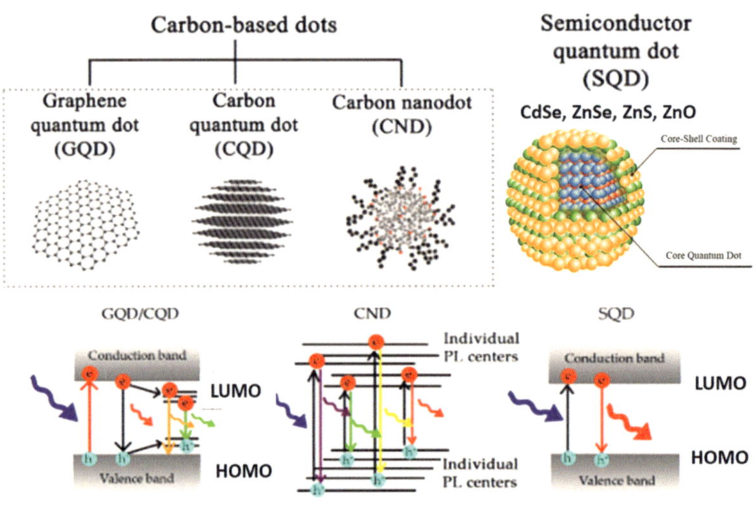

Figure 6.
Different types of quantum dots nanomaterials.

4.5 Polymer-based nanosensors

Polymers are a class of materials that consist basically of carbon and hydrogen forming a monomer (single basic unit of polymer). Chemical oxidation or electrochemical polymerization of these monomers forming polymers [42]. Conducting polymers represent a significant subclass of polymers that are recognized as essential functional organic materials for the development of next-generation electronic and optical devices [43]. Nanostructured conducting polymers featuring high surface area, small dimensions, and unique physical properties have been widely used to build various sensor devices [43, 44]. Conducting polymers has been of great scientific and technological importance since their discovery in 1970, because of their electrical, mechanical, and optical properties. In recent decades, conducting polymers have gained prominence as compelling materials for the development of electrochemical sensors. One of the primary advantages of biosensors based on conducting polymers, compared to other existing methods, is their ability to demonstrate enhanced response characteristics and heightened sensitivity to minor changes in their environment. This unique responsiveness makes conducting polymers particularly valuable in the design of advanced sensing applications [19]. Polyaniline, poly(3,4-ethylenedioxythiophene (PEDOT), and polyacetylene are among conducting polymers that have been the subject of a great number of studies due to their unique properties [45, 46]. Due to their unique changeable optical and electrochemical properties, conducting polymers such as polyaniline and its nanocomposite have revalorized the application of conjugating polymer in the field of radiation sensors and gamma detection [47, 48].

4.5.1 Hydrogels based nanosensors

Hydrogels are a subclass of polymeric materials, which are characterized by their polymeric network structure due to their ability to gelatinize when present in aqueous solution. This forms the chemical or physical crosslinking of individual polymer chains [42]. Hydrogels serve as highly effective fixed substrates for fluorescent sensors due to their exceptional light transmittance, customizable dimensions, and stable three-dimensional (3D) network architecture. The significant tunability of hydrogels allows for flexible packaging and modification of fluorescence sensors, catering to the need for portable and user-friendly detection systems. Furthermore, the 3D porous structure of hydrogels facilitates the diffusion of target analytes while providing a liquid sensing environment that enhances the sensitivity of fluorescence signals. This combination improves the accuracy and reliability of detection outcomes. These advantageous properties position hydrogel-based nanosensors as promising candidates for applications in food safety and environmental monitoring [49]. Researchers implanted five different hydrogel-encapsulated fluorescent nanosensors in experimental animals to monitor the inflammatory responses to determine the favorable design rules for hydrogel encapsulation that minimize adverse responses. Depending on this application, nanosensors were employed as injectable liquids or encapsulated in a hydrogel matrix as solid implants to monitor the inflammation [50, 51]. They found that hydrogels with higher crosslinking density allowed faster resolution of acute inflammation than lower ones [51].

4.6 Nanozymes

Nanozymes are nanomaterials that exhibit enzyme-like catalytic activities, which can be exploited for the development of colorimetric or electrochemical biosensors. Researchers have reported the use of nanozyme-based sensors for the detection of periodontitis-related biomarkers, such as hydrogen peroxide and myeloperoxidase, in clinical samples [52].

Liang and Qian [53] reviewed the application of nanozymes in detecting clinical biomarkers. The review highlights the integration of nanozymes with colorimetric, fluorescent, and electrochemical methods to create sensitive and precise biosensors. These advancements are particularly relevant for periodontitis detection, where early and accurate identification of biomarkers such as alkaline phosphatase (ALP) and interleukin-1beta (IL-1β) is crucial for effective treatment [53].

4.7 Ceramics-based nanosensors

Ceramics are categorized as inorganic and nonmetallic materials that play a crucial role in our everyday lives. A significant number of ceramics, including both oxides and non-oxides, are currently synthesized from polymer precursors. Typically, ceramics possess an amorphous or nanocrystalline structure, offering remarkable structural stability, resistance to oxidation, creep resistance, and favorable mechanical properties at elevated temperatures, along with good dielectric characteristics. However, they also exhibit a fundamental drawback: their susceptibility to fracture and the necessity of high-temperature processes for the fabrication of integrated substrates [42]. Advancements in modern technology have facilitated the creation of innovative ceramic materials that possess enhanced chemical, physical, and mechanical properties. These characteristics render them highly suitable for use in dental and orthopedic implants, where performance and biocompatibility are critical [54]. There is an electromechanical coupling in electroelastic materials. When they are subjected to an electric field, mechanical deformation occurs and produces electrical charges under mechanical stress. Due to these excellent properties, electroelastic materials are widely used in the mechanical and electrical engineering fields such as electromechanical transducers, ultrasonic generators, actuators, and sensors. Piezomagnetic, magneto-electro-elastic (MEE), electrostrictive and magnetostrictive materials are other kinds of smart materials, which have various practical applications in intelligent structures and many engineering applications. These days, due to the development of new technologies, new materials, and also in nanotechnology, piezoelectric and MEE nanomaterials and their nanostructures have captured significant attention in various fields such as fabricating electronic chipsets, microelectromechanical system (MEMS), nanoelectromechanical system (NEMS) and biosensors [55].

4.8 MXene-based nanosensors

MXene represents a novel class of two-dimensional materials characterized by their composition of transition metal carbides, carbonitrides, and nitrides, following the general formula Mn + 1XnTX. In this formula, M denotes an early transition metal, including elements such as scandium (Sc), titanium (Ti), vanadium (V), and chromium (Cr), while X signifies either carbon or nitrogen. The term TX refers to surface termination groups, which may include hydroxyl (OH), oxygen (O), or fluorine (F) functionalities [56]. Since the discovery of the first MXene in 2011, the scientific community had raised an idea among "the future of nanotechnology" [57]. The majority of MXenes and MXene-derived nanomaterials exhibit a combination of advantageous properties, including high volumetric capacitance, antibacterial activity, electrochromic behavior, exceptional electronic conductivity, and optical transparency [57]. MXenes' flexible surface chemistry and exceptional conductivity make them appealing for use in sensor applications. Many reviews offer opportunities for the field and place current developments in MXene-based sensors into perspective. It explains the appealing characteristics and operational principles of MXene-based sensors made from pure MXene or an MXene/polymer nanocomposite. The significance of altering MXenes' surface in order to increase their polymer affinity and create robust, self-healing sensors is outlined. The difficulties of several cutting-edge sensor fabrication techniques are explored. Emerging applications of MXene-based sensors including moisture, motion, gas, and humidity detection as well as pressure distribution mapping are critically examined. The food industry highlights the potential uses of MXene-based sensors for monitoring food ingredients [58, 59].

4.9 MOF-based sensing nanomaterials

Metal–organic frameworks (MOFs) have garnered significant attention from researchers due to their remarkably high specific surface areas, open and interconnected microporous structures, substantial content of transition metals, and the potential for post-synthesis customization and modification. Additionally, MOFs can exhibit semiconductor-like properties, further enhancing their appeal for various applications. As of now, over 20,000 distinct MOFs have been designed and synthesized, reflecting the extensive interest and ongoing advancements in this field [60]. In 2022, researchers successfully synthesized MOF-5/CoNi2S4, a hybrid nanostructure characterized by its low cost, safety, and bioactivity, using a one-pot high-gravity protocol. This innovative approach facilitated the efficient development of the nanostructure, highlighting its potential for various applications in fields such as catalysis and environmental remediation [61].

4.10 Upconversion nanoparticles for sensing applications

Upconversion is an optical phenomenon that entails the transformation of lower-energy photons into higher-energy photons. This process has been the subject of extensive research since the mid-1960s and has found numerous applications in optical devices. A key distinction between upconversion nanoparticles and other nanomaterials is their ability to emit visible light when exposed to near-infrared irradiation. This near-infrared excitation results in reduced autofluorescence, minimized scattering and absorption, and enhanced penetration depth in biological samples, making these nanoparticles particularly valuable for imaging and sensing applications in biological contexts [62]. The investigation of nanosensors capable of accurately diagnosing hypoxic conditions holds significant promise for assessing the malignancy of tumors and facilitating the implementation of more effective personalized treatment strategies. In this context, we present the design and synthesis of nanosensors that can selectively and reversibly measure hypoxia levels in both *in vitro* and *in vivo* settings. This advancement may enhance our understanding of tumor biology and improve therapeutic outcomes [63].

However, nanomaterial-based biosensors have shown promising results in the detection of oral cancer, but there are still several challenges that need to be addressed. These include the need for further optimization of sensor performance, validation of clinical utility, and the development of user-friendly, point-of-care devices for widespread adoption. Additionally, the long-term stability, biocompatibility, and scalability of these nanobiosensor systems need to be thoroughly investigated.

5. Integrated nanobiosensor platforms

To further enhance the diagnostic capabilities, researchers have developed integrated nanobiosensor platforms that combine multiple nanomaterials and detection techniques. These platforms leverage the synergistic effects of different nanomaterials to achieve improved sensitivity, selectivity, and multiplexing capabilities for the simultaneous detection of various caries-related biomarkers

- Nanoparticle-based targeting and detection of microcavities: Jones et al. [64] developed a novel bio-based nanoparticle system to detect active caries in vitro. These nanoparticles are made from cationic fluorescein-labeled food-grade starch, which fluoresces under a standard dental curing light. The nanoparticles selectively bind to active carious lesions, illuminating them and facilitating early diagnosis. This innovative approach demonstrated the potential for significant improvements in dental diagnostics [64]. Mady and Alzayyat [65] discussed the application of nanotechnology in the early detection of microcavities. They emphasized the importance of precise diagnosis prior to cavitation to facilitate targeted preventive interventions, including the application of fluorides and sealants [65].

- Nanomaterials-based electrochemical biosensors: Eissa et al. [66] provided an editorial on the integration of nanomaterials into electrochemical biosensors. These sensors demonstrate significant potential across various applications due to their low cost, ease of use, and miniaturization capabilities [66].

6. Challenges and future perspectives

While nanomaterial-based sensors have shown promising results in the detection of hazardous compounds diagnosis of diseases or infection-causative pathogens, and continuous monitoring of changes in certain environments, there are still several challenges that need to be addressed. These include the need for further optimization of sensor performance, validation of field utility, and the development of user-friendly, point-of-care devices for widespread adoption. Additionally, the long-term stability, biocompatibility, and scalability of these nanosensor systems need to be thoroughly investigated.

In the dental field, future research should focus on the development of more robust and reliable nanobiosensor platforms, the identification of novel caries-related biomarkers, and the integration of these sensors with advanced data analysis and machine learning algorithms to improve diagnostic accuracy and personalized treatment approaches. Collaborative efforts between researchers, clinicians, and industry partners will be crucial in translating these nanomaterial-based technologies from the laboratory to clinical practice, ultimately enhancing the early detection, management, and prevention of dental caries. Future research in this field should focus on the development of more robust and reliable nanobiosensor platforms, the identification of novel oral cancer biomarkers, and the integration of these sensors with advanced data analysis and machine learning algorithms to improve diagnostic accuracy and personalized treatment approaches. Collaborative efforts between researchers, clinicians, and industry partners will be crucial in translating these nanomaterial-based technologies from the laboratory to clinical practice, ultimately enhancing the early detection, management, and prevention of oral cancer.

DOI: http://dx.doi.org/10.5772/intechopen.1006836

7. Conclusion

In conclusion, nanobiosensors represent a transformative technology with extensive potential across diverse fields, particularly in medical and dental applications. From their meticulous preparation involving the functionalization of nanomaterials to their deployment in versatile sensing platforms, these devices have demonstrated remarkable capabilities in detecting a wide array of biomarkers with high specificity and sensitivity. The integration of nanotechnology with biosensing has enabled the development of innovative diagnostic tools that offer rapid, non-invasive, and precise detection of diseases. In dentistry, for instance, nanobiosensors have shown promise in early caries detection, monitoring periodontal health, and ensuring the success of dental implants. As research progresses, the refinement and optimization of nanobiosensor technologies will undoubtedly pave the way for their broader implementation, potentially revolutionizing diagnostic and therapeutic practices in healthcare.

Author details

Wessam S. Omara[1*], Maram A. AlGhamdi[2] and Noha Taymour[2]

1 Materials Science Department, Institute of Graduate Studies and Research, Alexandria University, Alexandria, Egypt

2 Department of Substitutive Dental Sciences, College of Dentistry, Imam Abdulrahman Bin Faisal University, Dammam, Saudi Arabia

*Address all correspondence to: wessam.omara@alexu.edu.eg

References

[1] Omara W, Elfeky SA. Effect of gold/amine nanoparticles on polyaniline electrochemical sensitivity to formaldehyde. Recent Patents on Nanotechnology. 2016;**10**:157-164

[2] Chandra P, Singh J, Singh A, Srivastava A, Goyal RN, Shim YB. Gold Nanoparticles and nanocomposites in clinical diagnostics using electrochemical methods. Journal of Nanoparticles. 2013;**2013**(1):535901

[3] Caudill ER et al. Wall teichoic acids govern cationic gold nanoparticle interaction with gram-positive bacterial cell walls. Chemical Science. 2020;**11**(16):4106-4118. DOI: 10.1039/c9sc05436g

[4] Karapa A, Kokkinos C, Fielden PR, Baldock SJ, Goddard NJ, Economou A. Talanta gold nanoparticle-modified sustainable plastic sensor chip for voltammetric monitoring of Hg (II). Talanta. 2023;**265**(April):124850. DOI: 10.1016/j.talanta.2023.124850

[5] Xian Y, Hu Y, Liu F, Xian Y, Wang H, Jin L. Glucose biosensor based on Au nanoparticles-conductive polyaniline nanocomposite. Biosensors & Bioelectronics. 2006;**21**(10):1996-2000. DOI: 10.1016/j.bios.2005.09.014

[6] Tamer U, Seçkin Aİ, Temur E, Torul H. Fabrication of biosensor based on polyaniline/gold Nanorod composite. International Journal of Electrochemistry. 2011;**2011**:1-7. DOI: 10.4061/2011/869742

[7] Galińska EM, Zagórski J. Brucellosis in humans--etiology, diagnostics, clinical forms. Annals of Agricultural and Environmental Medicine. 2013;**20**(2):233-238

[8] Ganganboina AB, Doong RA. Graphene quantum dots decorated gold-polyaniline nanowire for Impedimetric Detection of carcinoembryonic antigen. Scientific Reports. 2019;**9**(1):1-11. DOI: 10.1038/s41598-019-43740-3

[9] Suvarnaphaet P, Pechprasarn S. Graphene-based materials for biosensors: A review. Sensors (Switzerland). 2017;**17**(10):2161. DOI: 10.3390/s17102161

[10] Detection S. Recent advances in polymer-based biosensors for food. Polymers (Basel). 2023;**15**:1-21

[11] Radhapyari K, Kotoky P, Das MR, Khan R. Graphene-polyaniline nanocomposite based biosensor for detection of antimalarial drug artesunate in pharmaceutical formulation and biological fluids. Talanta. 2013;**111**:47-53. DOI: 10.1016/j.talanta.2013.03.020

[12] Omara W, Abdel-Latif SA, Azzazy HME, Abdel-Kader NS. Exploring polyaniline Nanofilaments for enhanced optical recognition of Lead in water: An integrated approach of experimental and theoretical studies. Applied Organometallic Chemistry. 2024;**38**(5):1-10. DOI: 10.1002/aoc.7439

[13] Xing YD, Li LY, Ai XC, Fu LM. Polyaniline-coated upconversion nanoparticles with upconverting luminescent and photothermal conversion properties for photothermal cancer therapy. International Journal of Nanomedicine. 2016;**11**:4327-4338. DOI: 10.2147/IJN.S97441

[14] Steffens C, Leite FL, Bueno CC, Manzoli A, Herrmann PSDP. Atomic force microscopy as a tool applied to nano/biosensors. Sensors (Switzerland).

2012;**12**(6):8278-8300. DOI: 10.3390/s120608278

[15] Wei D, Ivaska A. Electrochemical biosensors based on polyaniline. Chemia Analityczna. 2006;**51**:839-852

[16] da Silva AC, Córdoba de Torresi SI. Advances in conducting, biodegradable and biocompatible copolymers for biomedical applications. Frontiers in Materials. 2019;**6**(May):1-9. DOI: 10.3389/fmats.2019.00098

[17] Huber F, Lang HP, Backmann N, Rimoldi D, Gerber C. Direct detection of a BRAF mutation in total RNA from melanoma cells using cantilever arrays. National Nanotechnology. 2013;**8**(2):125-129. DOI: 10.1038/nnano.2012.263

[18] Zdarta J et al. Chitin-lignin material as a novel matrix for enzyme immobilization. Marine Drugs. 2015;**13**(4):2424-2446. DOI: 10.3390/md13042424

[19] Gerard M, Chaubey A, Malhotra BD. Application of conducting polymers to biosensors. Biosensors and Bioelectronics. 2002;**17**(5):345-359. DOI: 10.1016/S0956-5663(01)00312-8

[20] Kasas S, Longo G, Dietler G. Mechanical properties of biological specimens explored by atomic force microscopy. Journal of Physics D: Applied Physics. 2013;**46**(13):133001. DOI: 10.1088/0022-3727/46/13/133001

[21] Callister DG, Rethwisch WD. Materials Science and Engineering: An Introduction, 9th edn. 2013;9

[22] Mekuye B, Abera B. Nanomaterials: An overview of synthesis, classification, characterization, and applications. Nano Selenium. 2023;**4**(8):486-501. DOI: 10.1002/nano.202300038

[23] Xu W, Qing X, Liu S, Chen Z, Zhang Y. Manganese oxide nanomaterials for bacterial infection detection and therapy. Journal of Materials Chemistry B. 2022;**10**(9):1343-1358. DOI: 10.1039/d1tb02646a

[24] Rajesh JJ. Preparation and Characterization of Polyaniline Thin Films for Sensor Applications. The International Research Journal of Science and Engineering. 2017;**4**(2):51-55. DOI: 10.1016/S0042-207X(00)00182-2

[25] El-khawaga AM, Zidan A, El-mageed AIAA. Preparation methods of different nanomaterials for various potential applications: A review. Journal of Molecular Structure. 2023;**1281**(135148):1-21. DOI: 10.1016/j.molstruc.2023.135148

[26] Abid N et al. Synthesis of nanomaterials using various top-down and bottom-up approaches, influencing factors , advantages , and disadvantages: A review. Advances in Colloid and Interface Science. 2022;**300**(December):102597. DOI: 10.1016/j.cis.2021.102597

[27] De Freitas LF, Varca GHC, Batista JGDS, Lugão AB. An overview of the synthesis of gold nanoparticles using radiation technologies. Nanomaterials. 2018;**8**(11):939. DOI: 10.3390/nano8110939

[28] Lu Z, Lu N, Xiao Y, Zhang Y, Tang Z, Zhang M. Metal-nanoparticle-supported Nanozyme-based colorimetric sensor Array for precise identification of proteins and Oral bacteria. ACS Applied Materials & Interfaces. 2022;**14**:11156-11166. DOI: 10.1021/acsami.1c25036

[29] Fritea L, Banica F, Costea TO, Moldovan L, Dobjanschi L, Muresan M, et al. Metal nanoparticles and carbon-based nanomaterials for improved

performances of electrochemical (Bio) sensors with biomedical applications. Materials. 2021;**14**(21):6319. DOI: 10.3390/ma14216319

[30] Devices M et al. Leaders in Pharmaceutical Business Intelligence. 2016, 1-5

[31] Madkour LH. Applications of gold nanoparticles in medicine and therapy. Pharmacy and Pharmacology International Journal. 2018;**6**(3):157-174. DOI: 10.15406/ppij.2018.06.00172

[32] Kim CY, Shaban SM, Cho SY, Kim DH. Detection of periodontal disease marker with geometrical transformation of Ag Nanoplates. Analytical Chemistry. 2023;**95**(4):2356-2365. DOI: 10.1021/acs.analchem.2c04327

[33] Slimani Y, Hannachi E. Magnetic nanosensors and their potential applications. In: Nanosensors for Smart Cities. Elsevier; 2020. pp. 143-155. DOI: 10.1016/B978-0-12-819870-4.00009-8

[34] Mostufa S, Yari P, Rezaei B, Xu K, Wu K. Flexible magnetic field nanosensors for wearable electronics: A review. ACS Applied Nano Materials. 2023;**6**(15):13732-13765. DOI: 10.1021/acsanm.3c01936

[35] Xu C, Akakuru OU, Zheng J, Wu A. Applications of iron oxide-based magnetic nanoparticles in the diagnosis and treatment of bacterial infections. Frontiers in Bioengineering and Biotechnology. 2019;7(Jun):141. DOI: 10.3389/fbioe.2019.00141

[36] Zhang T, Xu Z, Liu X, Liu L, Jiang S, Zhang Z, et al. Activated silver nanoparticle-based platform for specific capture of Porphyromonas gingivalis in human saliva. Sensors and Actuators B: Chemical. 2024;**403**:135171. DOI: 10.1016/j.snb.2023.135171

[37] Speranza G. Carbon nanomaterials: Synthesis, functionalization and sensing applications. Nanomaterials. 2021;**11**(4):967. DOI: 10.3390/nano11040967

[38] Kumar V, Shukla SK, Thakur N. Carbon-based materials approach for environmental sensing. In: Functionalized Nanomaterials Based Devices for Environmental Applications. Elsevier; 2021. pp. 77-106. DOI: 10.1016/B978-0-12-822245-4.00015-5

[39] Fu C, Brand HS, Bikker FJ. The applications of carbon dots in oral health: A scoping review. Oral Diseases. 2024;**30**(4):1861-1172. DOI: 10.1111/odi.14702

[40] Alftaikhah SA, Issrani R, Alnasser M, Almutairi HA, Khattak O, Iqbal A, et al. Salivary Biomarkers in Periodontitis: A Scoping Review. Cureus. 2023;**15**(12):1-9. DOI: 10.7759/cureus.50207

[41] Xue J, Dong H, Ji L, Wang Y, Zhang J. Peptide-functionalized ZnSe:Mn quantum dots as fluorescent probes for accurate localization of hidden dental lesion sites. ACS Applied Nano Materials. 2023;**6**(15):14431-14438. DOI: 10.1021/acsanm.3c02446

[42] Hussain CM, Thomas S. Handbook of Polymer and Ceramic Nanotechnology. 2021

[43] Yoon H. Current trends in sensors based on conducting polymer nanomaterials. Nanomaterials. 2013;**3**(3):524-549. DOI: 10.3390/nano3030524

[44] Saylan Y, Kılıç S, Denizli A. Biosensing applications of molecularly imprinted-polymer-based nanomaterials. Processes. 2024;**12**(1):177

[45] Omara W, Amin R, Elhaes H, Ibrahim M, Elfeky SA. Preparation and

characterization of novel polyaniline nanosensor for sensitive detection of formaldehyde. Recent Patents on Nanotechnology. 2015;**9**(3):195-203

[46] Arshak K, Velusamy V, Korostynska O, Oliwa-stasiak K, Adley C. Conducting polymers and their applications to biosensors: Emphasizing on foodborne pathogen Detection. IEEE Sensors Journal. 2009;**9**(12):1942-1951

[47] Kane MC, Lascola RJ, Clark EA. Investigation on the effects of beta and gamma irradiation on conducting polymers for sensor applications. Radiation Physics and Chemistry. 2010;**79**(12):1189-1195. DOI: 10.1016/j.radphyschem.2010.07.012

[48] Maksoud MIAA, Youssef MA, Mohamady M, Aly MI, Ashour AH. Gamma radiation-induced synthesis of organoclays based polyaniline and ilmenite clay minerals for cesium ions removal from aqueous solutions. Separation and Purification Technology. 2023;**305**(October):122434. DOI: 10.1016/j.seppur.2022.122434

[49] Zhu J, Yu N, Bao C, Shi H, Li Q, Dai K. Upconversion-based intelligent dual-mode hydrogel nanosensor for visual quantitative detection of formaldehyde. Chemical Engineering Journal. 2023;**480**(December):2024. DOI: 10.1016/j.cej.2023.148106

[50] Taymour N, Fahmy AE, Gepreel MA, Kandil S, El-Fattah AA. Improved mechanical properties and bioactivity of silicate based bioceramics reinforced poly (ether-ether-ketone) nanocomposites for prosthetic dental implantology. Polymers. 2022;**14**(8):1632. DOI: 10.3390/polym14081632

[51] Lee MA, Jin X, Muthupalani S, Bakh NA, Gong X, Strano MS. In-Vivo fluorescent nanosensor implants based on hydrogel-encapsulation: investigating the inflammation and the foreign-body response. Journal of Nanobiotechnology. 2023;**21**(1):133. DOI: 10.1186/s12951-023-01873-8

[52] Shamsabadi A, Haghighi T, Carvalho S, Frenette LC, Stevens MM. The nanozyme revolution: Enhancing the performance of medical biosensing platforms. Advanced Materials. 2024;**36**(10):2300184. DOI: 10.1002/adma.202300184

[53] Liang D, Wang Y, Qian K. Nanozymes: Applications in clinical biomarker detection. Interdisciplinary Medicine. 2023;**1**(4):e20230020. DOI: 10.1002/inmd.20230020

[54] Rivera-muñoz EM. Hydroxyapatite-Based Materials: Synthesis and Characterization. 2014. DOI: 10.5772/19123

[55] Jandaghian AA, Rahmani O. Applications of smart ceramics in nano/micro sensors and biosensors. In: Advanced Ceramics for Energy and Environmental Applications. CRC Press; 2021. pp. 331-362. DOI: 10.1201/9781003005155-16

[56] Li W, Huang Y, Liu Y, Tekell MC, Emma D. Nano today three dimensional nanosuperstructures made of two-dimensional materials by design : Synthesis , properties , and applications. Nano Today. 2019;**29**:100799. DOI: 10.1016/j.nantod.2019.100799

[57] Perera AAPR, Madhushani KAU, Punchihewa BT, Kumar A. MXene-Based Nanomaterials for Multifunctional Applications. 2023

[58] Riazi H, Taghizadeh G, Soroush M. MXene-based nanocomposite sensors. ACS Omega. 2021;**6**(17):11103-11112. DOI: 10.1021/acsomega.0c05828

[59] Duan H, Tang SY, Goda K, Li M. Enhancing the sensitivity and stability of electrochemical aptamer-based sensors by AuNPs@ MXene nanocomposite for continuous monitoring of biomarkers. Biosensors and Bioelectronics. 2024;**246**:115918. DOI: 10.1016/j.bios.2023.115918

[60] Showman MS, Abd AM, Aziz E, Yahya R. Efficient photocatalytic degradation of petroleum oil spills in seawater using a metal - organic framework (MOF). Scientific Reports. 2022;**0123456789**:1-11. DOI: 10.1038/s41598-022-26295-8

[61] Rabiee N, Fatahi Y, Ahmadi S., Abbariki N, Ojaghi A. Bioactive Hybrid Metal-Organic Framework (MOF) - Based Nanosensors for Optical Detection of Recombinant SARS-CoV-2 Spike Antigen. 2022;825. DOI: 10.1016/j.scitotenv.2022.153902

[62] Chen J, Zhao JX. Upconversion nanomaterials: Synthesis, mechanism, and applications in sensing. Sensors. 2012;**12**(3):2414-2435. DOI: 10.3390/s120302414

[63] Liu J, Liu Y, Bu W, Bu J, Sun Y, Du J, et al. Ultrasensitive nanosensors based on upconversion nanoparticles for selective hypoxia imaging in vivo upon near-infrared excitation. Journal of the American Chemical Society. 2014;**136**(27):9701-9709. DOI: 10.1021/ja5042989

[64] Jones NA, Chang SR, Troske WJ, Clarkson BH, Lahann J. Nanoparticle-based targeting and detection of microcavities. Advanced Healthcare Materials. 2017;**6**(1):1600883. DOI: 10.1002/adhm.201600883

[65] Mady S, Alzayyat N. Nanotechnology in targeting and detection of microcavities. In: Nanotechnology in Conservative Dentistry. Woodhead Publishing; 2021. pp. 29-47. DOI: 10.1016/B978-0-323-90282-3.00007-0

[66] Eissa S, Tlili C, Mounir BA, Kanoun O. Nanomaterials-based electrochemical biosensors. Frontiers in Bioengineering and Biotechnology. 2022;**10**:1091592

Chapter 9

Iron Oxide Nanoparticles Application in Toxicity Therapeutics of CNS Disorders Indicated by Molecular MRI

Lei Zhao, Tao Peng and Shawn Y. Wu

Abstract

Nanomedicine holds promise in the process of diagnosis and treatment settings of neurotoxic conditions in many central nervous system (CNS) disorders, including neurodegenerative diseases and other non-tumor diseases herein, especially when using iron oxide nanoparticles for imaging biomarker development in magnetic resonance imaging (MRI) at the molecular level. The status and perspectives in the field of recent utilization of superparamagnetic iron oxide nanoparticles (SPIOs) and/or ultra-small superparamagnetic iron oxide (USPIOs) for theranostics evaluation are introduced, focusing on Alzheimer's disease (AD), Parkinson's disease (PD), and others based on the Aß cascade hypothesis and/or the oligomeric amyloid hypothesis as well as hyperphosphorylated tau fibrils leading to neurotoxicity in AD, the propagation hypothesis of neurotoxic misfolded α-synuclein in PD, and other mechanism in CNS disorders, including on neuroinflammation. In this chapter, we demonstrate the role of iron oxide nanoparticles as nano-theranostic agents in molecular neuroimaging with MR scans for CNS disorders, including neurodegenerative diseases, stroke, traumatic brain injury (TBI), and other non-tumor diseases. Some methods of doing imaging analysis in MR imaging data are provided. The introduction of the fundamental mechanism of nanoparticle-initiated neurotoxicity would seem out of the aim of our description.

Keywords: neurotoxicity, nanomedicine, molecular imaging, MR imaging, iron oxide nanoparticles, theranostics

1. Introduction

Besides the situation that liposomes, niosomes, micelles nanoparticles, nanocapsules, microparticles, dendrimers, colloidal gold, gold nanoshells, cyclodextrins, superparamagnetic particles, and carbon nanotubes are some of the drug delivery systems and nanocarriers for several diseases, such as neurodegenerative diseases [1–4], the small-sized iron-oxide-based contrast agents (CAs) could also be regarded as the largely employed nanomedicine applications in the central nervous system (CNS) [5, 6]. Magnetic iron oxide nanoparticles, also called IONPs or MNPs, are the most promising probes among various types of nanoparticles (NPs) for MR imaging approaches to pathological diagnosis, theranostics, and more specifically, nano-theranostics, which has become one of the critical research endeavors in this century. These included magnetic separation, drug and gene delivery, targeted therapeutics, cancer hyperthermia, and molecular imaging strategies for a large number of conditions.

Many inflammatory diseases in the CNS that are non-tumorous, such as neurodegenerative diseases, diabetes, and atherosclerosis, also require therapeutic approaches that offer high efficacy with minimal drug side effects. These conditions have been associated with neurotoxicity, which is closely related to the mechanisms of a specific disease and is caused by exposure to substances used in chemotherapy, radiation treatment, drug therapies, and organ transplants, as well as exposure to heavy metals such as lead, mercury, certain foods and food additives, pesticides, industrial and/or cleaning solvents, cosmetics, and some naturally occurring substances [7]. For example, the main sources of neurotoxicity in Alzheimer's disease (AD) pathogenesis are amyloid beta oligomers (AβOs) and free radicals [8, 9]. Hyperphosphorylated tau (P-tau) aggregates into paired helical filaments, which in turn accumulate into masses within nerve cells as neurofibrillary tangles and develop dystrophic neurites associated with Aβ plaques [10]. The exact mechanism that facilitates the development of senile plaques and neurofibrillary tangles (NFTs) in AD remains unclear. Besides the amyloid cascade hypothesis, which is currently the most widely accepted AD etiology, the oligomeric amyloid hypothesis has gained significant attention, offering the potential for definitive diagnosis without solely relying on postmortem histological findings of plaques. Like other neurodegenerative diseases, such as Parkinson's disease (PD) and multiple sclerosis (MS) [11], AD is accompanied by neuroinflammation [12–15], which is marked by an increase in activated microglia surrounding Aβ plaques [16–20]. Cerebrovascular dysfunction at the neurovascular unit (NVU), including the blood-brain barrier (BBB) breakdown, disrupted cerebral blood flow (CBF), hypoperfusion, oligemia, impaired vascular brain-to-blood clearance, and blood-to-brain transport of Aβ, contributes to disease pathophysiology [21–24]. The application of recent genome-wide and -omics approaches has provided novel insights into the critical molecular pathways of neurodegenerative diseases, relating to aggregation and accumulation of misfolded proteins, mitochondrial dysfunction, oxidative stress, oxidatively damaged DNA and impaired DNA repair, apoptosis, impaired autophagy-lysosomal activities, inflammation and microglia activation, perturbation of vesicle trafficking and synapse dysfunction, RNA processing and protein degradation pathways, as well as epigenetic deregulation of gene expression, all of which indicate the heterogeneous nature of those conditions [21–26]. Even though recent consensus criteria for the use of peripheral and imaging biomarkers improved the specificity and sensitivity of *in vivo* diagnosis [27, 28], significant limitations remain in longitudinally monitoring the progression or treatment of AD. In another example, biomarkers of small vessel ischemic strokes need further development, with the potential association between elevated plasma levels of intercellular adhesion molecule (ICAM) and small vessel disease being indicated by *in vivo* brain MRI through the observation of white matter hyperintensity progression in stroke [29, 30]. Given that the delivery of many potentially therapeutic neuroprotectants and diagnostic compounds to brain tissue is restricted by the BBB, magnetic iron oxide nanoparticles, which can readily cross the BBB without compromising its integrity, hold promise for use in the treatment of stroke and other CNS disorders.

As known, the use of molecular probes for the visualization and investigation of the structure, function, characteristics, and processes of cells via MR scans in living organisms is referred to as molecular MR imaging. In the field of diagnosing neurotoxicity by molecular MR imaging, particularly within the era of nanomedicine, this unique tool is used to noninvasively observe cell trafficking *in vivo* when combined with iron oxide nanoparticles (IONPs). Molecular MR imaging has garnered interest in the utilization of IONPs due to their enhanced magnetic properties. The superparamagnetic IONPs become magnetized in the presence of an external magnetic field and demagnetize once being without the magnetic field. This property of the IONPs is advantageous when using them in biological systems as MRI contrast agents.

Magnetic iron oxide nanoparticles, such as superparamagnetic iron oxide nanoparticles (SPIOs) and ultra-small superparamagnetic iron oxide (USPIOs), can be engineered to load drugs, contrast agents, and target moieties for cellular or intracellular components. These have emerged as potential alternatives for tracking and treating human diseases in the CNS, including neurodegenerative disorders, stroke, TBI, etc. For example, prototype imaging of the mouse model and isolated human brain tissue provided the clinical potential of MR imaging of synaptotoxic oligomers using targeted nanostructure probes [9].

In this chapter, we demonstrate the role of IONPs as nano-theranostic agents in molecular MRI for the treatment of CNS disorders, including neurodegenerative diseases, stroke, TBI, and other non-tumor diseases. The introduction of the fundamental mechanism of nanoparticles-initiated neurotoxicity would seem out of the aim of our description.

2. MR scan and IONPs for characterizations of neurodegenerative diseases at the molecular level

Progressive nervous system dysfunction due to the degeneration of selected neurons in the CNS is the general character of neurodegenerative diseases, including a heterogeneous group of either hereditary or sporadic conditions, some of which, the most well-known ones are AD and other dementias, PD, amyotrophic lateral sclerosis (ALS), and Huntington's disease (HD) [22, 30]. In AD, the most common subtype of dementia, the hyperphosphorylated tau protein, a major component of neurofibrillary tangles (NFTs), may play a critical role in disease progression. These NFTs, together with a major protein component of amyloid plaques—β-amyloid (Aβ peptides), which are generated from the parent molecule amyloid precursor protein (APP)—result in multiple Aβ forms of varying amino acid lengths (including the most neurotoxic species, Aβ1-42) and aggregate into oligomers and fibrils [31, 32]. The accumulation of small Aβ aggregates ('oligomers') is possibly an important factor in early disease stages of tau hyperphosphorylation and accumulation, oxidative stress, and inflammation. The hallmarks of AD and neuropathological features include extra-cellular Aβ plaques, intraneuronal hyperphosphorylated tau [33–35], oxidative stress, and neuroinflammation. Investigators believe abnormal Aβ metabolism that leads to high levels of toxic Aβ oligomers and combines with oxidative stress and inflammation would form an AD pathogenic cycle of neurodegeneration. The initiating steps of the neurodegeneration are hypothesized to begin decades before clinical diagnosis, though these changes remain in debate; meanwhile, diagnostic imaging has shown Aβ accumulation to start 20 years or more before the first clinical signs of AD [36]. Aβ accumulation reduces cognitive performance, particularly when Aβ is deposited within the temporal region [37]; reduces neurite outgrowth, and cerebral atrophy [38]; and has been associated with impaired synaptic function [39]. Mainly, there were positive relationships between synaptic/neuronal loss and NFTs load with cerebral atrophy and cognitive decline, which was also associated with cerebral Aβ load to some degree. Small oligomers of Aβ have also been found to be especially toxic to neurons with membrane damage, also causing Ca^{2+} leakage and oxidative damage, disruptions to insulin signaling pathways and synaptic function, and mitochondrial damage. Aβ-induced changes have been mentioned to occur early in the disease process, so the clinical interventions to interrupt the production of Aβ/Aβ oligomers or facilitate their removal from the CNS are regarded as the desirable ones. An approach that delays the onset of dementia by 5 years could reduce almost 50% of its incidence [40], while modeled projections suggest that delaying the onset of dementia by even one year could reduce the worldwide burden of cases in people over 60 years by approximately 10% [41]. Early preclinical prevention therapy, which may influence the accumulation or clearance of cerebral Aβ and tau pathology and/or reduce oxidative stress and chronic inflammation, would be highly significant for the reduction of AD prevalence as it slows or reverses these pathological changes. To this end, molecular imaging may hold promise in the early detection of AD and evaluation of the intervention effects.

2.1 Different types of magnetic iron oxide nanoparticles (IONPs)

So far, there is no U.S. Food and Drug Administration (FDA)-approved drug or carrier system with disease-modifying potential for Alzheimer's disease (AD); only treatments for the symptomatic management of AD are available. The challenges faced by the AD research community, as aforementioned, include transport-related issues, particularly those involving the blood-brain barrier (BBB); limited understanding of the basic pathology of AD, which hinders the development of more effective drug therapies (as AD involves several overlapping phenomena, such as inflammation, iron dysregulation, oxidative damage, cholesterol metabolism, and the "amyloid plaque" hypothesis) [42]; and the lack of early detection tools.

Nowadays, brain imaging of plaques in humans is conducted using positron emission tomography (PET), which is regarded as the current imaging "gold standard" for *in vivo* quantification of AD (Aβ) brain pathology [43, 44]. In 2012, the U.S. FDA approved the first commercially available radiotracer, ^{18}F-florbetapir (Amyvid), for PET imaging of amyloid plaques in the brain. However, this modality is limited by low spatial resolution, preventing the visualization of individual plaques. Other drawbacks of PET are high cost, limited availability, and exposure to radiation. Additionally, β-amyloid PEt alone may be insufficient to distinguish dementia syndromes that share overlapping β-amyloid pathology, such as dementia with Lewy bodies (LBs) and vascular dementia. In contrast, MRI does not involve ionizing radiation, provides superior spatial resolution, and is widely available in the clinical setting. Advanced MR imaging techniques at high and ultra-high magnetic fields may aid in a better understanding of the nature and progression of neurodegenerative diseases and have shown promise in detecting pathological changes corresponding to different stages. Molecular MR imaging with the application of IONPs also plays an important role in the development of therapeutics for AD and other neurodegenerative diseases through nanotechnologies. While cell-based therapies have been identified as novel treatment approaches for many kinds of diseases, studies [45–47] have shown that single, living, highly phagocytic large cells, such as macrophages or human endothelial cells, can be tracked over time by molecular MR imaging. Sophisticated techniques target iron oxide particles to specific molecular structures because they can be delivered intravenously without being specifically taken up by phagocytotic cells, such as the utilization of antibodies to recognize specific receptor molecules on the cell surface. Being in a variety of different sizes and coatings, SPIOs typically consist of a crystalline maghemite (Fe_2O_3) or magnetite (Fe_3O_4) iron core surrounded by a polymeric shell, and all of their varieties have been successfully used to label different immune, stem, or tumor cells, such as those from the smallest particles (VSOP) around 7 nm in diameter to the largest particles (MPIO, microparticles of iron oxide) being from 0.9 μm upward in diameter. The crystalline structure of the particles generates a strong resultant magnetic moment in a magnetic field and a large magnetic susceptibility, leading to local distortions of the magnetic field in the scanner and shortened relaxation times. When the transverse relaxation time is substantially reduced, it can be detected as a pronounced negative contrast in T_2^*-weighted gradient echo (GE) sequence use. IONPs can also be used as Aβ-targeted contrast agents (CAs) for AD by functionalizing the particle surface with moieties that selectively bind to Aβ plaques (i.e., using antibodies). They may also be engineered to bind Aβ peptides as probes, enabling researchers to visualize them via MR imaging and making this modality a feasible early detection tool [5, 48].

Ideal NPs should consist of unique magnetic nanomaterial or fluorescent active compounds to be observed noninvasively and for being specific to early-stage AD biomarkers to be packaged into biocompatible nano-vesicles that are decorated with a detecting ligand while many of them have been investigated for their delivery of diagnostic machinery across the BBB [49, 50]. For example, radiolabeled Aβ40 peptide can be imaged *in vivo,* forming the basis for its use in diagnostic and theranostic modalities related to AD. To improve its delivery across the BBB, targeting agents such as mAb, fusion proteins, and natural diamines are linked to these peptides [28]. Some of these targeting strategies have been used in combination with NPs. To detect amyloid plaques at early stages of the disease and to diminish the acquisition times of *in vivo* imaging, several studies have proposed the specific use of IONPs under MR imaging [51, 52]. The hypointense effect exhibited by these particles in T_2 and T_2^* images provides greater contrast in MRI, making this a good method for the early diagnosis of AD. Functionalized SPIOs with antibodies against Aβ (1–40) and Aβ (1–42) demonstrated specific labeling of both peptides in an *in vitro* study [53]. Anti-Aβ monoclonal antibody BAM10 was grafted onto γ-Fe_2O_3 nanoparticles in another study, showing interaction with $A\beta_{40}$ fibrils and targeting Aβ plaques *ex vivo* in rat brains when detected by MR imaging [54]. These IONPs present encouraging results in animal models, while their utilization in humans remains limited [55]. A promising approach [56] targeting Aβ oligomers at the instigation of early memory loss stages used antibody-laden 12–16 nm Fe_3O_4 magnetic nanostructures (MNS) stabilized with nitro-dopamine and polyethylene glycol (PEG). This AβO–MRI probe showed significant clinical potential for identifying synapto-toxic oligomers in the 5xFAD mouse model following intranasal delivery, thus providing a better candidate for an MRI contrast platform for earlier AD diagnosis.

Wadghiri et al. proposed the method using both Gd (DTPA) complexes and monocrystalline iron oxide nanoparticles (MION) coupled to Aβ (1–40) peptides [57]. The penetration inside the brain required transient BBB opening via mannitol injection. Wengenack's group illustrated that their in-house-developed anti-Ab antibody, IgG4.1, which binds to cerebral amyloid angiopathy (CAA), deposits in arterioles of APP transgenic mice and could effectively traverse the BBB and specifically target Aβ after producing IgG4.1-MIONs conjugated to the functionalized carboxylic acid groups of the phospholipids coating MIONs. Their MR imaging in *ex vivo* mouse brains through IgG4.1 nanoparticles with a 31-cm horizontal bore 9.4-T magnet detected the presence of CVA deposits in penetrating parenchymal cortical arterioles. This method is expected to improve the manifestations of AD that coexist with CAA and holds the potential for diagnosing and treating the condition (**Figure 1**) [58, 59]. Then, a putative CA was used where the BBB crossing properties were improved by modifying the Aβ moiety in a fully synthetic approach, creating Aβ derivatives with an attachment to gadolinium complexes. These strategies used to vectorize NPs involved the adsorption of antibodies or proteins [57, 60] onto the SPIO surface and suffered from difficulty in controlling the reproducibility of the noncovalent grafting as well as the orientation of the targeting moieties. Another strategy was based on streptavidin-biotin chemistry [61–64].

Several covalent conjugation strategies have also been developed that involve reactions with amino, carboxylic, aldehyde, or thiol groups [65]. There is an option based on the generation of aldehydes on a polysaccharidic coating, such as dextran, through an oxidative conjugation process [65, 66]. Regarding maghemite USPIO functionalized with either Rhodamine or Congo Red to mark Aβ (1-40) fibrils, it was shown that USPIO linked to a DDNP-carboxyl derivative was capable of binding to Aβ plaques, as proven by fluorophotometry. These NPs did not cross the BBB, but after mannitol

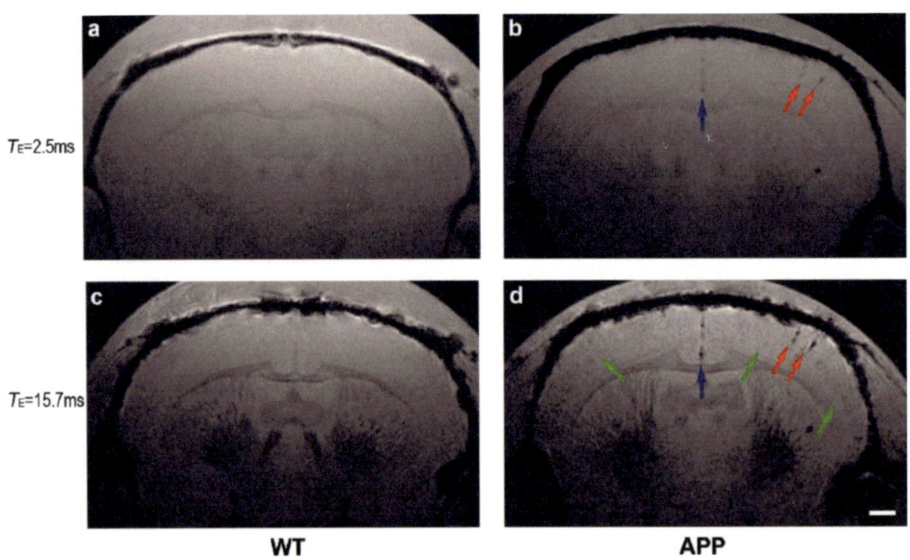

Figure 1.
Magnetic resonance images of wild-type (WT) (A, C) and amyloid precursor protein (APP) (B, D) mouse brains ex vivo after infusion of IgG4.1 nanoparticles into the external carotid artery. In the long-TE images (C, D), amyloid plaques in the APP brain appear as slightly dark foci in the cortex (D, green arrows), whereas the WT brain (C) is free of amyloid plaques and any dark foci in the cortex. In both the short- (A, B) and long-TE images, IgG4.1 nanoparticle-labeled cerebrovascular amyloid (CVA) deposits in cortical vessels in the APP brain appear as very dark foci (red arrows) on the infused side due to the strong paramagnetic effect of the iron oxide core (B, D). Cortical vessels in the WT brain are not visible because of the absence of CVA or nanoparticle binding in the vessels. The dark foci along the center of the cortex in the APP mouse brain are the arterioles that contain CVA and are labeled with IgG4.1 nanoparticles (blue arrows, B, D). Scale bar = 500 km (D). (Reprinted with permission from Poduslo et al. [58]).

injection, a T_2^* signal loss *in vivo* was apparent in the transgenic mice model of AD [67, 68]. USPIOs conjugated to Aβ1-42 peptides were used to detect amyloid plaque deposition and could identify AD mouse models [69]. When using a polyethylene glycol (PEG) coating, the BBB crossing was improved by femoral vein injection via 1-Ethyl-3-[3-dimethylaminopropyl] carbodiimide hydrochloride (EDC)/NHS linker to conjugate to the USPIO surface for a less toxic fragment of amyloid β peptides in transgenic mice, so as to replace co-injection of mannitol [70]. Investigators expressed the concern that the chemicals used in the additional steps might introduce cytotoxic effects, or that the former USPIO study presented in 2013 may form aggregates [71]. Curcumin-conjugated SPIOs (<100 nm) were stabilized by PEG-PLA as a block copolymer in a series of studies. Without using permeability-enhancing agents (such as mannitol) but with the NPs i.v. injection, Aβ plaques were detected *ex vivo* in AD mice brains. The mechanism of transport through the BBB proposed by the investigators was transcytosis, possibly influenced by the presence of PEG on the surface of NPs. Although PEG might interact with BBB endothelium cells and initiate transcytosis of NPs through the BBB, PEG coating of nanoparticles has been known to increase blood circulation time, because NPs have been found in the cytoplasm of vascular endothelial cells [72, 73].

Moreover, the USPIO application is also related to neuroinflammation study, becoming relatively popular in utilization due to its long blood half-life and implying a positive attribute for applications in the measurement of cerebral perfusion changes. Similar to acute cerebral ischemia and others [74, 75], AD is accompanied by neuroinflammation [12, 14, 15, 76], of which a central event is NF-kappaB (NF-κB) activation. This was a common feature of many neurodegenerative diseases (such as Huntington's, Parkinson's, or stroke); particularly AD, this could be predominantly found in neurons and glial cells in Aβ plaque surrounding areas, with obvious increased manifestation surrounding the Aβ plaques in the presence of activated microglia [16–20]. The pathological hallmarks of AD (Aβ and hyperphosphorylated tau) may induce NF-κB activation via various mechanisms, causing the expression of a large variety of pro-inflammatory molecules, such as cytokines and chemokines that are in part responsible for the neurotoxicity. Microglia have been implicated as active contributors to neuron damage in neurodegenerative diseases in which the overactivation of the resident innate immune cells in the brain and dysregulation of microglia might result in disastrous and progressive neurotoxic consequences, while their overactivity can amplify neuroinflammation and associated neurotoxicity. Though the uptake of SPIO and USPIO in mononuclear cells causes a transient oxidative stress, it does not affect *in vitro* cell survival, proliferation, mobility, and phagocytic properties even at high iron loads [77, 78]. Sillerud et al. [79] recently developed iron-platinum (Fe-Pt) NPs that target ionized calcium-binding adapter molecule 1 (Iba-1) on activated cerebral microglia to directly demonstrate neuroinflammation formation. The same group [80] previously found that anti-AβPP (amyloid-β protein precursor) antibodies targeting plaques *in vivo* had not been reported in literature until their treatment of the transgenic mice of AD with anti-AβPP-targeted SPIOs enhanced the conspicuity of the MRI-detected plaques, with the features including recognition by anti-human AβPP and Aβ antibodies, fluorescent staining with thioflavin-S, and the presence of activated microglia as detected by immunofluorescence. In 2013, investigators also reported the 9.4-T scan study on antibody-conjugated SPIOs. The designed CAs improved the negative contrast of brain slices after being injected by the CAs and the mice sacrificed, compared to those without any CA use, indicating that the anti-APP conjugated SPIOs crossed the BBB [81].

2.2 Applications of IONPs in neurodegenerative diseases

Solid advances have been made in understanding AD pathology. Nevertheless, pharmaceutical companies are repeatedly failing to develop new effective treatments for AD. Until now, pharmacotherapy of AD offered primarily symptomatic benefits but did not retard disease progression. Efforts are still needed to make effective treatment of AD. Some promising drugs are in development, such as enzymes to be used for decreasing Aβ production, some substances that increase Aβ clearance, immunotherapy, enzymatic degradation of Aβ, inhibition of Aβ aggregation, antioxidants, anti-inflammatory drugs, statins, etc. [82, 83]. A powerful approach to showing potential to treat and even cure some of the most common diseases of the CNS is gene therapy, which provided some preclinical efficacy in animal models of AD and held promise in clinical applications but with its main hurdle for AD, such as the diffusion of the lesions in the brain. AD impacts a few anatomical subregions of the brain that are involved in learning and memory. Therefore, adeno-associated virus (AAV)-based gene therapy strategies for AD should be based on safe neurosurgical protocols capable of efficiently targeting early affected regions through improved delivery methods, as well as the serotype application with good diffusion properties (anterograde and retrograde transport). The development of modified AAV vectors that can efficiently cross the BBB for neurodegenerative diseases, also MS (based on its corresponding preclinical model experimental autoimmune encephalomyelitis—EAE observations), should improve such strategies in the future [83–86]. The first gene therapy clinical trials that have been adopted for AD were based on intracerebral delivery of AAV2 encoding nerve growth factor (NGF). A first report of 10 patients with mild to moderate AD who received bilateral injections (total of six deliveries) into the basal forebrain region (into the nucleus basalis of Meynert) with doses escalating from 1.2×10^{e10} vg to 1.2×10^{e11} vg (clinical trial NCT00087789) showed the safety and tolerability of the procedure for at least 2 years post-injection. The trial also displayed persistent bioactivity of NGF in autopsy tissues, with a trophic response to NGF and lower cognitive decline [87]. A multicentric Phase II (clinical trial NCT00876863) consisting of 49 patients is ongoing, but results are not yet available [88].

Among these approaches, the inhibition of Aβ assembly has been considered the primary therapeutic strategy for this neurodegenerative disease. To date, it is the first mechanism that has received the most attention [89]. Degenerative diseases, such as AD [90–92] and PD, have shown a favorable response to cell therapy [93–95], which benefited TBI [96] and other CNS diseases/damages.

All the systems that were considered as treatment of AD must possess the capacity of crossing the BBB. Many studies have described the effect of nanoparticulate systems on Aβ fibrillation [97]. Fibril formation occurs by means of nucleation-dependent kinetics, where the formation of a critical nucleus is the key determining step, after which fibrillation proceeds rapidly. Stem cell-based therapies were regarded as a promising approach that could compensate for the massive and progressive neuronal and synaptic loss. In rat and mouse models of AD, improvements in learning and memory deficits have been shown through the transplantation of human neural stem cells (NSCs) overexpressing choline acetyltransferase [98, 99]. In an AD model with amyloid precursor protein (APP)/presenilin 1 (PS1) mouse, bilateral transplantation of NSCs into the hippocampus demonstrated the ability to integrate into the circuits, replacing some of the damaged/dead neurons; to promote brain plasticity (increasing synaptogenesis and long-term potentiation); as well as to reduce inflammation by regulating microglia and astrocyte activation, culminating in improvements in spatial learning and memory function, despite no alteration in terms of Aβ pathology were observed [100, 101]. Cell replacement in AD might also be achieved through the induction of endogenous NSCs. Both strategies face several challenges, including functional integration, recruitment, modulation of NSCs, microenvironmental control, etc. The application of biomaterials can be a useful tool to overcome some of these challenges [102].

There are a limited number of reports on the interaction of SPIOs with amyloidogenic proteins, with the majority of those focused on insulin amyloid fibril formation. One report showed the effect of IONPs on Aβ fibrillation, the effect of dextran polymer coating charge and thickness (i.e., a single polymer layer or a double polymer layer) on fibrillation kinetics [103]. A dual-surface area-dependent effect was observed depending on the surface coating charge, with lower concentrations of SPIOs inhibiting fibrillation, whereas higher concentrations enhanced the rate of Aβ fibrillation whose kinetics can be affected by the thickness and surface charge of the coating layer of SPIOs. Positive SPIOs promoted fibrillation at significantly lower particle concentrations than those of either the negatively charged or essentially uncharged. It was suggested that a negatively charged or uncharged surface coating of SPIOs be preferentially designed for *in vivo* medical imaging applications. Because the nanoparticles may influence the β-amyloid fibrillation process [104], the effect of the physicochemical properties of SPIOs deserves to be extensively illustrated, along with their charges, concentrations, and the coating, including their highest magnetic property on the fibrillation of β-amyloid and α-synuclein, at least being as carrier in treatment of AD, PD and other conditions in the CNS [105].

Based on controlled differentiation of NSCs to neurons and long-term monitoring in real time, investigators would achieve their successful applications. Zhang et al. [106] reported the synthesis of poly(2-hydroxyethyl methacrylate)-retinoic acid-poly(carboxybetaine) cell-penetrating peptide (PHEMA-RA-PCB-CPP) polymer to construct traceable, simple-component NPs for AD therapy. The polymer is self-assembled into NPs to encapsulate SPIOs in the hydrophobic region and formed complexes with small interfering (si) RNA against SOX9 (siSOX9, to down-regulate in advance the SOX9 protein expression of NSC maintenance and glial fate determination) to be of theranostic properties that depict the tracking of NPs *in vivo* due to the SPIOs, as well as the enhancement of neuronal differentiation due to the dual release of siSOX9 in an initial phase (inhibiting translation of Sox9) and retinoic acid in a later phase (a well-described promotor of neurogenesis). Based on the feature of siRNA as a potential agent for the down-regulation process via RNA interference (RNAi), which is a novel gene regulatory mechanism by activating a sequence-specific RNA degradation process (PTGS), while its short plasma half-life and poor membrane penetrability reduce the therapeutic efficiency, SPIO has been used to combine with the above approach so as to develop the charge-reversible zwitterionic poly(carboxybetaine) (PCB)-based NPs via protonation in an acidic environment. Because of their superior cellular uptake and temporally controlled release ability, the polymeric NPs efficiently controlled the neurons from NSCs treated with the NPs that ameliorated neurologic changes in a double transgenic mouse model (2x-tg) of AD and rescued memory deficits. In the system, MR imaging and SPIO with high r^2 value in the NPs in real-time monitored the transplantation site and migration of the NSCs. Lately, the same group synthesized positively charged polyprodrug amphiphiles PCB-Se-Se-simvastatin that could load hydrophobic SPIO simultaneously and absorb negative let-7b antisense oligonucleotide, then construct traceable co-delivery NPs that could attenuate behavior deficits and simultaneously track the NSCs in the

2xTg-AD mice model [107]. These polymeric NPs have the potential for theranostic applications for neurodegenerative diseases with their excellent tracing and therapeutic ability, although they need to improve the stability in the physiological environment prior to the application.

On the other hand, the intraperitoneal administration of curcumin-loaded poly-D, L-lactide-co-glycolic acid (PLGA) NPs, which is an FDA-approved NP for drug delivery due to its excellent biocompatibility and biodegradability, resulted in higher levels of both proliferating cells (BrdU-positive cells) and new neurons (BrdU/NeuN-double positive cells) in the subgranular zone (SGZ) of a rat model of AD through the injection of Aβ1-42 in the hippocampus. Also, the curcumin-loaded NPs treatment culminated in the amelioration of learning and memory deficits of AD model rats by enhancing endogenous neurogenesis through activation of the Wnt/β-catenin pathway. While similar effects have been observed with bulk curcumin, the use of NPs reduced the amount of compound needed to obtain the expected physiological response by 40 times [108]. These results raised a topic of high interest in the potential of NPs application in the transport and controlled release of drugs for neurodegenerative disorders. Also, Cheng et al. [109] developed a curcumin-conjugated SPIO stabilized by amphiphilic block copolymer (PEG-PLA) to have the ability to bind both Aβ plaques using Tg2576 transgenic mice, the clinically approved CA for MRI [110, 111], and the natural product extracted from turmeric, the root of the *Curcuma longa* plant, which has been demonstrated safe in a human clinical trial without adverse chronic effect during 6 months of high-dose consumption. Regarding the details of other nanoparticle technology besides the use of IONPs for stem cell therapy in neurodegenerative disorders, the interested reader is referred to the literature [112].

PD is the second most common neurodegenerative disorder after AD [113]; its neuropathological hallmark is the presence of intracellular inclusions known as LBs, which are composed mainly of α-synuclein (α-syn) proteins. It is physiologically necessary for α-synuclein-mediated SNARE-complex assembly for synaptic function, whereas α-synuclein at high doses pathologically misfolds into neurotoxic forms. Accordingly, α-syn has been shown to inhibit synaptic vesicle exocytosis in transfected midbrain dopaminergic neuron cultures in a dose-response manner. Toxic α-syn fibrils were also shown to impair the retrograde transport of BDNF-signaling endosomes [114]. Much evidence suggests that pathological α-syn can induce dopaminergic neuron death through various mechanisms (when their approximately 50% is lost in the substantia nigra (SN) pars compacta to the striatum causing PD), such as by inducing oxidative stress and inhibiting protein degradation. Their degradation subsequently causes loss of striatal dopamine (DA) projecting to the caudate-putamen region. The debilitating symptoms of PD commonly are tremors, rigidity, balance impairment, and slowness of movement. The long-term use of the DA precursor, levodopa, is limited by the development of severe motor side effects and a significant abating of therapeutic efficacy, though it provides superlative efficacy in PD's early stages and is regarded as the "gold standard" treatment. Therefore, there remains a major unmet clinical application for the development of effective neuroprotective, neurorestorative, or neuroreparatory therapies for this condition. PD, with its feature of relatively selective loss of dopaminergic neurons in the nigrostriatal pathway, may be regarded as a candidate for reparative cell therapies, also with DA neurons progressively lost in the SN representing as a hallmark but replaced through direct cell transplantation into the brain. This approach has been developed, validated, and clinically assessed using DA neuron-rich fetal ventral mesencephalon grafts, which have been shown to survive and reinnervate the denervated brain after transplantation, restoring motor function after transplantation. However, despite long-term symptomatic relief in some patients, significant limitations, including poor graft survival and the impact this has on the number of fetal donors required, have prevented this therapy from being more widely applied. There is potential for injectable biomaterial scaffolds to improve the delivery, engraftment, and survival of these grafts for cell adhesion, growth, and immune shielding in the brain through the provision of a supportive microenvironment [115].

Like in other neurodegenerative diseases, including AD [13–15] as well as MS, PD is accompanied by neuroinflammation [11, 14], which is manifested as microglia activation, astrocyte gliosis, and cytokine release. This exacerbates the neurodegenerative process and finally leads to neuronal death. Postmortem analysis reports also show neurotoxic microglia and astrocytes surrounding degenerated neurons in the SN. In various animal models of PD, microglia activation and astrocyte response can also be illustrated. Microglia activation promotes the release of reactive oxygen species (H_2O_2, O_2^-, NO) and toxic cytokines (TNF-α, IL-1β, IL-6, IFN-γ), then gives rise to oxidative stress and mitochondrial dysfunction, and finally leads to neuronal damage. Astrocytes also play critical roles in the initiation and progression of PD [65]. Microglia and other microenvironments in the brain may trigger the functional changes in astrocytes that produce plenty of neurotoxic factors and lack of growth factors for the neurons, eventually leading to neuronal death [28, 66].

To understand the physiology and efficacy of therapeutics for PD, especially cell-based therapy, it is necessary to develop accurate *in vivo* monitoring strategies. With the advent of molecular imaging technologies, *in vivo,* real-time monitoring of transplanted cell fate has become a reality. Functional neuroimaging techniques such as PET and functional MR imaging (fMRI) make it possible to evaluate *in vivo* specific neural circuits in the brain before and after the administration of a neurotoxin in PD models. FMRI studies using SPIOs and cerebral blood volume (CBV) measures [116] provide an optimal temporal and spatial resolution even in deep structures of the brain. For neuronal activation in anesthetized animals of PD, pharmacological challenges using selective agonists and antagonists of a neurotransmitter allow characterization of *in vivo* dynamic changes in cerebral function. Many reports have shown the use of amphetamine to induce DA release in the nigrostriatal system and evaluated the changes in CBV with fMRI [117–119]. The loss of DA cells induced by 6-Hydroxydopamine (6-OHDA), which was the first chemical agent used for its specific neurotoxic effects on catecholaminergic pathways, was accompanied by a parallel decrease in the CBV response to both amphetamine and CFT in rats [118]. Moreover, as in the PET studies mentioned in the literatures [117, 120], the results could demonstrate restoration of DA function after transplantation of fetal DA cells or embryonic stem cells using fMRI with an amphetamine challenge. Dopaminergic activity in grafts and hosts may be assessed using PET, which cannot directly measure the function of nondopaminergic cells. Also, the high cost and unavailability of PET in many hospital/clinical facilities call for the development of more accessible, versatile methods of cell-based therapy monitoring. At the same time, there were several advantages for molecular MR imaging over radionuclide imaging for monitoring cell-based therapy, such as superior delineation of morphology, no exposure to radiation, as well as the possibility of monitoring transplanted cells over long periods of time. Cell tracking with MRI is a relatively young field that aims to detect and monitor cells in living organisms. Cellular MRI combines the ability to obtain high-resolution MRI data with the use of magnetic contrast agents for labeling specific cells, thereby enhancing their detectability. Most cell tracking studies have used SPIOs to label cells. Different studies have reported the different cell types' successful applications in their detection by MRI when labeled with iron oxide agents at levels without negatively impacting cell viability, phenotype, or specific functions [121]. Many publications reported the use of IONPs to label or track stem cells, mainly focusing on experimental models of CNS disease, showing examples like imaging studies of iron-labeled neural stem cells transplanted into an embryonic mouse brain [122], iron-labeled oligodendroglial progenitors transplanted into the ventricles of a neonatal, myelin deficient rat brain [123], and iron-labeled human stem cells transplanted into the striatum of parkinsonian rats [124]. MRI-based diagnosis in transgenic mouse models of PD, HD, and ALS have also recently been studied [125–127].

The long-term effects of cell-based therapy on PD are unsatisfactory due to drug adaptation and disease progression. Both are current targets for stem cell therapies, especially HD, since there are no effective pharmaceutical treatments for it (although some drugs may lessen the symptoms somewhat). Investigators are developing several stem cell therapies for both diseases. Animal models are being employed to track and monitor the stem cells used in order to better understand these diseases and therefore optimize the development of treatments [126, 128]. Moraes et al. set up a rat model for investigating whether the neuroprotective effects of MSCs can be an effective treatment in HD with the MSCs labeled and with Feridex for monitoring via MRI [129]. Their results showed that the MSCs migrated away from the site of injection toward lesions caused by the disease, demonstrating a neuroprotective effect. Strategies to improve MRI cell tracking also include genetically engineering the graft cells to enhance the function of intracellular iron-regulatory proteins that act as endogenous CAs [130], and iron nanoparticle uptake by cells leads to subsequent compartmentalization in secondary endosomes within the cell cytoplasm, with the presence of intracellular iron causing a distortion in the magnetic field, appearing as regions of low signal intensity in areas containing iron-labeled cells, also facilitating labeled cell detection. SPIO and genetic MRI contrast enhancement methods were still regarded as being in the early stages of development while offering tremendous potential for clinical application in cell-based therapy.

Gene therapies have been shown as promising alternatives besides deep brain stimulations (DBS) and cell therapies [131]. While lysosomal hydrolase glucocerebrosidase activity showed reduced in PD, increasing it by gene transfer in mouse and rat models of PD indicated neuroprotective effects against dopaminergic neuron degeneration [132]. As for the developing non-viral gene delivery systems for brain-targeted gene therapy, Niu et al. [133] used polymer-coated IONPs to deliver plasmid DNA (pDNA) encoding a short hairpin RNA (shRNA) into 1-methyl-4-phenyl-1,2,3,6-tetrahydropyridine (MPTP) mice and improved motor dysfunction in this PD model by reversing α-synuclein–mediated elevation of apoptotic markers. AAV-mediated overexpression of the transcription factors Nurr1 and Foxa2 also indicated the protection of these dopaminergic neurons from Parkinson degenerative processes [134]. Gene therapy may be a good choice for PD therapy because it can rectify the mutations in some genes, such as genetic variability in *SNCA,* which influences the α-syn level, as the major component of LBs. Studies on small interfering RNA (siRNA) against *SNCA* showed this down-regulated the level of α-syn and had a recovering effect on PD [135]. Furthermore, the production and accumulation of Aβ were reduced by siRNA against β-secretase (BACE1) in AD [136, 137]. While selective small molecule drugs and peptides can be used to inhibit aggregation, increase aggregation clearance, and decrease aggregation spreading of α-syn that is degraded through the autophagylysosomal pathway and ubiquitin-proteasome system under physiological conditions, the intracellular clearance pathways are impaired, promoting the accumulation of proteins and ultimately leading to neurodegeneration. Increasing autophagy was shown to be a possible successful neuroprotection strategy for treating PD. Small autophagy-activating molecules, such as rapamycin, hold promise for rapid translation to patients [138]. It is known that neuroinflammation, microglia activation, and secretory neurotoxic activities in the brain are the causing factors of PD [128]; especially, samples from the brains of PD patients can be found with a reduction in the level of catalase, causing oxidative stress and degeneration of dopaminergic neurons [139]. Catalase, in this regard, is a potent anti-Parkinson drug because the molecule catalase can deactivate one million free radicals per second in a cycle of catalytic catalase [140]. A series of studies have depicted that AAV2 can be delivered by convection-enhanced delivery (CED) over a wide dose range without harmful effects on the brain, most importantly in humans [141, 142], *indicating* AAV2 to be a safer option for transduction of neurons in the brain.

Progressive amelioration of functional deficits, recovery of DA levels, and regrowth of fibers to the striatal neuropil was observed in the presence of high expression of glial-derived neurotrophic factor (GDNF), which was a potent biologic that had neurotrophic and neuroprotective effects when delivered to the degenerated striatum, being effective in ameliorating symptoms in parkinsonian nonhuman primates (NHP) using a gene therapy approach [143, 144], further strengthening the preliminary results that AAV2-GDNF delivered by CED was safe and effective in an NHP model of advanced PD [141, 145, 146]. Achieving clinically significant effects in PD patients needs a validated treatment platform to ensure viral vector delivery to the chosen target area and accurately predict the areas of subsequent GDNF expression. The study from several collaborative institutes [143] was unique so far among the preclinical literature in that the initiation of neurotrophic factor gene delivery was delayed at least 3 months following 1-methy-4-phenyl-1,2,3,6-tetrahydropyridine (MPTP) administration in NHP experiments, a timeframe during which MPTP lesioning of the nigrostriatal pathway became complete, ensuring a direct assessment of neuronal repair comparable to future assessments of the neurorestorative potential of GDNF in PD patients. The results suggest that clinical investigation of AAV2-GDNF in PD patients is warranted and, in combination with the current reports, illustrates how the protocol should be clinically implemented.

Importantly, the MRI-guided methodologies in development have been used for these research works, and by including Gd-loaded liposomes (GDL) with AAV, real-time MR imaging and tracking of CED of viral vectors to the three different regions of the NHP brain (corona radiata, putamen, and thalamus) was achieved [147]. An FDA-approved, interventional MRI targeting system (iMRI) for achieving precise and real-time CED in planned AAV2-GDNF gene therapy trials for PD patients was then modeled in NHP [148]. This system predicts the anatomic distribution within the target structure of subsequent expression of GDNF (other than α-Syn) in a series of studies. A phase Ib trial (NCT01973543) was the first use of iMRI and extended into targets in a gene therapy trial for a neurodegenerative disorder and adopted the ClearPoint platform to deliver a gene for amino acid decarboxylase (AADC) carried by AAV2 to the putamen in medically refractory PD. A parallel study using the same technique was also enrolling patients at the NIH using AAV2 to deliver a gene encoding GDNF (NCT01621581), with both trials utilizing bilaterally mounted Smart Frames to allow for simultaneous infusions in bilateral hemispheres [149, 150]. Also, the methodologies of intracranial co-transplantation of mesencephalic NSCs stably overexpressing human GDNF (GDNF-mNSCs) together with fetal DA neurons were applied in a preclinical study [151] and showed that SPIO-based imaging was capable of therapeutic effects assessment, monitoring motor functional improvement in PD with so many perspectives.

3. Molecular MR imaging with IONPs for other CNS lesions

Better understanding of the pathological conditions of the nervous system might be achieved by the advances in nanotechnology and the development of formulations that could enhance the therapy of neurological diseases [152]. To do surface modification for making the NPs functionalized, even conjugated with transporters of the BBB, has been proposed to enhance their transport efficiency into the brain. Recent developments in both MRI acquisition methods and magnetic iron oxide nanoparticle synthesis and post-synthesis modifications have led to significant improvements in SPIO detection sensitivity and *in vivo* biodistribution. IONPs application studies include the diseases in neuroscience mentioned above, but also others like stroke, TBI, MS, etc. [11, 29, 66, 96, 153] described in this section.

3.1 For stroke studies

As one of the leading causes of death in the world, ischemic stroke has limitations in treatment options due to the therapeutic time window. Nowadays the only effective treatment identified to date is early (<4.5 h) thrombolysis and intravenous (i.v.) administration of recombinant tissue plasminogen activator (rtPA), despite considerable research efforts in treatment options [154]. In several randomized clinical trials, endovascular treatment has been shown to improve functional outcomes for selected patients with acute ischemic stroke. Thrombectomy with stent retrievers is regarded as standard care for acute ischemic stroke patients with a proximal large vessel occlusion in the anterior circulation. Both of the treatments with short therapeutic time windows limited their clinical use. Many studies have demonstrated the critical role of neuroimaging in the evaluation of stroke onset, processing, and post-ischemic brain injury as well as BBB permeability [155–158]. To prudently select patients potentially benefiting from reperfusion therapy, being as the most convincing treatment in acute ischemic stroke, is extremely important for the goal to minimize reperfusion injuries in clinical practice while being the similar case when demanding insights from basic research. Neuroimaging is thereby a bridge translating basic research results to clinical applications, especially in stroke, and bi-directionally clinical imaging findings may in turn help to apprehend underlying disease mechanisms.

Small amount of research on nanoparticles has focused on brain imaging for the CNS diseases, though those have been studied extensively on the frequent conditions of neuro-oncology. After facing the main challenge of the BBB and getting more understanding of the leaky mechanism of it, previously developed nanoparticles that target tumors according to the enhanced permeability and retention (EPR) effect could also be applied to brain diseases [159, 160]. What can now be encapsulated in NPs, such as statins or candesartan, belonged to those drugs that showed promising results *in vitro* through endocytotic or transcytotic routes to cross the BBB [161, 162]. When conjugated with NPs, the labeled antibodies of CNS injury biomarkers such as S100 calcium-binding protein B (S100B), vascular cell adhesion molecule (VCAM), glial fibrillary acidic protein (GFAP), etc., can be quickly detected using a CT scan or MR scan. For the purpose of circulating NPs conjugated with blood clotting biomarkers in an effort of early diagnosis or even the disruption of a molecular cascade that leads to thrombosis, their capability to endure degradation could be applied [163]. Bhatia and the team made steps toward the diagnostic use of NPs when the way of constant monitoring for coagulation was provided by conjugating thrombin-sensitive peptide substrates on the surface of the NPs to detect the changes of thrombin levels in the circulatory system [164]. Because the BBB represents a challenge for molecular imaging of the brain, CAs should be able to cross the BBB to reach parenchymal targets. Nowadays there seems to be no CA able to cross the BBB in sufficiently large quantities to induce significant signal changes in MR imaging. Molecular imaging of the brain is therefore limited to endothelial proteins (such as VCAM-1, ICAM-1, E-selectin, and P-selectin), also to diseases with compromised BBB (such as stroke, MS, or TBI), and to cells capable of crossing the BBB (such as monocytes) [29, 30, 165]. Investigators in the Netherlands performed molecular MR imaging in the subacute phase of stroke of ICAM-1, one of those that mediate the diapedesis of neutrophils (among the first cells to adhere to the cerebrovasculature after stroke), playing different roles as adhesion molecules, with VCAM-1, which is mainly implicated in the diapedesis of monocytes and lymphocytes. MPIO-αICAM-1 was applied in their study to depict the overexpression of ICAM-1 by activated endothelial cells 24 hours after ischemic onset in a transient mechanical occlusion of the middle cerebral artery (MCAO) mouse model [165]. The same group then conducted a longitudinal molecular neuroimaging observations with MR scans of ICAM-1 at 1, 2, 3, 7, and 21 days post-stroke [166], showing that MPIO-αICAM-1 binding was detectable at days 1, 2, and 3 in the lesion core, and at day 2 also in the border zone of the lesion and contralesional tissue. The findings had been regarded as being in line with the results of MPIOs-αVCAM1–enhanced imaging with multiparametric MR scan [74, 167], "inflammatory penumbra," and the presence of cerebrovascular inflammation being perilesional areas expressing VCAM-1 after pMCAO, which was recruited by the ischemic core during the subacute phase. These suggested an imaging criterion to select patients for subacute anti-inflammatory treatments. Meanwhile, the MR imaging with MPIO-αICAM-1 illustrated the expression of ICAM-1 by some leucocyte subtypes [166], suggesting that ICAM-1-targeted MPIOs could be applied for *in vivo* tracking of leucocytes besides showing ICAM-1 expression on vascular endothelium at different stages after stroke. Therefore, the clinically applicable MPIO holds promise in MRI-based diagnosis of neuroinflammation and identification of anti-inflammatory targets in acute stroke patients. These above plus others [168–170], have depicted molecular MRI applications to get valuable insights into inflammatory events involved in cerebral pathology in ischemic stroke and other neurological disorders (such as MS).

Pathophysiological events occur after ischemic stroke, including ionic imbalance [171], neuroinflammation [172, 173], excitotoxicity [174], and activity of microglia, all contributing to neuronal death and being the target of gene therapy. Main points related to that are a variety of endogenous proteins are upregulated by ischemic stroke, indicating the strategies of modulation of these gene expressions so as to provide neuroprotection, neurogenesis, and angiogenesis. By using a 9.4-T animal MR system with an in-plane resolution of 90–120 μm/pixel, investigators have developed a gene transcript-targeted MRI (GT-tMRI) technique [175, 176] in which oligonucleotide-conjugated SPIOs and an MRI-based version of *in situ* hybridization mapping were applied in a forebrain ischemia model using bilateral carotid occlusion for 60 min (BCAO-60), known to stimulate the growth of new vasculature (neovascularization) by recruiting pericytes (which express nestin) and expansion of

GFAP-expressing glial cells active in gliogenesis in the CNS. Through labeling target-specific CAs to small DNAs (18–26 nt in length), investigators have recently shown that the MR-CAs enter the vascular endothelia by caveolae and then are transported through the BBB and glial end-foot to the rough endoplasmic reticulum of specific cells where mRNAs are located in a novel model of ocular ischemia in living mice. This modality exhibits near-cellular resolution with signal specificity and sensitivity using clinical or high-field-strength magnets. The results have further demonstrated the robustness of the technique for detecting specialized cells associated with the progression of brain diseases (as shown through measures of oxidative stress and neuroinflammatory responses) and/or recovery (as evidenced by gliogenesis, angiogenesis, and neurogenesis). These also included pericyte-specific CA developed using phosphorothioate-modified antisense oligodeoxynucleotides (ODNs), which were short segments of DNA binding to complementary mRNA sequences to stop translation of encoded proteins for *in vivo* reporting and quantitative characterization of vascular pericytes in the region of BBB injury after cerebral ischemia in mice. MR contrast agent, which was designed to generate bioactivated and theranostic agents, could be used in complex physiological conditions, such as to report enzymatic activities, gene expression, metalloproteinases (MMPs), as well as signal transduction and intracellular messengers, at least showing the expression of mmp-9 mRNA from the global cerebral ischemia (GCI)-induced brain damage region associated with hyperintensity in diffusion-weighted imaging (DWI). A recent study also showed that Xenon (Xe)-liposomes in combination with late rtPA treatment developed a noticeable reduction in MMP9 immunoreactivity in the MCAO model, predominantly observed within the microvessel walls, and reduced rtPA-induced hemorrhage [2]. Noninvasive imaging may have promise in these sorts of research when applied for the monitoring of neuroinflammation. Investigations provided longitudinal data as evidence for additional biomarkers of angiogenesis-associated pericytes *in vivo* with target-specific CAs after noninvasive delivery to noninvasively reveal neural progenitor cells during vascularization in the subventricular zone (SVZ). Those can be based on delivery methods for some brain probes specific to actin mRNA, nestin mRNA, etc., with SPIOs conjugation, which was ever successfully applied to living mouse brains MR imaging for their correlation with histology (**Figure 2**) [175–177]. Uptake using intracerebroventricular delivery is robust, and it is most suitable for the utilizations in preclinical testing, also for the applications in antisense strategies for gene knockdown, gene therapies, and stem cell tracking.

Gene therapy is not only an alternative therapeutic option for ischemic stroke; it is also broadly defined as a genetic-based material (e.g., DNA or RNA) delivering an approach to specific cells to modify their gene expression patterns or treating diseases at the genetic level. Antisense ODNs have already been used in the treatment of experimental ischemic stroke as early as 1993 [178]. Selectively reducing the expression of NMDA (N-methyl-D-aspartate) receptors by antisense ODNs could prevent neurotoxicity elicited by NMDA *in vitro* and reduce the cerebral infarct size following experimental occlusion of the cerebral artery in rats. The GT-tMRI technique has been extended to test the hypothesis that a neuroprotective biologic encoded in the complimentary DNA (cDNA) carried by a viral vector could be delivered at multiple time points using eye drops to the BCAO-60 model of C57blcak6 mice [179]. By encoding human granulocyte colony-stimulating factor (hG-CSF) in a replication-deficient, self-complementary AAV vector (scAAV2-CMV-hG-CSF), it may elicit a continuous supply of endogenous G-CSF (a cytokine that stimulates growth and differentiation of myeloid precursors), passing through the BBB by receptor-mediated endocytosis

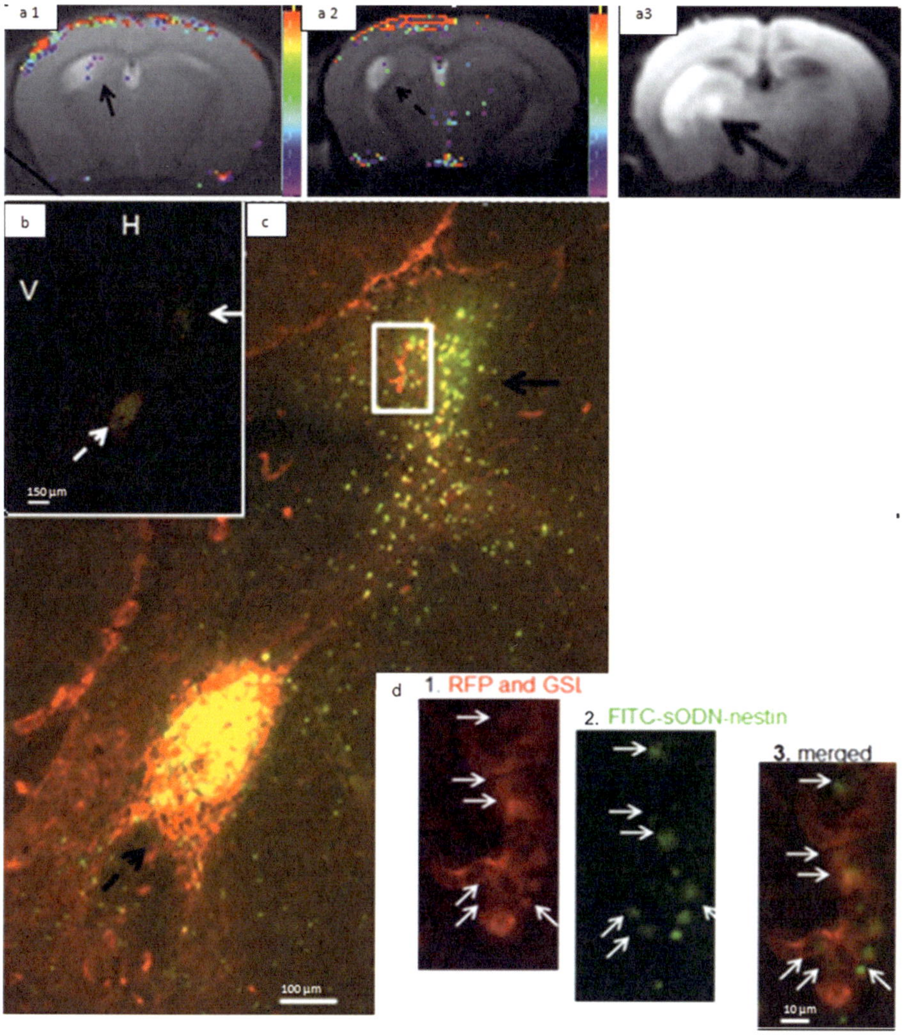

Figure 2.
SPION-nestin was administered, respectively, in weeks 5 and 10 to the mouse, which was acquired by the deltaR$_2^*$ map by GT-tMRI 24 hours after SPION-nestin administration (a1, a2). A few ROIs were observed as no more than two pixels (0.007 mm^3 per voxel) and scattered throughout the brain for at least 7 brain slices (or 3.5 mm). Except in the proximity of the esophagus, trachea, and ear canals and the surface of the skull in the proximity of the sagittal sinus due to potential artifacts of air and hair, two of these ROIs (arrows) were located at the region of abnormal water diffusion (ever demonstrated by DWI-MRI two days after BCAO (c3). MRI-cell typing was validated by FITC-sODN-nestin transfection during week 12. Only two matching ROIs with FITC were identified by histology (b). Actin-derived RFP-producing cells on endothelia (GSL-I positive stain, arrows) retained FITC-sODN-nestin (c). Two cohorts of pericytes (arrows, two to three pixels) were apparent near the ventricle (V) and below the hippocampus (H). Enlarged views of the boxed area were shown (d).

to provide autocrine neuroprotection in response to brain damage and leading to a reduction in infarct volume. This monitoring system will be regarded as central to evaluating the therapeutic efficacy of exogenous cDNA in the central nervous system. Furtherly, investigators have validated hG-CSF cDNA expression using translatable MRI in living brains, being potential for great impact in the area of experimental gene therapy in animal models of heart attack, stroke, AD, PD, and ALS.

Only a few studies have evaluated the utility of USPIOs in stroke with MRI. The first *in vivo* study of phagocyte-labeling with intravenously injected USPIO (Ferumoxtran-10) methodology in mice model of permanent occlusion of the distal middle cerebral artery (pMCAO) was reported in 2007 to have therapeutic implications in the assessment of the potential neurotoxicity of activated microglia/ macrophages [180]. Henning et al. performed a preclinical study providing proof of principle that this type of imaging may be useful in the development of future neuroprotective strategies [181]. A group in Germany investigated the uptake of iron by cells of the mononuclear phagocyte system that can cause them to adopt a pro-inflammatory M1 phenotype [182]. Another study provided evidence that the administration of ferumoxytol at a dose sufficient for perfusion mapping after stroke did not cause significant preloading of peripheral monocytes/macrophages or affect the inflammatory response to stroke [183]. Preliminary studies in patients with stroke reported heterogeneous patterns of USPIO enhancement after its injection, which could be due to different forms of accumulation (intra- and extra-cellular) and variation in timings of CA injection, as well as of post-contrast MRI acquisition [184]; and according to Farell et al. [185], demonstrating that ferumoxytol images show signal changes on T_1WI and T_2WI in different types of inflammatory lesions, as well as different enhancement patterns compared to gadolinium-based contrast agent (GBCA)-enhanced scans. The later findings had indications of ferumoxytol image application in stroke monitoring. Actually, USPIO enhancement was present in the early subacute phase (1–4 days) in 3 out of 9 and in 1 out of 5 patients, respectively, in the only two studies reporting results at this time point [186]. In these studies, USPIO uptake was not related to BBB damage. In contrast to this uniform response at the late stage, enhancement at earlier stages revealed a heterogeneous pattern.

In the stroke research field, many *in vivo* animal model studies and limited human studies have been performed to address stroke-associated pathophysiological processes using nanotechnological compounds (examples, see **Table 1**), showing opportunities to pursue for translation of nanotechnology applications with MRI into actual tools for human clinical practice.

3.2 For MS, TBI, and other studies

Like in the case of MS, observing monocyte infiltration by imaging for multiple small lesions in the CNS pathologies remains challenging due to smaller-sized lesions and, consequently, an accumulation of a low number of labeled monocytes. Although some examples have been documented elsewhere [187], it is difficult to use in vitro labeling strategies for monocytes while cell trafficking in human neuroinflammation has been performed following direct i.v. INOPs injection. Engberink et al. [188] used magnetoelectroporation (MEP) for monocyte labeling with iron oxide particles injected into rats so that EAE was induced. The results demonstrated successful implementation for *in vivo* monocyte migration in rats with EAE. Their methods for targeting monocyte recruitment in the inflammatory response could facilitate bench-to-bedside translation of novel therapeutics. Both preclinical and clinical studies have clearly shown that INOPs provide additional information for MS and TBI beyond Gd-based assessments of BBB breakdown that are relevant to disease pathophysiology and therapy effects [189]. A longitudinal study comparing USPIO enhancement and quantitative susceptibility mapping (QSM) may lead to better interpretations of QSM variations, which are essential for understanding the mechanisms of MS and for the development and monitoring of new treatments [190]. During the subacute

Nanoparticle types	Study subjects	Study objectives	Reference
Ultrasmall superparamagnetic iron oxide particles (USPIO)	Human	A single-center, open-labeled, clinical phase II study to test the potential for USPIO-enhanced versus conventional GBCA-enhanced MRI to image macrophages in human ischemic stroke lesions.	[170]
	Human	To investigate 10 patients who presented with acute anterior circulation stroke through USPIO administered on day 6 and signal alterations in 9/10 patients on day 9 T_1-weighted imaging for studying on hematogenous macrophage as the main target.	[186]
Ultrasmall superparamagnetic iron oxide particles (USPIO)	Mouse	To monitor phagocytic activity by intravenously injected USPIO for 72 hours after focal cerebral ischemia in mice.	[180]
Ultrasmall superparamagnetic iron oxide particles (USPIO)	Rat	To demonstrate a method of the iron labeling of bone marrow-derived macrophages and the imaging findings of their activity after tMCAO, and whether so-called “CNS-resident macrophages” are essential for the initiation and maintenance of the inflammatory response through antigen presentation and cytokine expression.	[181]
Micro-sized particles of iron oxide (MPIOs) conjugated with specific antibodies (P-selectin, ICAM-1, VCAM-1)	Mouse	Through MPIOs-αP-selectin being able to identify and delineate transient ischemic attacks (TIAs) 24 hours after induction to provide evidence of whether the method could discriminate TIAs from epilepsy and migraine-mimicking TIAs in the clinical setting.	[74]
	Mouse	About the feasibility of the approach of targeted CAs to detect the neuroinflammatory biomarker ICAM-1 after stroke and a sort of most efficient imaging procedure; To determine the diagnostic potential of ICAM-1-targeted MPIO for in vivo MRI of vascular ICAM-1 expression and leukocyte infiltration at different stages after stroke.	[165, 166]
	Mouse	To assess the spatiotemporal evolution of the cerebrovascular inflammation occurring after ischemic and hemorrhagic strokes using MRI after injection of MPIOs-αVCAM-1.	[167]
Iron oxide microparticle (SPIO) (conjugated phosphorothioate oligodeoxynucleotide)	Mouse	To demonstrate the use of modified INOPs to be able to target cellular mRNAs and detect active transcriptions of specific mRNAs using antisense imaging agents (e.g., phosphorothioate oligodeoxynucleotide).	[175]

Table 1.
MR imaging examples with IONPs for potential applications in ischemic stroke.

time window after primary TBI, neuroinflammation is also an important pathological mechanism. Maximum USPIO uptake in a murine TBI model occurred, reflecting macrophage infiltration between 66 and 72 hours post-injury [191]. Foley et al. found a similar time window in a study on MPIO, with the intra-macrophagic localization of the NP, which was confirmed by *ex vivo* MR microscopy [192]. The combination of Gd- and USPIO-enhanced MRI allowed for monitoring the time course of BBB breakdown and cellular inflammatory response in rabbits with mild TBI [193], revealing a mismatch between these two mechanisms. GBCA-enhancement occurred on the first day, reached 100% of the subjects after 48 h, and peaked at 72 h when 90% of the lesions were also enhancing at USPIO; then it progressively decreased five days after injury while USPIO enhancement tended to expand at these later time points.

Cell therapy holds great promise in the treatment of several neurological disorders, including stroke, spinal cord injury (SCI), and ALS, in addition to neurodegenerative diseases. Yet the mechanisms underlying the improved functional recovery of ischemic stroke animal model subjected to NSC therapy and potential candidates for cell therapy are still unclear. Inflammatory and immune responses mediate secondary injury after acute cerebral ischemia, whose main inflammatory regulators are microglia in focal brain tissue. Intravenous administration of NSCs in a rat stroke model can reduce OX-42$^+$ microglia and MPO$^+$ (myeloperoxidase) neutrophil infiltration into brain lesions, promoting neuroprotection, and also attenuate both cerebral and splenic activations of TNF-α, IL-6, and NF-κB [194]. Elevated transcription factor (TF) proteins (such as NF-κB) and microglial activation are also known associations in ischemia-reperfusion injury. However, a relevant beneficial aspect of stroke-induced neuroinflammation is the induction of a strong regenerative response that leads to robust expansions of endogenous NSCs [195]. These NSCs can be attracted to the site of the lesion by various inflammation-associated cytokines such as stromal cell-derived factor-1, tumor necrosis factor-alpha, and interferon-γ [196–199]. In the inflamed region of the brain, the accumulation of USPIOs-labeled monocytes was detected by MRI, and the delay between injection and imaging was usually long (about 24 hrs), limiting the interest of such CAs in the acute phases. The relative amount of CAs eventually reaching parenchyma also directly depended on the number of monocytes entering the brain, limiting the sensitivity of this method. The possibility of false-positive findings in the case of BBB leakage is another limitation because USPIO readily crosses the BBB after stroke as well as in active plaques in MS investigation [200]. Another report [201] showed that NSCs after transplantation into mouse striatum can incorporate SPIOs for MR imaging, as demonstrated by immunostaining using anti-dextran antibodies, Prussian blue assays, and transmission electron microscopy, but without impairment of cellular viability, proliferation, migration, differentiation, and neurosphere formation. Results indicated that some neurons and astrocytes can carry SPION due to the observation of SPIO-labeled NSCs showing dextran-positive dots in these cells. This corresponds to a study in which human NSCs labeled with SPIOs and transplanted into the neonatal, adult, or injured rodent brain (Feridex) were grown without adverse effects on differentiation, long-term survival, or migration of the grafted labeled NSCs. Ferritin can be introduced into proliferating (stem) cells as a nMRI reporter gene using viral vectors, rendering them detectable for MR imaging in a mouse study of KU Leuven [202]. The same group also showed further quantitative evaluation of ferritin-labeled endogenous NSCs in rodent brains using MRI bias field correction [203]. Endogenous NSCs migration from the SVZ to the injured site was confirmed by MRI visualization using micron-sized IONPs, followed by histological staining. Progress in this field has also been made with the pipeline development of a monoclonal antibody binding to endogenous neural progenitors in a brain tumor mice model, coupled to magnetic glyconanoparticles allowing for their detection by MR imaging [204]. To track iPSC-derived neural progenitors, another method has been developed using MRI-based imaging and multimodal imaging (with PET) [205]. In the first report [206] on the longitudinal mouse MR imaging for quantitative evaluation of T_2^* time at the TBI site, investigators used absolute bias correction for up to 21 days after transplantation of dual labeled (red fluorescent dye PKH26 and USPIO) MSCs, providing a protocol for *in vivo* imaging of stem cell tracking and therapies for different types of neurodegenerative diseases. A study showed a hypointense signal on day 3 after intravenous transplantation of human NSCs (hNSCs) in a rat focal ischemia model [207]. Translational studies of systemically administered USPIO (Sinerem/Combidex) for neuroinflammation MRI demonstrated in animal models have evolved into pilot studies in patients with ischemic stroke, generally indicating that the infiltration of blood-borne monocytes in stroke lesions is a delayed event [184, 185, 208, 209].

During the application of IONPs in tracking the cells for regenerative medicine with cell therapy, the stem cells can take them up via common cellular pathways such as endocytosis, phagocytosis, or antibody-mediated uptake (the former two are common routes of entry into stem cells). Specific antibodies can also be attached to the surface of IONPs to induce target specificity. Based on the application of a magnetic field, investigators have successfully created an area for the mesenchymal stem cells (MSCs) tagged with poly-lysine-coated iron oxide NPs to migrate to the site of action for the treatment of SCI in rats [210].

4. Multimodal MR imaging techniques application combined with nanomedicine

During the disclosure of the nanotheranostics processing in CNS disorders, MR technology applications, data collection, and software utilization for imaging analysis have made solid contributions to magnetic iron oxide nanoparticles' function of detecting and monitoring neurotoxicity, along with the development of more efficient and safer theranostics systems.

MRI is uniquely suited for *in vivo* visualization of SPIOs in target organs based on the strong local magnetic field inhomogeneities by iron oxide crystal generating hypointense areas in T_2WI or T_2^*WI as shown in AD, PD studies [67, 68, 184, 206, 211–213]. Iron-oxide labeled cells implantation has also been imaged longitudinally *in vivo* using high-resolution MRI [214]. Investigations suggested that combined with relaxation time mapping techniques, especially T_2 and T_2^*, multiparametric MRI and susceptibility measurement could be an efficient way to measure both NPs concentration as well as dispersed and cell-internalized SPIO fractions *in vivo* regardless of whether the particles were dispersed or clustered, to gain insight into tissue function or detect and quantify molecular imaging paradigms, with the basic and advanced MR techniques and different acquisition sequences [215]. Regarding quantitative analysis of SPIOs, an ALS study from Willenbrock et al. [127] has shown the lowest detectable number of SPIO-MTH53A cells to be 1 × 105 with a 1.0-T MR scanner versus 2.5 × 104 with a 7.0-T field strength; and the *in vitro* data indicated that the field strength is not directly inversely proportional to the detectable cell number, establishing the feasibility of basic concepts for hUCBC-labeling and MRI protocol.

In a study [216] of 51 patients with symptoms of PD using multiple sequences of susceptibility-weighted imaging (SWI), T_2^* maps, T_2-WI, and FLAIR-SPACE imaging, the results indicated that SWI contrast correlated well with iron concentration in the brain. SWI data yielded significantly increased values in PD patients through voxel-wise analysis versus other Parkinsonism in the bilateral, right-dominant thalamus [217]. A significant increase of the physiological low SWI signal in the left SN of PD patients was demonstrated by additional ROI analysis that was performed due to the inaccuracy of automatic spatial normalization in the infratentorial region in relation to the small size of SN and RN versus the non-significant trend in the contralateral SN, being consistent with previous studies [218]. The asymmetry might be related to Parkinsonism being clinically more prominent on the right in many PD patients (12 right versus 4 left dominant Parkinsonism). Regarding SWI application, QSM has been used to study abnormal iron accumulation in various brain regions in MS and the magnetic susceptibility of WM lesions, and by inference, iron-containing microglia dynamically evolve in MS patients, providing a wide window into chronic inflammatory activity in established non-enhancing lesions even to be used for treatment monitoring [190]; specifically, the QSM-hyperintense rim of a WM lesion corresponds to iron in pro-inflammatory activated microglia [219]. QSM may also have a broad utility in other disorders characterized by chronic microglial activation, such as AD, providing the potential of simple, yet robust, noninvasive, and challenge-free oxygen consumption measurements, with further development significantly indicating the imaging strategies of a wide range of diseases [220].

Dynamic contrast-enhanced MR imaging (DCE-MRI), as the most widely used technique to investigate the degree of BBB breakdown that triggers entry into the brain of blood-derived neurotoxic molecules and cells causing vascular pathology in brain parenchyma and direct neuron damage, plus other biomarkers, have been used in AD, PD, HD, ALS, MS, HIV-1-associated dementia, and chronic traumatic encephalopathy [221–224]. These features and other changes in BBB function (increased permeability and changes in brain perfusion, including CBF dysregulation and reduction) belonged to a cascade of pathogenic events, at least in AD. The factors to develop neurodegeneration and dementia include neurotoxic effects by increased Aβ accumulation in brain parenchyma and impaired clearance, which can be promoted by inadequate cerebral perfusion while impairing endothelial function and neurovascular coupling and making CBF reduced (supporting the vascular-mediated neurodegeneration hypothesis) [225, 226]. CBV and other parameters (for example, cerebral metabolic rate of oxygen consumption ($CMRO_2$), micro bleeding events, white matter connectivity, white matter lesions (WML), and regional brain atrophy, etc.) may also be measured by MR methods to evaluate neurovascular dysfunction in early dementia and AD.

CBV maps are usually obtained using advanced perfusion modality dynamic susceptibility contrast (DSC) -MRI as well as through a bolus administration of a GBCAs and acquisition of the time course of the MR signal [227]; while the CBV accuracy estimates by DSC are dependent on the intravascular localization of the CA and the leaky BBB situation. USPIO (Ferumoxytol, such as the FDA-approved compound Feraheme), whose larger size (than Gd-based CAs) confines it to the intravascular space and has the potential to measure CBV more accurately in stroke [183] or other diseases [228]. With the long half-life of USPIO in the blood, some alternative approaches [229, 230], for example, steady-state methods, can be developed for the measurement of CBV with higher resolution and/or signal to noise ratio (SNR), even other parameters, such as acquiring both DSC and steady-state CBV (or SS-CBV) maps in volunteers and comparing the quantitative values at different doses and spatial resolutions. Corresponding maps were derived using SS_CBV(%) = $100 \times \text{delta}R_2{}^*{}_{\text{tissue}} / (\text{delta}R_2{}^*{}_{\text{sag.sinus}})^{0.5}$, based on the expected linear relationship between $\text{delta}R_2{}^*$ and CBV in tissues and a quadratic relationship between $\text{delta}R_2{}^*$ and CA concentration in the sagittal sinus, while $\text{delta}R_2{}^*$ were attained from literature (= contrast-enhanced $R_2{}^*$ value - baseline $R_2{}^*$ value) [231–236].

Moreover, a proof of concept introduced by the recently developing machine learning (ML)-based approach to estimation of the perfusion parameters in DSC MR imaging for clinical stroke research obtained better robustness to noise with the use of FDA-approved software (for example, Olea Sphere from Olea medical); it may allow clinical decision-making to be more useful for neurovascular perfusion assessment and treatment selection in acute stroke at least [237, 238]. Earlier, Crimi et al. [239] reported that MS patients had been classified according to three major different lesion patterns for the nature of inflammatory lesions, and the finding could be useful for more accurate prognosis and thus for more suitable treatments at an early stage. An NIH group [240] developed temporal similarity perfusion (TSP) maps, together with a model-free, iterative process for their generation with no need for the arterial input function (AIF) compared to traditional time-to-peak (TTP) and mean-transit time (MTT) maps, to assess the clinical value for the detection of perfusion deficits. The proposed methods operated without the need during the generation of TSP maps. There have already been applications of convolutional neural network (CNN)-based architecture in DCE-MRI [241].

For quantifying exogenous iron contents, particularly IONPs-based CAs, is of important scope for the labeled stem cell tracking and microvasculature imaging or CBV estimation. Besides R_2 or R_2^* relaxation rates achieved IONPs concentration by the measurement in the GRE sequence, the ultrashort echo (UTE) sequence may have more advantages than the former. Hong et al. demonstrated that through a UTE sequence with the 3D IR-Cones technique in detail, for R_1 mapping, IONPs with concentrations up to about 45 mM were reliably quantified [242].

In addition, the novel MR techniques above, combined with advanced post-processing procedures, have been used in neuroimaging studies on magnetic iron oxide nanoparticles. In AD research, with USPIO-PEG-A$\beta_{1\text{-}42}$ injected transgenic and USPIO-alone-injected transgenic mice, the regional differences in T_2^* values were examined with voxel-based morphometry (VBM) [69, 70] using statistical parametric mapping (SPM) extensively applied in MR imaging as well as in PET [243, 244], showing the correlation of the amyloid plaque distribution with histology. The results from voxel-based analysis (VBA) supported the notion that USPIO-PEG-A$\beta_{1\text{-}42}$ facilitates their utilization in longitudinal studies for amyloid plaque detection but no need to get permeability of the BBB increased through co-injection of a specific CA. In other diseases, such as temporal lobe epilepsy (TLE), VBM can also be involved in the assessment of whole-brain tissue class volumes to find correlations between material-specific cognitive dysfunction and localized gray matter loss [245], and it also has potential in combined application with the proposed tools of the support vector machine (SVM) classifier (for hippocampus shape analysis) [246] and supervised learning models used for classification and regression analysis based on hippocampal volume and shape features.

5. Conclusion and future prospects

Nanomedicine has so much potential for many CNS disorders' theranostics, allowing monitoring of biodistribution and therapeutic activity simultaneously based on magnetic iron oxide nanoparticles conjugated with biomaterials/agents delivering. The CNS was ever considered an immune-privileged organ missing a lymphatic system and was protected from the peripheral blood cells by the BBB. However, the observation is clear now that the BBB can respond to soluble factors and plasma proteins and interconnect with the peripheral immune system cells, developing neuroimmune system interactions, being feasible of setting up the concept that neuroinflammation leads to AD pathology, even contributing to many diseases. Recent advanced MR imaging applications have shed more light on the CNS disease pathology at the cellular or even molecular level with IONPs utilization, and when combined with PET, the multimodalities could have greatly impacted the design of nanomedicine to target specific cell populations in the CNS and to validate various antineurotoxic interventions for the CNS disorders. Despite the study results elsewhere [247] showing IONPs can induce cell damage by violating membrane integrity, degrading DNA, and initiating oxidative stress in the cell, even the side effects at the organism level, IONPs have been considered the most biocompatible ones due to the participation of iron in metabolic processes.

Molecular MR imaging would be much more used for evaluation of nanoparticles as drug delivery vehicles in preclinical studies, for example, there would be more application exploration of stable viral vector-mediated marking of endogenous stem cells and their progeny with a reporter gene for MR imaging. GT-tMRI technique mentioned above may serve as a promising drug delivery system and a functional and noninvasive tool to assess all the elements of the NVU in conjunction with pericytes, vascular smooth muscle cells from the vessel wall, neurons, glial cells, etc. Meanwhile, altered inflammation and cytokine signaling pathways associated with neurodegenerative diseases can be implicated by single-cell transcriptional profiling of endothelial cells which may differentiate healthy aging from multiple neurodegenerative conditions [248]. Importantly, these could also be translated and implemented in clinical applications. Other nanoprobes, for example, ADx-001, based on the new Gd construct, is currently being conducted in a first-in-human Phase 1 clinical trial (clinical trial NCT05453539) as the first agent for patients with suspected AD through *in vivo* Aβ molecular MR imaging after performing comprehensive preclinical safety evaluations in small and large animal models [249]. Molecular MR imaging studies with liposome-based Gd CAs were of robust perspectives as well in AD and stroke [2, 147, 250]. In addition, multimodality utilization strategy with novel devices, for example, ultrasound-mediated technologies for drug release [251], could have more advantages in gene therapy and others for CNS applications. Regarding cell transplantation and gene therapy strategies for treating patients with PD, for instance, they could provide robust imaging biomarkers to improve the therapeutic efficacy or diagnosis efficiency in the clinical settings. They also have perspectives, if with the collaboration of some appropriate PET tracers' utilization, in wider scope applications in not only the treatment of neurodegenerative diseases and neuropsychiatric disorders [252, 253], but now also in interventions of neurotoxicity induced by common drugs of abuse, including amphetamines, cocaine, opiates, and alcohol.

Moreover, additional but promising applications arise from machine learning techniques [237–239], which is a branch of artificial intelligence, together with radiomics-based models [254, 255], without being limited to only one modality imaging run or only in oncology studies, will play an important role in CNS disorders. With the extraction of abundant imaging features and analysis of large amounts of quantitative data, advanced imaging biomarkers could be more helpful in various CNS pathophysiologies.

Acknowledgements

Authors of this chapter sincerely would like to thank Ian Liu, a master's degree candidate in Brown University, USA, for his editing efforts in the text.

Conflict of interest

There is no conflict of interest during the design and editing of this chapter.

Author details

Lei Zhao[1], Tao Peng[2] and Shawn Y. Wu[1,3]*

1 Northern Medical Institute of Health, Middletown, NY, USA

2 Department of Internal Medicine, UT Health Science Center Houston, Houston, TX, USA

3 Department of Diagnostic Imaging, Northern Medical Institute of Health, Middletown, NY, USA

*Address all correspondence to: shengywu58@gmail.com

References

[1] Mornet S, Vasseur S, Grasset F, Duguet E. Magnetic nanoparticle design for medical diagnosis and therapy. Journal of Materials Chemistry. 2004;**14**:2161-2175. DOI: 10.1039/B402025A

[2] Peng T, Booher K, Moody MR, Yin X, Aronowski J, McPherson DD, et al. Enhanced cerebroprotection of xenon-loaded liposomes in combination with rtPA thrombolysis for embolic ischemic stroke. Biomolecules. 2023;**13**(8):1256. DOI: 10.3390/biom13081256

[3] Sun C, Lee JSH, Zhang MQ. Magnetic nanoparticles in MR imaging and drug delivery. Advanced Drug Delivery Reviews. 2008;**60**:1252-1265. DOI: 10.1016/j.addr.2008.03.018

[4] Padmanabhan P, Kumar A, Kumar S, Chaudhary RK, Gulyás B. Nanoparticles in practice for molecular imaging applications: An overview. Acta Biomaterialia. 2016;**41**:1-16. DOI: 10.1016/j.actbio.2016.06.003

[5] Shapiro EM, Skrtic S, Sharer K, Hill JM, Dunbar CE, Koretsky AP. MRI detection of single particles for cellular imaging. Proceedings of the National Academy of Sciences of the United States of America. 2004;**101**:10901-10906. DOI: 10.1073/pnas.0403918101

[6] Wu S. Iron oxide nanoparticles imaging tracking by MR advanced techniques: Dual contrast approaches. In: Hashim A, editor. Smart Nanoparticles Technology. InTechOpen; 2012. pp. 149-172. DOI: 10.5772/33648

[7] Slikker W Jr, Sobotka TJ. Current and future approaches to neurotoxicity risk assessment. Annals of the New York Academy of Sciences. 1997;**825**:406-418. DOI: 10.1111/j.1749-6632.1997.tb48452.x

[8] Nazem A, Mansoori GA. Nanotechnology for Alzheimer's disease detection and treatment. Insciences Journal. 2011;**1**:169-193. DOI: 10.5640/insc.0104169

[9] Kayed R, Head E, Thompson JL, McIntire TM, Milton SC, Cotman CW, et al. Common structure of soluble amyloid oligomers implies common mechanism of pathogenesis. Science. 2003;**300**(5618):486-489. DOI: 10.1126/science.1079469

[10] Goedert M, Klug A, Crowther RA. Tau protein, the paired helical filament and Alzheimer disease. Journal of Alzheimer's Disease. 2006;**9**(3):195-207. DOI: 10.3233/JAD-2006-9S323

[11] Kim YS, Joh TH. Microglia, major player in the brain inflammation: Their roles in the pathogenesis of Parkinson's disease. Experimental & Molecular Medicine. 2006;**38**(4):333-347. DOI: 10.1038/emm.2006.40

[12] Bornemann KD, Wiederhold KH, Pauli C, Ermini F, Stalder M, Schnell L, et al. Abeta-induced inflammatory processes in microglia cells of APP23 transgenic mice. The American Journal of Pathology. 2001;**158**(1):63-73. DOI: 10.1016/s0002-9440(10)63945-4

[13] Granic I, Dolga AM, Nijholt IM, van Dijk G, Eisel UL. Inflammation and NF-kappaB in Alzheimer's disease and diabetes. Journal of Alzheimer's Disease. 2009;**16**:809-821. DOI: 10.3233/JAD-2009-0976

[14] Khandelwal PJ, Herman AM, Moussa CE. Inflammation in the early stages of neurodegenerative pathology. Journal of Neuroimmunology. 2011;**238**:1-11. DOI: 10.1016/j.jneuroim.2011.07.002

[15] Couturier J, Paccalin M, Morel M, Terro F, Milin S, Pontcharraud R, et al. Prevention of the β-amyloid peptide-induced inflammatory process by inhibition of double-stranded RNA-dependent protein kinase in primary murine mixed co-cultures. Journal of Neuroinflammation. 2011;**8**:72. DOI: 10.1186/1742-2094-8-72

[16] Mrak RE, Griffin WST. Glia and their cytokines in a progression of neurodegeneration. Neurobiology of Aging. 2005;**26**:349-354. DOI: 10.1016/j.neurobiolaging.2004.05.010

[17] El Khoury J, Luster AD. Mechanisms of microglia accumulation in Alzheimer's disease: Therapeutic implications. Trends in Pharmacological Sciences. 2008;**29**:626-632. DOI: 10.1016/j.tips.2008.08.004

[18] Schlachetzki JC, Hüll M. Microglial activation in Alzheimer's disease. Current Alzheimer Research. 2009;**6**:554-563. DOI: 10.2174/156720509790147179

[19] Graeber MB. Changing face of microglia. Science. 2010;**330**:783-788. DOI: 10.1126/science.1190929

[20] Burguillos MA, Deierborg T, Kavanagh E, Persson A, Hajji N, Garcia-Quintanilla A, et al. Caspase signalling controls microglia activation and neurotoxicity. Nature. 2011;**472**:319-324. DOI: 10.1038/nature09788

[21] Montagne A, Barnes SR, Sweeney MD, Halliday MR, Sagare AP, Zhao Z, et al. Blood-brain barrier breakdown in the aging human hippocampus. Neuron. 2015;**85**(2):296-302. DOI: 10.1016/j.neuron.2014.12.032

[22] Zlokovic BV. Neurovascular pathways to neurodegeneration in Alzheimer's disease and other disorders. Nature Reviews. Neuroscience. 2011;**12**(12):723-738. DOI: 10.1038/nrn3114

[23] Zlokovic BV, Deane R, Sagare AP, Bell RD, Winkler EA. Low-density lipoprotein receptor-related protein-1: A serial clearance homeostatic mechanism controlling Alzheimer's amyloid β-peptide elimination from the brain. Journal of Neurochemistry. 2010;**115**(5):1077-1089. DOI: 10.1111/j.1471-4159.2010.07002.x

[24] Montagne A, Nation DA, Pa J, Sweeney MD, Toga AW, Zlokovic BV. Brain imaging of neurovascular dysfunction in Alzheimer's disease. Acta Neuropathologica. 2016;**131**(5):687-707. DOI: 10.1007/s00401-016-1570-0

[25] Marttinen M, Takalo M, Natunen T, Wittrahm R, Gabbouj S, Kemppainen S, et al. Molecular mechanisms of Synaptotoxicity and Neuroinflammation in Alzheimer's disease. Frontiers in Neuroscience. 2018;**12**:963. DOI: 10.3389/fnins.2018.00963

[26] Migliore L, Uboldi C, Di Bucchianico S, Coppedè F. Nanomaterials and neurodegeneration. Environmental and Molecular Mutagenesis. 2015;**56**(2):149-170. DOI: 10.1002/em.21931

[27] Dubois B, Feldman HH, Jacova C, Dekosky ST, Barberger-Gateau P, Cummings J, et al. Research criteria for the diagnosis of Alzheimer's disease: Revising the NINCDS-ADRDA criteria. Lancet Neurology. 2007;**6**:734-746. DOI: 10.1016/S1474-4422(07)70178-3

[28] Jack CR Jr, Albert MS, Knopman DS, McKhann GM, Sperling RA, Carrillo MC, et al. Introduction to the recommendations from the National Institute on Aging-Alzheimer's association workgroups on diagnostic guidelines for Alzheimer's disease.

Alzheimer's & Dementia. 2011;**7**:257-262. DOI: 10.1016/j.jalz.2011.03.004

[29] Markus HS, Hunt B, Palmer K, Enzinger C, Schmidt H, Schmidt R. Markers of endothelial and hemostatic activation and progression of cerebral white matter hyperintensities: Longitudinal results of the austrian stroke prevention study. Stroke. 2005;**36**:1410-1414. DOI: 10.1161/01.STR.0000169924.60783.d4

[30] Przedborski S, Vila M, Jackson-Lewis V. Neurodegeneration: What is it and where are we? The Journal of Clinical Investigation. 2003;**111**:3-10. DOI: 10.1172/JCI200317522

[31] Ittner LM, Götz J. Amyloid-β and tau – A toxic pas de deux in Alzheimer's disease. Nature Reviews. Neuroscience. 2011;**12**:67-72. DOI: 10.1038/nrn2967

[32] Ittner LM, Ke YD, Delerue F, Bi M, Gladbach A, van Eersel J, et al. Dendritic function of tau mediates amyloid-beta toxicity in Alzheimer's disease mouse models. Cell. 2012;**142**:387-397. DOI: 10.1016/j.cell.2010.06.036

[33] Kosik KS, Joachim CL, Selkoe DJ. Microtubule-associated protein tau (tau) is a major antigenic component of paired helical filaments in Alzheimer disease. Proceedings of the National Academy of Sciences of the United States of America. 1986;**83**:4044-4048. DOI: 10.1073/pnas.83.11.4044

[34] Glenner GG, Wong CW. Alzheimer's disease: Initial report of the purification and characterization of a novel cerevrovascular amyloid protein. Biochemical and Biophysical Research Communications. 1984;**120**:885-890. DOI: 10.1016/j.bbrc.2012.08.020

[35] Masters CL, Multhaup G, Simms G, Pottgiesser J, Martins RN, Beyreuther K. Neuronal origin of a cerebral amyloid: Neurofibrillary tangles of Alzheimer's disease contain the same protein as the amyloid of plaque cores and blood vessels. The EMBO Journal. 1985;**4**:2757-2763

[36] Villemagne VL, Rowe CC. Long night's journey into the day: Amyloid-beta imaging in Alzheimer's disease. Journal of Alzheimer's Disease. 2013;**33**(Suppl. 1):S349-S359. DOI: 10.3233/JAD-2012-129034

[37] Chételat G, Villemagne VL, Pike KE, Ellis KA, Bourgeat P, Jones G, et al. Australian imaging biomarkers and lifestyle study of ageing (AIBL) research group. Independent contribution of temporal beta-amyloid deposition to memory decline in the pre-dementia phase of Alzheimer's disease. Brain. 2011;**134**:798-807. DOI: 10.1093/brain/awq383

[38] Chételat G, Villemagne VL, Villain N, Jones G, Ellis KA, Ames D, et al. Accelerated cortical atrophy in cognitively normal elderly with high beta-amyloid deposition. Neurology. 2012;**78**:477-484. DOI: 10.1212/WNL.0b013e318246d67a

[39] Haass C, Selkoe DJ. Soluble protein oligomers in neurodegeneration: Lessons from the Alzheimer's amyloid beta-peptide. Nature Reviews. Molecular Cell Biology. 2007;**8**:101-112. DOI: 10.1038/nrm2101

[40] Vickland V, Morris T, Draper B, et al. Modelling the Impact of Interventions to Delay the Onset of Dementia in Australia. Report for Alzheimers Australia. Sydney, Australia: Alzheimer's Australia Inc; 2012

[41] Johnson E, Brookmeyer R, Ziegler-Graham K. Modeling the effect of Alzheimer's disease on mortality.

International Journal of Biostatistics. 2007;**3**(Article 13):12-21

[42] Fonseca-Santos B, Gremião MPD, Chorilli M. Nanotechnology-based drug delivery systems for the treatment of Alzheimer's disease. International Journal of Nanomedicine. 2015;**10**:4981-5003. DOI: 10.2147/IJN.S87148

[43] Marcus C, Mena E, Subramaniam RM. Brain PET in the diagnosis of Alzheimer's disease. Clinical Nuclear Medicine. 2014;**39**:e413-e422. DOI: 10.1097/RLU.0000000000000547

[44] Attems J, Jellinger KA. The overlap between vascular disease and Alzheimer's disease–lessons from pathology. BMC Medicine. 2014;**12**:206. DOI: 10.1186/s12916-014-0206-2

[45] Zhang Z, van den Bos EJ, Wielopolski PA, de Jong-Popijus M, Bernsen MR, Duncker DJ, et al. In vitro imaging of single living human umbilical vein endothelial cells with a clinical 3.0-T MRI scanner. Magma. 2005;**18**(4):175-185. DOI: 10.1007/s10334-005-0108-6

[46] Heyn C, Bowen C, Rutt B, Foster P. Detection threshold of single SPIO-labeled cells with FIESTA. Magnetic Resonance in Medicine. 2005;**53**:312-320. DOI: 10.1002/mrm.20356

[47] Heyn C, Ronald JA, Mackenzie LT, MacDonald IC, Chambers AR, Rutt BK, et al. In vivo magnetic resonance imaging of single cells in mouse brain with optical validation. Magnetic Resonance in Medicine. 2006;**55**:23-29. DOI: 10.1002/mrm.20747

[48] Grenier N, Brader P. Principles and basic concepts of molecular imaging. Pediatric Radiology. 2011;**41**(2):144-160. DOI: 10.1007/s00247-010-1835-z

[49] Giljohann DA, Mirkin CA. Drivers of biodiagnostic development. Nature. 2009;**462**:461-464. DOI: 10.1038/nature08605

[50] Azria D, Blanquer S, Verdier J, Belamie E. Nanoparticles as contrast agents for brain nuclear magnetic resonance imaging in Alzheimer's disease diagnosis. Journal of Materials Chemistry B. 2017;**5**:7216-7237. DOI: 10.1039/C7TB01599B

[51] Rocha S. Targeted drug delivery across the blood brain barrier in Alzheimer's disease. Current Pharmaceutical Design. 2013;**19**:6635-6646. DOI: 10.2174/1381612811319990613

[52] McLachlan SJ, Morris MR, Lucas MA, Fisco RA, Eakins MN, Fowler DR, et al. Phase I clinical evaluation of a new iron oxide MR contrast agent. Journal of Magnetic Resonance Imaging. 1994;**4**:301-307. DOI: 10.1002/jmri.1880040313

[53] Yang CC, Yang SY, Chieh JJ, Horng H, Hong CY, Yang HC, et al. Biofunctionalized magnetic nanoparticles for specifically detecting biomarkers of Alzheimer's disease in vitro. ACS Chemical Neuroscience. 2011;**2**:500-505. DOI: 10.1021/cn200028j

[54] Skaat H, Corem-Slakmon E, Grinberg I, Last D, Goez D, Mardor Y, et al. Antibody-conjugated, dual-modal, near-infrared fluorescent iron oxide nanoparticles for antiamyloidgenic activity and specific detection of amyloid-β fibrils. International Journal of Nanomedicine. 2013;**8**:4063-4076. DOI: 10.2147/IJN.S52833

[55] Salerno M, Porqueras DSD. Alzheimer's disease: The use of contrast agents for magnetic resonance imaging to detect amyloid beta peptide inside the brain. Coordination Chemistry

Reviews. 2016;**327**:27-34. DOI: 10.1016/j.ccr.2016.04.018

[56] Viola KL, Sbarboro J, Sureka R, De M, Bicca MA, Wang J, et al. Towards non-invasive diagnostic imaging of early-stage Alzheimer's disease. Nature Nanotechnology. 2015;**10**:91-98. DOI: 10.1038/nnano.2014.254

[57] Wadghiri YZ, Sigurdsson EM, Sadowski M, Elliott JI, Li Y, Scholtzova H, et al. Detection of Alzheimer's amyloid in transgenic mice using magnetic resonance microimaging. Magnetic Resonance in Medicine. 2003;**50**:293-302. DOI: 10.1002/mrm.10529

[58] Poduslo JF, Hultman KL, Curran GL, Preboske GM, Chamberlain R, Marjanska M, et al. Targeting vascular amyloid in arterioles of Alzheimer disease transgenic mice with amyloid beta protein antibody-coated nanoparticles. Journal of Neuropathology and Experimental Neurology. 2011;**70**:653-661. DOI: 10.1016/j.addr.2024.115219

[59] Poduslo JF, Curran GL, Peterson JA, McCormick DJ, Fauq AH, Khan MA, et al. Design and chemical synthesis of a magnetic resonance contrast agent with enhanced in vitro binding, high blood-brain barrier permeability, and in vivo targeting to Alzheimer's disease amyloid plaques. Biochemistry. 2004;**43**:6064-6075. DOI: 10.1021/bi0359574

[60] Weissleder R, Lee AS, Khaw BA, Shen T, Brady TJ. Antimyosin-labeled monocrystalline iron oxide allows detection of myocardial infarct: MR antibody imaging. Radiology. 1992;**182**(2):381-385. DOI: 10.1148/radiology.182.2.1732953

[61] Jung HI, Kettunen MI, Davletov B, Brindle KM. Detection of apoptosis using the C2A domain of synaptotagmin I. Bioconjugate Chemistry. 2004;**15**(5):983-987. DOI: 10.1021/bc049899q

[62] Artemov D, Mori N, Okollie B, Bhujwalla ZM. MR molecular imaging of the Her-2/neu receptor in breast cancer cells using targeted iron oxide nanoparticles. Magnetic Resonance in Medicine. 2003;**49**(3):403-408. DOI: 10.1002/mrm.10406

[63] Bulte JWM, Hoekstra Y, Kamman RL, Magin RL, Webb AG, Briggs RW, et al. Specific MR imaging of human lymphocytes by monoclonal antibody-guided dextran-magnetite particles. Magnetic Resonance in Medicine. 1992;**25**(1):148-157. DOI: 10.1002/mrm.1910250115

[64] Go KG, Bulte JWM, De Ley L, The TH, Kamman RL, Hulstaert CE, et al. Our approach towards developing a specific tumour-targeted MRI contrast agent for the brain. European Journal of Radiology. 1993;**16**(3):171-175. DOI: 10.1016/0720-048X(93)90064-T

[65] Sun EY, Josephson L, Kelly KA, Weissleder R. Development of nanoparticle libraries for biosensing. Bioconjugate Chemistry. 2006;**17**(1):109-113. DOI: 10.1021/bc050290e

[66] Sharifi S, Seyednejad H, Laurent S, Atyabi F, Saei AA, Mahmoudi M. Superparamagnetic iron oxide nanoparticles for in vivo molecular and cellular imaging. Contrast Media & Molecular Imaging. 2015;**10**:329-355. DOI: 10.1002/cmmi.1638

[67] Zhou J, Fa H, Yin W, Zhang J, Hou C, Huo D, et al. Synthesis of superparamagnetic iron oxide nanoparticles coated with a DDNP-carboxyl derivative for in vitro magnetic resonance imaging of Alzheimer's disease. Materials Science

& Engineering. C, Materials for Biological Applications. 2014;**37**:348-355. DOI: 10.1016/j.msec.2014.01.005

[68] Zhang D, Fa HB, Zhou JT, Li S, Diao XW, Yin W. The detection of β-amyloid plaques in an Alzheimer's disease rat model with DDNP-SPIO. Clinical Radiology. 2015;**70**(1):74-80. DOI: 10.1016/j.crad.2014.09.019

[69] Yang J, Wadghiri YZ, Hoang DM, Tsui W, Sun Y, Chung E, et al. Detection of amyloid plaques targeted by USPIO-Aβ1-42 in Alzheimer's disease transgenic mice using magnetic resonance microimaging. NeuroImage. 2011;**55**:1600-1609. DOI: 10.1016/j.neuroimage.2011.01.023

[70] Wadghiri YZ, Li J, Wang J, Hoang DM, Sun Y, Xu H, et al. Detection of amyloid plaques targeted by bifunctional USPIO in Alzheimer's disease transgenic mice using magnetic resonance microimaging. PLoS One. 2013;**8**(2):e57097. DOI: 10.1371/journal.pone.0057097

[71] Cheng KK, Chan PS, Fan S, Kwan SM, Yeung KL, Wang YX, et al. Curcumin-conjugated magnetic nanoparticles for detecting amyloid plaques in Alzheimer's disease mice using magnetic resonance imaging (MRI). Biomaterials. 2015;**44**:155-172. DOI: 10.1016/j.biomaterials.2014.12.005

[72] Cheng KK, Yeung CF, Ho SW, Chow SF, Chow AH, Baum L. Highly stabilized curcumin nanoparticles tested in an in vitro blood-brain barrier model and in Alzheimer's disease Tg2576 mice. The AAPS Journal. 2013;**15**:324-336. DOI: 10.1208/s12248-012-9444-4

[73] Ku S, Yan F, Wang Y, Sun Y, Yang N, Ye L. The blood-brain barrier penetration and distribution of PEGylated fluorescein-doped magnetic silica nanoparticles in rat brain. Biochemical and Biophysical Research Communications. 2010;**394**(4):871-876. DOI: 10.1016/j.bbrc.2010.03.006

[74] Quenault A, de Lizarrondo SM, Etard O, Gauberti M, Orset C, Haelewyn B, et al. Molecular magnetic resonance imaging discloses endothelial activation after transient ischaemic attack. Brain. 2017;**140**(1):146-157. DOI: 10.1093/brain/aww260

[75] Hoyte LC, Brooks KJ, Nagel S, Akhtar A, Chen R, Mardiguian S, et al. Molecular magnetic resonance imaging of acute vascular cell adhesion molecule-1 expression in a mouse model of cerebral ischemia. Journal of Cerebral Blood Flow and Metabolism. 2010;**30**(6):1178-1187. DOI: 10.1038/jcbfm.2009.287

[76] Ramanan VK, Saykin AJ. Pathways to neurodegeneration: Mechanistic insights from GWAS in Alzheimer's disease, Parkinson's disease, and related disorders. American Journal of Neurodegenerative Disease. 2013;**2**:145-175. DOI: PMC3783830/

[77] Inglese M, Bester M. Diffusion imaging in multiple sclerosis: Research and clinical implications. NMR in Biomedicine. 2010;**23**(7):865-872. DOI: 10.1002/nbm.1515

[78] Henkelman RM, Stanisz GJ, Graham SJ. Magnetization transfer in MRI: A review. NMR in Biomedicine. 2001;**14**(2):57-64. DOI: 10.1002/nbm.683

[79] Tafoya MA, Madi S, Sillerud LO. Superparamagnetic nanoparticle-enhanced MRI of Alzheimer's disease plaques and activated microglia in 3X transgenic mouse brains: Contrast optimization. Journal of Magnetic Resonance Imaging. 2017;**46**(2):574-588. DOI: 10.1002/jmri.25563

[80] Sillerud LO, Solberg NO, Chamberlain R, Orlando RA, Heidrich JE, Brown DC, et al. SPION-enhanced magnetic resonance imaging of Alzheimer's disease plaques in AβPP/PS-1 transgenic mouse brain. Journal of Alzheimer's Disease. 2013;**34**(2):349-365. DOI: 10.3233/JAD-121171

[81] Chopra K, Misra S, Kuhad A. Current perspectives on pharmacotherapy of Alzheimer's disease. Expert Opinion on Pharmacotherapy. 2011;**12**:335-350. DOI: 10.1517/14656566.2011.520702

[82] Corbett A, Ballard C. New and emerging treatments for Alzheimer's disease. Expert Opinion on Emerging Drugs. 2012;**17**:147-156. DOI: 10.1517/14728214.2012.675327

[83] Mildner A, Schmidt H, Nitsche M, Merkler D, Hanisch UK, Mack M, et al. Microglia in the adult brain arise from Ly-6ChiCCR2+ monocytes only under defined host conditions. Nature Neuroscience. 2007;**10**:1544-1553. DOI: 10.1038/nn2015

[84] Davoust N, Vuaillat C, Cavillon G, et al. Bone marrow CD34+/B220+ progenitors target the inflamed brain and display in vitro differentiation potential toward microglia. The FASEB Journal. 2006;**20**:2081-2092. DOI: 10.1096/fj.05-5593com

[85] Ajami B, Bennett JL, Krieger C, McNagny KM, Rossi FM. Infiltrating monocytes trigger EAE progression, but do not contribute to the resident microglia pool. Nature Neuroscience. 2011;**14**:1142-1149. DOI: 10.1038/nn.2887

[86] Peters C, Steward CG. National Marrow Donor Program; international bone marrow transplant registry; Working party on inborn errors, European bone marrow transplant group. Hematopoietic cell transplantation for inherited metabolic diseases: An overview of outcomes and practice guidelines. Bone Marrow Transplantation. 2003;**31**:229-239. DOI: 10.1038/sj.bmt.1703839

[87] Rafii MS. Preclinical Alzheimer's disease therapeutics. Journal of Alzheimer's Disease. 2014;**42**:S545-S549. DOI: 10.3233/JAD-141482

[88] Tuszynski MH, Yang JH, Barba D, Hs U, Bakay RA, Pay MM, et al. Nerve growth factor gene therapy: Activation of neuronal responses in Alzheimer disease. JAMA Neurology. 2015;**72**:1139-1147. DOI: 10.1001/jamaneurol.2015.1807

[89] Brambilla D, Le Droumaguet B, Nicolas J, Hashemi SH, Wu LP, Moghimi SM, et al. Nanotechnologies for Alzheimer's disease: Diagnosis, therapy, and safety issues. Nanomedicine: Nanotechnology, Biology and Medicine. 2011;**7**:521-540. DOI: 10.1016/j.nano.2011.03.008

[90] Sanchez-Ramos J, Cimino C, Avila R, Rowe A, Chen R, Whelan G, et al. Pilot study of granulocyte-colony stimulating factor for treatment of Alzheimer's disease. Journal of Alzheimer's Disease. 2012;**31**:843-855. DOI: 10.3233/JAD-2012-120196

[91] Qiao L, Huang H, Muresanu DF. Clinical neurorestorative progress in Alzheimer's disease. Journal of Neurorestoratology. 2014;**5**:1-9. DOI: 10.2147/JN.S74143

[92] Seo SW, Lee JI, Kim CH, Chang JW, Roh JH, Oh W, et al. A phase I trial of parenchymal injection of umbilical cord stem cells in Alzheimer's disease. Alzheimer's & Dementia. 2013;**9**(4 Suppl):291. DOI: 10.1016/j.jalz.2013.05.590

[93] Ma Y, Tang C, Chaly T, Greene P, Breeze R, Fahn S, et al. Dopamine cell implantation in Parkinson's disease: Long-term clinical and (18)F-FDOPA PET outcomes. Journal of Nuclear Medicine. 2010;**51**:7-15. DOI: 10.2967/jnumed.109.066811

[94] Kefalopoulou Z et al. Long-term clinical outcome of fetal cell transplantation for Parkinson disease: Two case reports. JAMA Neurology. 2014;**71**:83-87. DOI: 10.1001/jamaneurol.2013.4749

[95] Chen L, Huang H, Duan W, Mao G. Clinical neurorestorative progress in Parkinson disease. Journal of Neurorestoratology. 2015;**3**:101-107. DOI: 10.2147/JN.S74144

[96] Huang H, Chen L, Huang H. Clinical neurorestorative progress in traumatic brain injury. Journal of Neurorestoratology. 2015;**3**:57-62. DOI: 10.2147/JN.S74486

[97] Estelrich J, Linazasoro G. Potential applications of nanotechnologies to neurodegenerative diseases (Alzheimer's and Parkinson's diseases). In: Pastorin G, Forcada J, Van der Werk A, editors. Controlled Release Systems: Advances in Nanobottles and Active Nanoparticles. USA: Pan Stanford Publishing; 2016. pp. 351-383

[98] Park D, Lee HJ, Joo SS, Bae DK, Yang G, Yang YH, et al. Human neural stem cells over-expressing choline acetyltransferase restore cognition in rat model of cognitive dysfunction. Experimental Neurology. 2012;**234**:521-526. DOI: 10.1016/j.expneurol.2011.12.040

[99] Park D, Yang YH, Bae DK, Lee SH, Yang G, Kyung J, et al. Improvement of cognitive function and physical activity of aging mice by human neural stem cells over-expressing choline acetyltransferase. Neurobiology of Aging. 2013;**34**:2639-2646. DOI: 10.1016/j.neurobiolaging.2013.04.026

[100] Zhang W, Wang GM, Wang PJ, Zhang Q, Sha SH. Effects of neural stem cells on synaptic proteins and memory in a mouse model of Alzheimer's disease. Journal of Neuroscience Research. 2014;**92**:185-194. DOI: 10.1002/jnr.23299

[101] Zhang Q, Wu H, Wang Y, Gu G, Zhang W, Xia R. Neural stem cell transplantation decreases neuroinflammation in a transgenic mouse model of Alzheimer's disease. Journal of Neurochemistry. 2016;**136**:815-825. DOI: 10.1111/jnc.13413

[102] Chen Y, Pan C, Xuan A, Xu L, Bao G, Liu F, et al. Treatment efficacy of NGF nanoparticles combining neural stem cell transplantation on Alzheimer's disease model rats. Medical Science Monitor. 2015;**21**:3608-3615. DOI: 10.12659/msm.894567

[103] Mahmoudi M, Quinlan-Pluck F, Monopoli MP, Sheinani S, Vali H, Dawson KA, et al. Influence of the physicochemical properties of superparamagnetic iron oxide nanoparticles on amyloid β protein fibrillation in solution. ACS Chemical Neuroscience. 2013;**4**:475-485. DOI: 10.1021/cn300196n

[104] Liao YH, Chang YJ, Yoshiike Y, Chang YC, Chen YR. Negatively charged gold nanoparticles inhibit Alzheimer's amyloid-β fibrillization, induce fibril dissociation, and mitigate neurotoxicity. Small. 2012;**8**(23):3631-3639. DOI: 10.1002/smll.201201068

[105] Javdani N, Rahpeyma SS, Ghasemi Y, Raheb J. Effect of superparamagnetic nanoparticles

coated with various electric charges on α-synuclein and β-amyloid proteins fibrillation process. International Journal of Nanomedicine. 2019;**14**:799-808. DOI: 10.2147/IJN.S190354

[106] Zhang R, Li Y, Hu B, Lu Z, Zhang J, Zhang X. Traceable nanoparticle delivery of small interfering RNA and retinoic acid with temporally release ability to control neural stem cell differentiation for Alzheimer's disease therapy. Advanced Materials. 2016;**28**:6345-6352. DOI: 10.1002/adma.201600554

[107] Li Y, Li Y, Ji W, Lu Z, Liu L, Shi Y, et al. Positively charged polyprodrug amphiphiles with enhanced drug loading and reactive oxygen species-responsive release ability for traceable synergistic therapy. Journal of the American Chemical Society. 2018;**140**:4164-4171. DOI: 10.1021/jacs.8b01641

[108] Tiwari SK, Agarwal S, Seth B, Yadav A, Nair S, Bhatnagar P, et al. Curcumin-loaded nanoparticles potently induce adult neurogenesis and reverse cognitive deficits in Alzheimer's disease model via canonical Wnt/beta-catenin pathway. ACS Nano. 2014;**8**(1):76-103. DOI: 10.1021/nn405077y

[109] Baum L, Lam CW, Cheung SK, Kwok T, Lui V, Tsoh J, et al. Six-month randomized, placebo-controlled, double-blind, pilot clinical trial of curcumin in patients with Alzheimer disease. Journal of Clinical Psychopharmacology. 2008;**28**:110-113. DOI: 10.1097/jcp.0b013e318160862c

[110] Wang YX, Xuan S, Port M, Idee J. Recent advances in superparamagnetic iron oxide nanoparticles for cellular imaging and targeted therapy research. Current Pharmaceutical Design. 2013;**19**:6575-6693. DOI: 10.2174/1381612811319370003

[111] Maurea S, Mainenti P, Tambasco A, Imbriaco M, Mollica C, Laccetti E, et al. Diagnostic accuracy of MR imaging to identify and characterize focal liver lesions: Comparison between gadolinium and superparamagnetic iron oxide contrast media. Quantitative Imaging in Medicine and Surgery. 2014;**4**:181-189. DOI: 10.3978/j.issn.2223-4292.2014.01.02

[112] Vissers C, Ming GL, Song H. Nanoparticle technology and stem cell therapy team up against neurodegenerative disorders. Advanced Drug Delivery Reviews. 2019;**148**:239-251. DOI: 10.1016/j.addr.2019.02.007

[113] Mhyre TR, Boyd JT, Hamill RW, Maguire-Zeiss KA. Parkinson's disease. Sub-Cellular Biochemistry. 2012;**65**:389-455. DOI: 10.1007/978-94-007-5416-4_16

[114] Tenenbaum L, Humbert-ClaudeGlial M. Cell line-derived neurotrophic factor gene delivery in Parkinson's disease: A delicate balance between neuroprotection, trophic effects, and unwanted compensatory mechanisms. Frontiers in Neuroanatomy. 2017;**11**:29. DOI: 10.3389/fnana.2017.00029

[115] Moriarty N, Parish CL, Dowd E. Primary tissue for cellular brain repair in Parkinson's disease: Promise, problems and the potential of biomaterials. The European Journal of Neuroscience. 2019;**49**(4):472-486. DOI: 10.1111/ejn.14051

[116] Mandeville JB, Marota JJ, Kosofsky BE, Keltner JR, Weissleder R, Rosen BR, et al. Dynamic unctional imaging of relative cerebral blood volume during rat forepaw stimulation. Magnetic Resonance in Medicine. 1998;**39**:615-624. DOI: 10.1002/mrm.1910390415

[117] Bjorklund LM, Sanchez-Pernaute R, Chung S, Andersson T,

Chen IY, Mc-Naught KS, et al. Embryonic stem cells develop into functional dopaminergic neurons after transplantation in a Parkinson rat model. Proceedings of the National Academy of Sciences of the United States of America. 2002;**99**:2344-2349. DOI: 10.1073/pnas.022438099

[118] Chen YI, Galpern WR, Brownell AL, Matthews RT, Bogdanov M, Isacson O, et al. Detection of dopaminergic neurotransmitter activity using pharmacologic MRI: Correlation with PET, microdialysis, and behavioral data. Magnetic Resonance in Medicine. 1997;**38**:389-398. DOI: 10.1002/mrm.1910380306

[119] Jenkins BG, Sanchez-Pernaute R, Brownell AL, Chen YCI, Isacson O. Mapping dopamine function in primates using pharmacologic magnetic resonance imaging. The Journal of Neuroscience. 2004;**24**(43):9553-9560. DOI: 10.1523/JNEUROSCI.1558-04.2004

[120] Chen YI, Brownell AL, Galpern W, Isacson O, Bogdanov M, Beal MF, et al. Detection of dopaminergic cell loss and neural transplantation using pharmacological MRI, PET and behavioral assessment. Neuroreport. 1999;**10**:2881-2886

[121] Li L, Jiang W, Luo K, Song H, Lan F, Wu Y, et al. Superparamagnetic iron oxide nanoparticles as MRI contrast agents for non-invasive stem cell labeling and tracking. Theranostics. 2013;**3**:595-615. DOI: 10.7150/thno.5366

[122] Tang H, Sha H, Sun H, Wu X, Xie L, Wang P, et al. Tracking induced pluripotent stem cells-derived neural stem cells in the central nervous system of rats and monkeys. Cellular Reprogramming. 2013;**15**:435-442. DOI: 10.1089/cell.2012.0081

[123] Bulte JW, Duncan ID, Frank JA. In vivo magnetic resonance tracking of magnetically labeled cells after transplantation. Journal of Cerebral Blood Flow and Metabolism. 2002;**22**:899-907. DOI: 10.1097/00004647-200208000-00001

[124] Stroh A, Boltze J, Sieland K, Hild K, Gutzeit C, Jung T, et al. Impact of magnetic labeling on human and mouse stem cells and their long-term magnetic resonance tracking in a rat model of Parkinson disease. Molecular Imaging. 2009;**8**:166-178. DOI: 10.2310/7290.2009.00017

[125] Liu XG, Lu S, Liu DQ, Zhang L, Zhang LX, Yu XL, et al. ScFv-conjugated superparamagnetic iron oxide nanoparticles for MRI-based diagnosis in transgenic mouse models of Parkinson's and Huntington's diseases. Brain Research. 2019;**15**(1707):141-153. DOI: 10.1016/j.brainres.2018.11.034

[126] Valera E, Masliah E. Therapeutic approaches in Parkinson's disease and related disorders. Journal of Neurochemistry. 2016;**139**(Suppl. 1):346-352. DOI: 10.1111/jnc.13529

[127] Willenbrock S, Knippenberg S, Meier M, Hass R, Wefstaedt P, Nolte I, et al. In vivo MRI of intraspinally injected SPIO-labelled human CD34+ cells in a transgenic mouse model of ALS. In Vivo. 2012;**26**(1):31-38

[128] Bartus RT, Kordower JH, Johnson EM Jr, Brown L, Kruegel BR, Chu Y, et al. Post-mortem assessment of the short and long-term effects of the trophic factor neurturin in patients with α-synucleinopathies. Neurobiology of Disease. 2015;**78**:162-171. DOI: 10.1016/j.nbd.2015.03.023

[129] Moraes L, Vasconcelos-dos-Santos A, Santana FC, Godoy MA,

Rosado-de-Castro PH, Jasmin A-PRL, et al. Neuroprotective effects and magnetic resonance imaging of mesenchymal stem cells labeled with SPION in a rat model of Huntington's disease. Stem Cell Research. 2012;**9**(2):143-155. DOI: 10.1016/j.scr.2012.05.005

[130] Goldhawk DE, Rohani R, Sengupta A, Gelman N, Prato FS. Using the magnetosome to model effective gene-based contrast for magnetic resonance imaging. Wiley Interdisciplinary Reviews. Nanomedicine and Nanobiotechnology. 2012;**4**:378-388. DOI: 10.1002/wnan.1165

[131] Hocquemiller M, Giersch L, Audrain M, Parker S, Cartier N. Adeno-associated virus-based gene therapy for CNS diseases. Human Gene Therapy. 2016;**27**:478-496. DOI: 10.1089/hum.2016.087

[132] Rocha EM, Smith GA, Park E, Cao H, Brown E, Hayes MA, et al. Glucocerebrosidase gene therapy prevents α-synucleinopathy of midbrain dopamine neurons. Neurobiology of Disease. 2015;**82**:495-503. DOI: 10.1016/j.nbd.2015.09.009

[133] Niu S, Zhang LK, Zhang L, Zhuang S, Zhan X, Chen WY, et al. Inhibition by multifunctional magnetic nanoparticles loaded with alpha-synuclein RNAi plasmid in a Parkinson's disease model. Theranostics. 2017;7:344-356. DOI: 10.7150/thno.16562

[134] Oh SM, Chang MY, Song JJ, Rhee YH, Joe EH, Lee HS, et al. Combined Nurr1 and Foxa2 roles in the therapy of Parkinson's disease. EMBO Molecular Medicine. 2015;7(5):510-525. DOI: 10.15252/emmm.201404610

[135] Helmschrodt C, Höbel S, Schöniger S, Bauer A, Bonicelli J, Gringmuth M, et al. Polyethylenimine nanoparticle-mediated siRNA delivery to reduce α-Synuclein expression in a model of Parkinson's disease. Molecular Therapy - Nucleic Acids. 2017;**9**:57-68. DOI: 10.1016/j.omtn.2017.08.013

[136] Kao SC, Krichevsky AM, Kosik KS, Tsai LH. BACE1 suppression by RNA interference in primary cortical neurons. The Journal of Biological Chemistry. 2004;**279**:1942-1949. DOI: 10.1074/jbc.M309219200

[137] Singer O, Marr RA, Rockenstein E, Crews L, Coufal NG, Gage FH, et al. Targeting BACE1 with siRNAs ameliorates Alzheimer disease neuropathology in a transgenic model. Nature Neuroscience. 2005;**8**:1343-1349. DOI: 10.1038/nn1531

[138] Dehay B, Bourdenx M, Gorry P, Przedborski S, Vila M, Hunot S, et al. Targeting α-synuclein for treatment of Parkinson's disease: Mechanistic and therapeutic considerations. Lancet Neurology. 2015;**14**:855-866. DOI: 10.1016/S1474-4422(15)00006-X

[139] Abraham S, Soundararajan CC, Vivekanandhan S, Behari M. Erythrocyte antioxidant enzymes in Parkinson's disease. The Indian Journal of Medical Research. 2005;**121**:111-115

[140] Surendran S, Rajasankar S. Parkinson's disease: Oxidative stress and therapeutic approaches. Neurological Sciences. 2010;**31**:531-540

[141] Johnston LC, Eberling J, Pivirotto P, Hadaczek P, Federoff HJ, Forsayeth J, et al. Clinically relevant effects of convection-enhanced delivery of AAV2-GDNF on the dopaminergic nigrostriatal pathway in aged rhesus monkeys. Human Gene Therapy. 2009;**20**:497-510. DOI: 10.1089/hum.2008.137

[142] Fiandaca M, Forsayeth J, Bankiewicz K. Current status of gene therapy trials for Parkinson's disease. Experimental Neurology. 2008;**209**:51-57. DOI: 10.1016/j.expneurol.2007.08.009

[143] Kells AP, Eberling J, Su X, Pivirotto P, Bringas J, Hadaczek P, et al. Regeneration of the MPTP-lesioned dopaminergic system after convection-enhanced delivery of AAV2-GDNF. The Journal of Neuroscience. 2010;**30**:9567-9577. DOI: 10.1523/JNEUROSCI.0942-10.2010

[144] Richardson RM, Kells AP, Rosenbluth KH, Salegio EA, Fiandaca MS, Larson PS, et al. Interventional MRI-guided putaminal delivery of AAV2-GDNF for a planned clinical trial in Parkinson's disease. Molecular Therapy. 2011;**19**:1048-1057. DOI: 10.1038/mt.2011.11

[145] Eberling JL, Kells AP, Pivirotto P, Beyer J, Bringas J, Federoff HJ, et al. Functional effects of AAV2-GDNF on the dopaminergic nigrostriatal pathway in parkinsonian rhesus monkeys. Human Gene Therapy. 2009;**20**(5):511-518. DOI: 10.1089/hum.2008.201

[146] Su X, Kells AP, Huang EJ, Lee HS, Hadaczek P, Beyer J, et al. Safety evaluation of AAV2-GDNF gene transfer into the dopaminergic nigrostriatal pathway in aged and parkinsonian rhesus monkeys. Human Gene Therapy. 2009;**20**(12):1627-1640. DOI: 10.1089/hum.2009.103

[147] Fiandaca MS, Varenika V, Eberling J, McKnight T, Bringas J, Pivirotto P, et al. Real-time MR imaging of adeno-associated viral vector delivery to the primate brain. NeuroImage. 2009;**47**(Suppl. 2):T27-T35. DOI: 10.1016/j.neuroimage.2008.11.012

[148] Han SJ, Bankiewicz K, Butowski NA, Larson PS, Aghi MK. Interventional MRI-guided catheter placement and real time drug delivery to the central nervous system. Expert Review of Neurotherapeutics. 2016;**16**(6):635-639. DOI: 10.1080/14737175.2016.1175939

[149] Safety Study of AADC Gene Therapy (VY-AADC01) for Parkinson's Disease (AADC). ClinicalTrials.gov Identifier: NCT01973543. Available from: https://clinicaltrials.gov/ct2/show/NCT01973543?term=NCT01973543&rank=1

[150] AAV2-GDNF for Advanced Parkinson s Disease. ClinicalTrials.gov Identifier: NCT01621581. Available from: https://clinicaltrials.gov/ct2/show/NCT01621581?term=NCT01621581&rank=1

[151] Deng X, Liang Y, Lu H, Yang Z, Liu R, Wang J, et al. Co-transplantation of GDNF-overexpressing neural stem cells and fetal dopaminergic neurons mitigates motor symptoms in a rat model of Parkinson's disease. PLoS One. 2013;**8**(12):e80880. DOI: 10.1371/journal.pone.0080880. eCollection 2013

[152] Gaudin A, Andrieux K, Couvreur P. Nanomedicines and stroke: Toward translational research. Journal of Drug Delivery Science and Technology. 2015;**30**:278-299. DOI: 10.1016/j.jddst.2015.07.018

[153] Bhaskar S, Tian F, Stoeger T, Kreyling W, de la Fuente JM, Grazú V, et al. Multifunctional nanocarriers for diagnostics, drug delivery and targeted treatment across blood-brain barrier: Perspectives on tracking and neuroimaging. Particle and Fibre Toxicology. 2010;7:3. DOI: 10.1186/1743-8977-7-3

[154] Hacke W, Kaste M, Bluhmki E, Brozman M, Dávalos A, Guidetti D, et al. Thrombolysis with alteplase 3 to 4.5 hours after acute ischemic stroke.

The New England Journal of Medicine. 2008;**359**(13):1317-1329. DOI: 10.1056/NEJMoa0804656

[155] Vidarsson L, Thornhill RE, Liu F, Mikulis DJ, Kassner A. Quantitative permeability magnetic resonance imaging in acute ischemic stroke: How long do we need to scan? Magnetic Resonance Imaging. 2009;**27**(9):1216-1222. DOI: 10.1016/j.mri.2009.01.019

[156] Bang OY, Saver JL, Alger JR, Shah SH, Buck BH, Starkman S, et al. Patterns and predictors of blood–brain barrier permeability derangements in acute ischemic stroke. Stroke. 2009;**40**(2):454-461. DOI: 10.1161/STROKEAHA.108.522847

[157] Neeman M. Functional and molecular MR imaging of angiogenesis: Seeing the target, seeing it work. Journal of Cellular Biochemistry. Supplement. 2002;**39**:11-17. DOI: 10.1002/jcb.10399

[158] Li L, Jiang Q, Zhang L, Ding G, Wang L, Zhang R, et al. Ischemic cerebral tissue response to subventricular zone cell transplantation measured by iterative self-organizing data analysis technique algorithm. Journal of Cerebral Blood Flow and Metabolism. 2006;**26**:1366-1377. DOI: 10.1038/sj.jcbfm.9600288

[159] Huang L, Liu Y. In vivo delivery of RNAi with lipid-based nanoparticles. Annual Review of Biomedical Engineering. 2011;**13**:507-530. DOI: 10.1146/annurev-bioeng-071910-124709

[160] Shiraishi K, Wang Z, Kokuryo D, Aoki L, Yokoyama M. A polymeric micelle magnetic resonance imaging (MRI) contrast agent reveals blood-brain barrier (BBB) permeability for macromolecules in cerebral ischemia-reperfusion injury. Journal of Controlled Release. 2017;**253**:165-171. DOI: 10.1016/j.jconrel.2017.03.020

[161] Sierra S, Ramos MC, Molina P, Esteo C, Vázquez JA, Burgos JS. Statins as neuroprotectants: A comparative in vitro study of lipophilicity, blood-brain-barrier penetration, lowering of brain cholesterol, and decrease of neuron cell death. Journal of Alzheimer's Disease. 2011;**23**:307-318. DOI: 10.3233/JAD-2010-101179

[162] So G, Nakagawa S, Morofuji Y, Hiu T, Hayashi K, Tanaka K, et al. Candesartan improves ischemia-induced impairment of the blood-brain barrier in vitro. Cellular and Molecular Neurobiology. 2015;**35**:563-572. DOI: 10.1007/s10571-014-0152-8

[163] Jickling GC, Sharp FR. Blood biomarkers of ischemic stroke. Neurotherapeutics. 2011;**8**:349-360. DOI: 10.1007/s13311-011-0050-4

[164] Lin KY, Kwong GA, Warren AD, Wood DK, Bhatia SN. Nanoparticles that sense thrombin activity as synthetic urinary biomarkers of thrombosis. ACS Nano. 2013;**7**:9001-9009. DOI: 10.1021/nn403550c

[165] Deddens LH, van Tilborg GA, van der Toorn A, van der Marel K, Paulis LE, van Bloois L, et al. MRI of ICAM-1 upregulation after stroke: The importance of choosing the appropriate target-specific particulate contrast agent. Molecular Imaging and Biology. 2013;**15**(4):411-422. DOI: 10.1007/s11307-013-0617-z

[166] Deddens LH, van Tilborg GAF, van der Marel K, Hunt H, van der Toorn A, Viergever MA, et al. In vivo molecular MRI of ICAM-1 expression on endothelium and leukocytes from subacute to chronic stages after experimental stroke. Translational Stroke

Research. 2017;**8**:440-448. DOI: 10.1007/s12975-017-0536-4

[167] Gauberti M, Montagne A, Marcos-Contreras OA, Le Béhot A, Maubert E, Vivien D. Ultra-sensitive molecular MRI of vascular cell adhesion molecule-1 reveals a dynamic inflammatory penumbra after strokes. Stroke. 2013;**44**(7):1988-1996. DOI: 10.1161/STROKEAHA.111.000544

[168] McAteer MA, Sibson NR, von Zur MC, Schneider JE, Lowe AS, Warrick N, et al. In vivo magnetic resonance imaging of acute brain inflammation using microparticles of iron oxide. Nature Medicine. 2007;**13**:1253-1258. DOI: 10.1038/nm1631

[169] van Kasteren SI, Campbell SJ, Serres S, Anthony DC, Sibson NR, Davis BG. Glyconanoparticles allow pre-symptomatic in vivo imaging of brain disease. Proceedings of the National Academy of Sciences of the United States of America. 2009;**106**:18-23. DOI: 10.1073/pnas.0806787106

[170] Saleh A, Schroeter M, Jonkmanns C, Hartung HP, Modder U, Jander S. In vivo MRI of brain inflammation in human ischaemic stroke. Brain. 2004;**127**(7):1670-1677. DOI: 10.1093/brain/awh191

[171] Leng T, Shi Y, Xiong ZG, Sun D. Proton-sensitive cation channels and ion exchangers in ischemic brain injury: New therapeutic targets for stroke? Progress in Neurobiology. 2014;**115**:189-209. DOI: 10.1016/j.pneurobio.2013.12.008

[172] Tuttolomondo A, Di Raimondo D, di Sciacca R, Pinto A, Licata G. Inflammatory cytokines in acute ischemic stroke. Current Pharmaceutical Design. 2008;**14**(33):3574-3589. DOI: 10.2174/138161208786848739

[173] Kim JY, Kawabori M, Yenari MA. Innate inflammatory responses in stroke: Mechanisms and potential therapeutic targets. Current Medicinal Chemistry. 2014;**21**(18):2076-2097. DOI: 10.2174/0929867321666131228205146

[174] Globus MY, Busto R, Martinez E, Valdes I, Dietrich WD, Ginsberg MD. Comparative effect of transient global ischemia on extracellular levels of glutamate, glycine, and gamma-aminobutyric acid in vulnerable and nonvulnerable brain regions in the rat. Journal of Neurochemistry. 1991;**57**(2):470-478. DOI: 10.1111/j.1471-4159.1991.tb03775.x

[175] Liu CH, Huang S, Cui J, Kim YR, Farrar CT, Moskowitz MA, et al. MR contrast probes that trace gene transcripts for cerebral ischemia in live animals. The FASEB Journal. 2007;**21**(11):3004-3015. DOI: 10.1096/fj.07-8203com

[176] Liu CH, You Z, Liu CM, Kim YR, Whalen MJ, Rosen BR, et al. Diffusion-weighted magnetic resonance imaging reversal by gene knockdown of matrix metalloproteinase-9 activities in live animal brains. The Journal of Neuroscience. 2009;**29**(11):3508-3517. DOI: 10.1523/JNEUROSCI.5332-08.2009

[177] Wu S, Ren J, Yang J, Liu C, Liu P. Multi-parameters evaluation on adult neurogenesis with angiogenesis by gene transcript-targeted MR imaging in preclinical cerebral ischemia-repair model. In: Scientific Informal Presentations — Molecular Imaging. Chicago, USA: Radiological Society of North America (RSNA); 2012

[178] Wahlestedt C, Golanov E, Yamamoto S, Yee F, Ericson H, Yoo H, et al. Antisense oligodeoxynucleotides to NMDA-R1 receptor channel protect

cortical neurons from excitotoxicity and reduce focal ischaemic infarctions. Nature. 1993;**363**(6426):260-263. DOI: 10.1038/363260a0

[179] Ren J, Chen YI, Liu CH, Chen PC, Prentice H, Wu JY, et al. Noninvasive tracking of gene transcript and neuroprotection after gene therapy. Gene Therapy. 2016;**23**:1-9. DOI: 10.1038/gt.2015.81

[180] Wiart M, Davoust N, Pialat JB, Desestret V, Moucharrafie S, Cho TH, et al. MRI monitoring of neuroinflammation in mouse focal ischemia. Stroke. 2007;**38**:131-137. DOI: 10.1161/01.STR.0000252159.05702.00

[181] Henning EC, Ruetzler CA, Gaudinski MR, Hu TC, Latour LL, Hallenbeck JM, et al. Feridex preloading permits tracking of CNS-resident macrophages after transient middle cerebral artery occlusion. Journal of Cerebral Blood Flow and Metabolism. 2009;**29**:1229-1239. DOI: 10.1038/jcbfm.2009.48

[182] Sindrilaru A, Peters T, Wieschalka S, Baican C, Baican A, Peter H, et al. An unrestrained proinflammatory M1 macrophage population induced by iron impairs wound healing in humans and mice. The Journal of Clinical Investigation. 2011;**121**:985-997. DOI: 10.1172/JCI44490

[183] Doyle KP, Quach LN, D'Arceuil HE, Buckwalter MS. Ferumoxytol administration does not alter infarct volume or the inflammatory response to stroke in mice. Neuroscience Letters. 2015;**0**:236-240. DOI: 10.1016/j.neulet.2014.10.041

[184] Saleh A, Schroeter M, Ringelstein A, Hartung HP, Siebler M, Modder U, et al. Iron oxide particle-enhanced MRI suggests variability of brain inflammation at early stages after ischemic stroke. Stroke. 2007;**38**:2733-2737

[185] Farell B, Hamilton B, Dosa E, Rimely E, Nasseri M, Gahramanov S, et al. Using iron oxide nanoparticles to diagnose CNS inflammatory diseases and PCNSL. Neurology. 2013;**81**(3):256-263. DOI: 10.1212/WNL.0b013e31829bfd8f

[186] Nighoghossian N, Wiart M, Cakmak S, Berthezene Y, Derex L, Cho TH, et al. Inflammatory response after ischemic stroke: A USPIO-enhanced MRI study in patients. Stroke. 2007;**38**:303-307. DOI: 10.1161/01.STR.0000254548.30258.f2

[187] Ugga L, Romeo V, Tedeschi E, Brunetti A, Quarantelli M. Superparamagnetic iron oxide nanocolloids in MRI studies of neuroinflammation. Journal of Neuroscience Methods. 2018;**310**:12-23. DOI: 10.1016/j.jneumeth.2018.06.008

[188] Oude Engberink RD, van der Pol SM, Walczak P, van der Toorn A, Viergever MA, Dijkstra CD, et al. Magnetic resonance imaging of monocytes labeled with ultrasmall superparamagnetic particles of iron oxide using magnetoelectroporation in an animal model of multiple sclerosis. Molecular Imaging. 2010;**9**:268-277. DOI: 10.2310/7290.2010.00016

[189] Maarouf A, Ferré JC, Zaaraoui W, Le Troter A, Bannier E, Berry I, et al. Ultra-small superparamagnetic iron oxide enhancement is associated with higher loss of brain tissue structure in clinically isolated syndrome. Multiple Sclerosis. 2016;**22**:1032-1039. DOI: 10.1177/1352458515607649

[190] Zhang Y, Gauthier SA, Gupta A, Comunale J, Chia-Yi Chiang G, Zhou D,

et al. Longitudinal change in magnetic susceptibility of new enhanced multiple sclerosis (MS) lesions measured on serial quantitative susceptibility mapping (QSM). Journal of Magnetic Resonance Imaging. 2016;**44**:426-432. DOI: 10.1002/jmri.25144

[191] Mishra SK, Kumar BS, Khushu S, Singh AK, Gangenahalli G. Early monitoring and quantitative evaluation of macrophage infiltration after experimental traumatic brain injury: A magnetic resonance imaging and flow cytometric analysis. Molecular and Cellular Neurosciences. 2017;**78**:25-34. DOI: 10.1016/j.mcn.2016.11.008

[192] Foley LM, Hitchens TK, Ho C, Janesko-Feldman KL, Melick JA, Bayir H, et al. Magnetic resonance imaging assessment of macrophage accumulation in mouse brain after experimental traumatic brain injury. Journal of Neurotrauma. 2009;**26**:1509-1519. DOI: 10.1089/neu.2008.0747

[193] Ouyang L, Zeng S, Zheng G, Lu GM. Early inflammatory response following traumatic brain injury in rabbits using USPIO- and Gd-enhanced MRI. BioMed Research International. 2016;**2016**:8431987. DOI: 10.1155/2016/8431987

[194] Lee ST, Chu K, Jung KH, Kim SJ, Kim DH, Kang KM, et al. Anti-inflammatory mechanism of intravascular neural stem cell transplantation in haemorrhagic stroke. Brain. 2008;**131**(Pt 3):616-629. DOI: 10.1093/brain/awm306

[195] Belmadani A, Tran PB, Ren D, Miller RJ. Chemokines regulate the migration of neural progenitors to sites of neuroinflammation. The Journal of Neuroscience. 2006;**26**:3182-3191. DOI: 10.1523/JNEUROSCI.0156-06.2006

[196] Imitola J, Raddassi K, Park KI, Mueller FJ, Nieto M, Teng YD, et al. Directed migration of neural stem cells to sites of CNS injury by the stromal cell-derived factor 1alpha/CXC chemokine receptor 4 pathway. Proceedings of the National Academy of Sciences of the United States of America. 2004;**101**:18117-18122. DOI: 10.1073/pnas.0408258102

[197] Robin AM, Zhang ZG, Wang L, Zhang RL, Katakowski M, Zhang L, et al. Stromal cell-derived factor 1alpha mediates neural progenitor cell motility after focal cerebral ischemia. Journal of Cerebral Blood Flow and Metabolism. 2006;**26**:125-134. DOI: 10.1038/sj.jcbfm.9600172

[198] Thored P, Arvidsson A, Cacci E, Ahlenius H, Kallur T, Darsalia V, et al. Persistent production of neurons from adult brain stem cells during recovery after stroke. Stem Cells. 2006;**24**:739-747. DOI: 10.1634/stemcells.2005-0281

[199] Widera D, Mikenberg I, Elvers M, Kaltschmidt C, Kaltschmidt B. Tumor necrosis factor alpha triggers proliferation of adult neural stem cells via IKK/NF-kappaB signaling. BMC Neuroscience. 2006;7:64. DOI: 10.1186/1471-2202-7-64

[200] Desestret V, Brisset JC, Moucharrafie S, Devillard E, Nataf S, Honnorat J, et al. Early-stage investigations of ultrasmall superparamagnetic iron oxide-induced signal change after permanent middle cerebral artery occlusion in mice. Stroke. 2009;**40**:1834-1841. DOI: 10.1161/strokeaha.108.531269

[201] Azevedo-Pereira RL, Rangel B, Tovar-Moll F, Gasparetto EL, Attias M, Zaverucha-do-Valle C, et al. Superparamagnetic iron oxide nanoparticles as a tool to track mouse

neural stem cells in vivo. Molecular Biology Reports. 2019;**46**:191-198. DOI: 10.1007/s11033-018-4460-9

[202] Vande Velde G, Rangarajan JR, Toelen J, Dresselaers T, Ibrahimi A, Krylychkina O, et al. Evaluation of the specificity and sensitivity of ferritin as an MRI reporter gene in the mouse brain using lentiviral and adeno-associated viral vectors. Gene Therapy. 2011;**18**(6):594-605. DOI: 10.1038/gt.2011.2

[203] Vande Velde G, Raman Rangarajan J, Vreys R, Guglielmetti C, Dresselaers T, Verhoye M, et al. Quantitative evaluation of MRI-based tracking of ferritin-labeled endogenous neural stem cell progeny in rodent brain. NeuroImage. 2012;**62**:367-380. DOI: 10.1016/j.neuroimage.2012.04.040

[204] Elvira G, Garcia I, Benito M, Gallo J, Desco M, Penades S, et al. Live imaging of mouse endogenous neural progenitors migrating in response to an induced tumor. PLoS One. 2012;7:e44466. DOI: 10.1371/journal.pone.0044466

[205] Ji B, Kaneko H, Minamimoto T, Inoue H, Takeuchi H, Kumata K, et al. Multimodal imaging for dreadd-expressing neurons in living brain and their application to implantation of iPSC-derived neural progenitors. The Journal of Neuroscience. 2016;**36**(45):11544-11558. DOI: 10.1523/JNEUROSCI.1279-16.2016

[206] Mishra SK, Khushu S, Singh AK, Gangenahalli G. Homing and tracking of iron oxide labelled mesenchymal stem cells after infusion in traumatic brain injury mice: A longitudinal in vivo MRI study. Stem Cell Reviews and Reports. 2018;**14**(6):888-900. DOI: 10.1007/s12015-018-9828-7

[207] Song M, Kim Y, Kim Y, Ryu S, Song I, Kim SU, et al. MRI tracking of intravenously transplanted human neural stem cells in rat focal ischemia model. Neuroscience Research. 2009;**64**(2):235-239. DOI: 10.1016/j.neures.2009.03.006

[208] Cho TH, Nighoghossian N, Wiart M, Desestret V, Cakmak S, Berthezène Y, et al. USPIO-enhanced MRI of neuroinflammation at the sub-acute stage of ischemic stroke: Preliminary data. Cerebrovascular Diseases. 2007;**24**:544-546. DOI: 10.1159/000111222

[209] Deddens LH, Van Tilborg GAF, Mulder WJM, De Vries HE, Dijkhuizen RM. Imaging Neuroinflammation after stroke: Current status of cellular and molecular MRI strategies. Cerebrovascular Diseases. 2012;**33**:392-402. DOI: 10.1159/000336116

[210] Hamasaki T, Tanaka N, Kamei N, Ishida O, Yanada S, Nakanishi K, et al. Magnetically labeled neural progenitor cells, which are localized by magnetic force, promote axon growth in organotypic cocultures. Spine. 2007;**32**:2300-2305. DOI: 10.1097/BRS.0b013e318154c651

[211] Artemov D. Molecular magnetic resonance imaging with targeted contrast agents. Journal of Cellular Biochemistry. 2003;**90**:518-524. DOI: 10.1002/jcb.10660

[212] Weissleder R, Stark DD, Engelstad BL, Bacon BR, Compton CC, White DL, et al. Superparamagnetic iron oxide: Pharmacokinetics and toxicity. AJR. American Journal of Roentgenology. 1989;**152**:167-173. DOI: 10.2214/ajr.152.1.167

[213] Bulte JW, Vymazal J, Brooks RA, Pierpaoli C, Frank JA. Frequency dependence of MR relaxation times. II. Iron oxides. Journal of Magnetic

Resonance Imaging. 1993;**3**:641-648. DOI: 10.2214/ajr.152.1.167

[214] Lee N, Choi Y, Lee Y, Park M, Moon WK, Choi SH, et al. Water-dispersible ferromagnetic iron oxide nanocubes with extremely high r2 relaxivity for highly sensitive in vivo MRI of tumors. Nano Letters. 2012;**12**:3127-3131. DOI: 10.1021/nl3010308

[215] Chamberlain R, Reyes D, Curran GL, Marjanska M, Wengenack TM, Poduslo JF, et al. Comparison of amyloid plaque contrast generated by T2-weighted, T2*-weighted, and susceptibility-weighted imaging methods in transgenic mouse models of Alzheimer's disease. Magnetic Resonance in Medicine. 2009;**61**(5):1158-1164. DOI: 10.1002/mrm.21951

[216] Rossi M, Ruottinen H, Elovaara I, Ryymin P, Soimakallio S, Eskola H, et al. Brain iron deposition and sequence characteristics in parkinsonism: Comparison of SWI, T2* maps, T2-weighted-, and FLAIR-SPACE. Investigative Radiology. 2010;**45**(12):795-802. DOI: 10.1097/RLI.0b013e3181ec9c96

[217] Haller S, Badoud S, Nguyen D, Barnaure I, Montandon ML, Lovblad KO, et al. Differentiation between Parkinson disease and other forms of parkinsonism using support vector machine analysis of susceptibility-weighted imaging (SWI): Initial results. European Radiology. 2013;**23**:12-19. DOI: 10.1007/s00330-012-2579-y

[218] Gupta D, Saini J, Kesavadas C, Sarma PS, Kishore A. Utility of susceptibility-weighted MRI in differentiating Parkinson's disease and atypical parkinsonism. Neuroradiology. 2010;**52**(12):1087-1094. DOI: 10.1007/s00234-010-0677-6

[219] Wisnieff C, Ramanan S, Olesik J, Gauthier S, Wang Y, Pitt D. Quantitative susceptibility mapping (QSM) of white matter multiple sclerosis lesions: Interpreting positive susceptibility and the presence of iron. Magnetic Resonance in Medicine. 2015;**74**(2):564-570. DOI: 10.1002/mrm.25420

[220] Acosta-Cabronero J, Williams GB, Cardenas-Blanco A, Arnold RJ, Lupson V, Nestor PJ. In vivo quantitative susceptibility mapping (QSM) in Alzheimer's disease. PLoS One. 2013;**8**:e81093. DOI: 10.1371/journal.pone.0081093

[221] Poduslo JF, Wengenack TM, Curran GL, Wisniewski T, Sigurdsson EM, Macura SI, et al. Molecular targeting of Alzheimer's amyloid plaques for contrast-enhanced magnetic resonance imaging. Neurobiology of Disease. 2002;**11**:315-329. DOI: 10.1006/nbdi.2002.0550

[222] Kisler K, Nelson AR, Montagne A, Zlokovic BV. Cerebral blood flow regulation and neurovascular dysfunction in Alzheimer disease. Nature Reviews. Neuroscience. 2017;**18**:419-434. DOI: 10.1038/nrn.2017.48

[223] Sweeney M, Sagare A, Zlokovic B. Blood-brain barrier breakdown in Alzheimer's disease and other neurodegenerative disorders. Nature Reviews. Neurology. 2018;**14**:133-150. DOI: 10.1038/nrneurol.2017.188

[224] Wang Y, Zhang R, Tao C, Xu Z, Chen W, Wang C, et al. Blood-brain barrier disruption and perivascular beta-amyloid accumulation in the brain of aged rats with spontaneous hypertension: Evaluation with dynamic contrast-enhanced magnetic resonance imaging. Korean Journal of Radiology. 2018;**19**(3):498-507. DOI: 10.3348/kjr.2018.19.3.498

[225] Popa-Wagner A, Buga A-M, Popescu B, Muresanu D. Vascular

cognitive impairment, dementia, aging and energy demand. A vicious cycle. Journal of Neural Transmission Vienna Austria. 1996;**122**(Suppl. 1):S47-S54. DOI: 10.1007/s00702-013-1129-3

[226] Erickson MA, Banks WA. Blood–brain barrier dysfunction as a cause and consequence of Alzheimer's disease. Journal of Cerebral Blood Flow & Metabolism. 2013;**33**:1500-1513. DOI: 10.1038/jcbfm.2013.135

[227] Barbier EL, Lamalle L, Decorps M. Methodology of brain perfusion imaging. Journal of Magnetic Resonance Imaging. 2001;**13**:496-520. DOI: 10.1002/jmri.1073

[228] Weinstein JS, Varallyay CG, Dosa E, Gahramanov S, Hamilton B, Rooney WD, et al. Superparamagnetic iron oxide nanoparticles: Diagnostic magnetic resonance imaging and potential therapeutic applications in neurooncology and central nervous system inflammatory pathologies, a review. Journal of Cerebral Blood Flow and Metabolism. 2010;**30**:15-35. DOI: 10.1038/jcbfm.2009.192

[229] Christen T, Ni W, Qiu D, Schmiedeskamp H, Bammer R, Moseley M, et al. High-resolution cerebral blood volume imaging in humans using the blood pool contrast agent ferumoxytol. Magnetic Resonance in Medicine. 2013;**70**(3):705-710. DOI: 10.1002/mrm.24500

[230] Christen T, Pannetier NA, Ni WW, et al. MR vascular fingerprinting: A new approach to compute cerebral blood volume, mean vessel radius, and oxygenation maps in the human brain. NeuroImage. 2014;**89**:262-270. DOI: 10.1016/j.neuroimage.2013.11.052

[231] Simonsen CZ, Ostergaard L, Vestergaard-Poulsen P, Rohl L, Bjornerud A, Gyldensted C. CBF and CBV measurements by USPIO bolus tracking: Reproducibility and comparison with Gd-based values. Journal of Magnetic Resonance Imaging. 1999;**9**:342-347. DOI: 10.1002/(SICI)1522-2586

[232] Bjornerud A, Johansson LO, Briley-Saebo K, Ahlstrom HK. Assessment of T1 and T2* effects in vivo and ex vivo using iron oxide nanoparticles in steady state-dependence on blood volume and water exchange. Magnetic Resonance in Medicine. 2002;**47**:461-471. DOI: 10.1002/mrm.10066

[233] Troprès I, Grimault S, Vaeth A, Grillon E, Julien C, Payen JF, et al. Vessel size imaging. Magnetic Resonance in Medicine. 2001;**45**:397-408. DOI: 10.1002/1522-2594

[234] Wang Y-XJ, Idée J-M. A comprehensive literatures update of clinical researches of superparamagnetic resonance iron oxide nanoparticles for magnetic resonance imaging. Quantitative Imaging in Medicine and Surgery. 2017;**7**:88-122. DOI: 10.21037/qims.2017.05.05

[235] Fernandez-Seara MA, Wehrli FW. Postprocessing technique to correct for background gradients in image-based R*(2) measurements. Magnetic Resonance in Medicine. 2000;**44**:358-366. DOI: 10.1002/1522-2594

[236] Kim YR, van Meer MP, Tejima E, Murata Y, Mandeville JB, Dai G, et al. Functional MRI of delayed chronic lithium treatment in rat focal cerebral ischemia. Stroke. 2008;**39**:439-447. DOI: 10.1161/STROKEAHA.107.492215

[237] McKinley R, Hung F, Wiest R, Liebeskind DS, Scalzo F. A machine learning approach to perfusion imaging with dynamic susceptibility contrast

MR. Frontiers in Neurology. 2018;**9**:717. DOI: 10.3389/fneur.2018.00717

[238] McKinley R, Häni L, Gralla J, El-Koussy M, Bauer S, Arnold M, et al. Fully automated stroke tissue estimation using random forest classifiers (FASTER). JCBFM. 2017;**37**(8):2728-2741. DOI: 10.1177/0271678X16674221

[239] Crimi A, Commowick O, Maarouf A, Ferré JC, Bannier E, Tourbah A, et al. Predictive value of imaging markers at multiple sclerosis disease onset based on gadolinium- and USPIO-enhanced MRI and machine learning. PLoS One. 2014;**9**:e93024. DOI: 10.1371/journal.pone.0093024

[240] Song S, Bokkers RPH, Luby M, Edwardson MA, Brown T, Shah S, et al. Temporal similarity perfusion mapping: A standardized and model free method for detecting perfusion deficits in stroke. PLoS One. 2017;**12**(10):e0185552. DOI: 10.1371/journal.pone.0185552

[241] Ulas C, TettehG TMJ, Armitage PA, Makin SD, Wardlaw JM, Davies ME, et al. Direct estimation of pharmacokinetic parameters from DCE-MRI using deep CNN with forward physical model loss. In: Frangi AF, Schnabel JA, Davatzikos C, Alberola-Lpez C, Fichtinger G, editors. Medical Image Computing and Computer Assisted Intervention - MICCAI. Cham: Springer International Publishing; 2018. pp. 39-47. DOI: 10.1007/978-3-030-00928-1_5

[242] Hong W, He Q, Fan S, Carl M, Shao H, Chen J, et al. Imaging and quantification of iron-oxide nanoparticles (IONP) using MP-RAGE and UTE based sequences. Magnetic Resonance Imaging. 2017;**78**(1):226-232. DOI: 10.1002/mrm.26371

[243] Ashburner J. Computational anatomy with the SPM software. Magnetic Resonance Imaging. 2009;**27**(8):1163-1174. DOI: 10.1016/j.mri.2009.01.006

[244] Dubois A, Hérard AS, Flandin G, Duchesnay E, Besret L, Frouin V, et al. Quantitative validation of voxel-wise statistical analyses of autoradiographic rat brain volumes: Application to unilateral visual stimulation. NeuroImage. 2008;**40**(2):482-494. DOI: 10.1016/j.neuroimage.2007.11.054

[245] Focke NK, Thompson PJ, Duncan JS. Correlation of cognitive functions with voxel-based morphometry in patients with hippocampal sclerosis. Epilepsy & Behavior. 2008;**12**(3):472-476. DOI: 10.1016/j.yebeh.2007.12.011

[246] Kohan Z, Azmi R. Hippocampus shape analysis for temporal lobe epilepsy detection in magnetic resonance imaging. In: Medical Imaging: Biomedical Applications in Molecular, Structural, and Functional Imaging. Vol. 9788. San Diego, CA: SPIE-Intl Soc Optical Eng; 2016. p. 97882T. DOI: 10.1117/12.2216936

[247] Park EJ, Oh SY, Lee SJ, Lee K, Kim Y, Lee BS, et al. Chronic pulmonary accumulation of iron oxide nanoparticles induced Th1-type immune response stimulating the function of antigen-presenting cells. Environmental Research. 2015;**143**:138-147. DOI: 10.1016/j.envres.2015.09.030

[248] Zhao L, Li Z, Vong JSL, Chen X, Lai HM, Yan LYC, et al. Pharmacologically reversible zonation-dependent endothelial cell transcriptomic changes with neurodegenerative disease associations in the aged brain. Nature Communications. 2020;**11**:4413. DOI: 10.1038/s41467-020-18249-3

[249] Badachhape AA, Working PK, Srivastava M, Bhandari P, Stupin IV,

Devkota L, et al. Pre-clinical dose-ranging efficacy, pharmacokinetics, tissue biodistribution, and toxicity of a targeted contrast agent for MRI of amyloid deposition in Alzheimer's disease. Scientific Reports. 2020;**10**:16185. DOI: 10.1038/s41598-020-73233-7

[250] Parekh P, Mu Q, Badachhape A, Bhavane R, Srivastava M, Devkota L, et al. A surrogate marker for very early-stage tau pathology is detectable by molecular magnetic resonance imaging. Theranostics. 2022;**12**:5504-5521. DOI: 10.7150/thno.72258

[251] Perolina E, Meissner S, Raos B, Harland B, Thakur S, Svirskis D. Translating ultrasound-mediated drug delivery technologies for CNS applications. Advanced Drug Delivery Reviews. 2024;**208**:115274. DOI: 10.1016/j.addr.2024.115274

[252] Su YY, Yang GF, Lu GM, Wu S, Zhang LJ. PET and MR imaging of neuroinflammation in hepatic encephalopathy. Metabolic Brain Disease. 2015;**30**(1):31-45. DOI: 10.1007/s11011-014-9633-1

[253] Quillard T, Libby P. Molecular imaging of atherosclerosis for improving diagnostic and therapeutic development. Circulation Research. 2012;**111**:231-244. DOI: 10.1161/CIRCRESAHA.112.268144

[254] Berenguer R, del Pastor-Juan MR, Canales-Vázquez J, Castro-García M, Villas MV, Mansilla Legorburo F, et al. Radiomics of CT features may be nonreproducible and redundant: Influence of ct acquisition parameters. Radiology. 2018;**288**:407-415. DOI: 10.1148/radiol.2018172361

[255] Aerts HJWL, Velazquez ER, Leijenaar RTH, Parmar C, Grossmann P, Carvalho S, et al. Decoding tumour phenotype by noninvasive imaging using a quantitative radiomics approach. Nature Communications. 2014;**5**:4006. DOI: 10.1038/ncomms5006